Nanotechnology
Synthesis to Applications

T0199368

Nanotechnology
Synthesis to Applications

Edited by
Sunipa Roy, Chandan Kumar Ghosh,
and Chandan Kumar Sarkar

CRC Press
Taylor & Francis Group
Boca Raton London New York

CRC Press is an imprint of the
Taylor & Francis Group, an **Informa** business

CRC Press
Taylor & Francis Group
6000 Broken Sound Parkway NW, Suite 300
Boca Raton, FL 33487-2742

First issued in paperback 2020

© 2018 by Taylor & Francis Group, LLC
CRC Press is an imprint of Taylor & Francis Group, an Informa business

No claim to original U.S. Government works

ISBN-13: 978-0-367-57302-7 (pbk)
ISBN-13: 978-1-138-03273-6 (hbk)

Visit the Taylor & Francis Web site at
http://www.taylorandfrancis.com

and the CRC Press Web site at
http://www.crcpress.com

Dedicated to our families for their love and support

Contents

Preface

Nano is a Greek word meaning "dwarf." One nanometer is 10^{-9} meter. In modern-day technology, the term "nano" has assumed great significance because physical and chemical properties of materials change significantly when their sizes are reduced to dimensions of 100 nm or less. Thus nanoscience and nanotechnology refer to the science and technology of materials and systems having dimensions in the range of 1–100 nm. In the field of electronics, there are immense possibilities of nanostructured semiconductor devices, some of which are demonstrated by different authors in this book.

This technological advancement also demanded understanding of the new device physics as the properties of materials changed significantly at such reduced dimensions. These dimensions are comparable to the electron wavelength of motion, and hence the device characteristics are governed by the confinement of the electron wave function, commonly referred to as quantum confinement effect. Quantum confinement effect due to changes in the size and shape of nanoparticles (NPs) can modify the energy bands of semiconductors and insulators. A brief description on electrical transport of nanostructure is given in this book.

Nano particles have attracted much interest in recent years by virtue of their unusual mechanical, electrical, optical, and magnetic properties. Due to their special properties, NPs are finding wide applications in all fields of engineering. Several synthetic approaches have been developed to control their morphology and synthesis.

Metal NPs are of particular interest in medical diagnostics because of their antibacterial activity and low toxicity to human cells. This antibacterial effect is further enhanced with the reduction in particle size. Biomedical devices must be biocompatible, which indicates nontoxicity of the reactants used in nanoparticle synthesis. The so-called "green synthesis" method suits well for this purpose. A novel green synthesis of NPs makes use of environmental-friendly, nontoxic, and safe reagents, which is discussed in Chapter 7.

Nanocomposites for energy harvesting are a latest area of research and contain many unexpected benefits in the field of technology. Keeping this in mind, a chapter has been written to present the matter from an experimental point of view.

Nanotechnology is a completely new science that came on the market a few years ago, though many are not fully conscious of its presence in daily life. The application of nanotechnology can be found in many everyday items. Thousands of commercial products incorporate nanomaterials. Though it is a relatively new science, its applications range from consumer goods to medicine to improving the environment.

In the present work, we have covered different areas of nanotechnology from energy harvesting to gas sensing. A systematic step-by-step approach has been adopted to facilitate the thinking of new researchers in the nanotechnology domain.

This edited volume bridges the gap between and presents the latest trends and updates in three topics of ever-increasing importance (nanogenerator, thin-film solar cell, and green synthesis of metallic NPs) in current and future society. The optical responses of an ensemble of NPs drastically differ from individual NPs due to collective resonance of free electrons on the surface. It provides the state-of-the-art as well as current challenges and advances in the sustainable preparation of novel metal NPs and their applications. Further, the global energy consumption is estimated to rise by more than 56% within the next 20 years. At present, most of the energy produced is from the combustion of fuels, such as natural gas and coal. To restrain global dependence on exhaustible natural resources and their hazardous effects, more technical and scientific research has been directed toward the renewable energies to reduce the cost of energy production. Though there are many solid-state devices to generate renewable energies, among them, piezoelectric nanogenerators, which convert mechanical energy, vibrational energy, and hydraulic energy into electrical energy, can be used as self-powered nanodevices that operate at very low power (nW to μW). Moreover, nanotechnology-enhanced, thin-film solar cells (which convert the energy of the sun into electricity) are a potentially important upcoming technology. With the dynamic development we currently see in the area of thin-film/ heterojunction-based solar cell research, this book enhances awareness in the area of renewable energy applications. Finally, this edited volume covers from the very basics to the more advanced, trendy developments, containing a unique blend of nano, green, and renewable energy.

The book consists of 15 chapters.

Chapter 1 provides an introduction to the concept of nanotechnology and its application.

Chapter 2 introduces the basic concept of quantum mechanics. This chapter starts with quantum theory of radiation, developed from Maxwell's theory for black body radiation. It is stressed that Maxwell's theory, developed on the basis of the wave nature of radiation, does not predict the spectral dependency of radiation. Quantum theory for radiation was introduced by Planck on the basis of the particle property of radiation, defined as a photon. Newton's law that predicts the position and momentum of any macroscopic object at any instant of time does not hold good for the quantum particle. Then Heisenberg's uncertainty principle illustrates the fact that position and momentum of any microscopic particle cannot be measured simultaneously. At this point, quantum theory was developed by Schrödinger on the basis of the wave nature of quantum particles. It is attributed that wave function (solution to Schrödinger's equation) contains all information regarding the quantum particle, and a few mathematical tools, defined as operators, have been introduced to extract measurable parameters. A few approximation methods are also discussed to find out wave function where direct solution of Schrödinger's equation is not possible.

Chapter 3 provides a brief introduction of crystal structure followed by wave function of electrons. The chemical potential of electrons originates from the following two, electron–lattice and electron–electron interaction. This chapter deals with Sommerfeld free electron theory neglecting these two interactions. But this does not differentiate metal, semiconductor, and insulator. Then, Bloch theory is discussed considering electron–lattice interaction. The concept of density of states describing number of electronic states per unit energy interval is introduced in this chapter. The operational principles of a few nanoscale devices have also been introduced in this chapter on the basis of wave function of any quantum particle and tunneling phenomena.

Chapter 4 discusses different synthesis techniques of nanomaterials, including top-down and bottom-up approaches. In this chapter, various characterization techniques have been discussed to identify the morphology and properties of nanomaterials. Native defects that persist in the film can be identified. Structural characterizations like XRD, FESEM, and EDX confirming the crystal structure and determining the crystallite size are discussed lucidly. The impurity present in nanomaterials has been discussed through FTIR spectroscopy. Moreover, the development of crystal growth techniques, such as MOCVD and MBE, is discussed in this chapter. Parallely, sophisticated patterning and fabrication tools, such as e-beam lithography and ion beam etching techniques, have enabled the reduction of conventional silicon VLSI components to the nanoscale.

Chapter 5 deals with electrical transport in nanostructured materials. It is briefly discussed that in a nanostructured material, the fundamental transport mechanism differs from that of the bulk. Hence, ballistic transport is designated for the nanostructured materials.

Chapter 6 provides information about the chemical, sonochemical, and radiolysis methods for the synthesis of noble metals such as silver, gold, and platinum NPs. A few physical methods have also been discussed briefly. It has been investigated that surface-to-volume ratio plays a fundamental role in determining the properties of NPs, which is dependent on morphology of the particles, that is, it can be stated that the shape of a NP also plays a crucial role in determining its different physical and chemical properties. Therefore, we have also introduced a few methods (template-assisted method, hydrothermal method, etc.) to prepare noble metal NPs with different shapes.

Chapter 7 demonstrates green synthesis methods of noble metal NPs. It is well known that the physical and chemical routes for the synthesis of noble metal NPs include various chemicals as starting materials, and the equipment used are very costly sometimes. The chemicals used in the process are carcinogenic as well. At this stage, green synthesis protocols could be a suitable alternative for the synthesis of noble metal NPs. In this chapter, we have elaborated a few green synthesis methods for the preparation of noble metal NPs. Their main advantage lies in the fact that they are environment friendly and very cost effective. In this context, various green methods, including different microorganisms (bacteria, fungi, algae) and plant (leaf, root, flower etc.) extract, have been briefly discussed.

Chapter 8 discusses various useful application opportunities of NPs, particularly noble metal NPs. It is already established that various types of dyes, from textile industries, are very harmful to the environment. This chapter deals with photocatalytic processes by which dyes can be degraded using noble metal NPs. This chapter also provides a brief mention about the detection of mercury and hydrogen peroxide by these noble metal NPs. These NPs also have potential in a few biological applications, and this chapter suggests some ideas on this matter. For example, antibacterial, antifungal, anticancer, and antiviral activities of noble metal NPs have been briefly discussed in this chapter along with their mechanism. The present chapter also illustrates its basic principle and opportunity of bioimaging of noble metal.

Chapter 9 provides a coherent coverage of the ever craved nanogenerators from fundamental materials, basic theory and principles of physics, scientific approach, and technological applications to have a full picture about the development of this technology.

Chapter 10 discusses the overall view of solar cell technology. High-efficiency solar cells and the physical principles of design, fabrication, characterization, and applications of novel photovoltaic devices are presented in this chapter. Solar panel manufacturing techniques are briefed, and solar photovoltaic power plant design is presented in Chapter 10.

Chapter 11 offers an approach to nanotechnology-based VOC (volatile organic compound), as VOCs are thought to be carcinogenic in high concentrations and after long-term exposure. Different types of nanostructures used in the detection of VOCs are summarized. The main attraction of this chapter is the introduction of different packaging technologies specifically used for gas sensing.

Chapter 12 illustrates graphene and its properties in nanoscale science. It behaves well enough in the field of nanoelectronics. A detailed study about its honeycomb lattice structure and its importance in the formation of Brillouin zone are presented in this chapter. The zero-bandgap semiconductor property of graphene has been elaborately discussed. The author concludes this chapter by discussing some application areas of graphene.

Chapter 13 attempts to help realize why a high-performance sub-100 nm gate transistor demands semiconductor nanotechnology of highest quality and how that technology might be used to realize right-first-time, manufacturing systems.

Chapter 14 presents high-mobility III–V semiconductors having significant transport advantages, which are extensively used as alternative channel materials for upcoming high-scaled devices. The III–V compound semiconductor binaries such as GaAs, GaN, InP; ternaries such as InGaAs, AlGaAs,

AlGaN, and AlInN; and quaternary InAlGaN, InAlGaAs, and GaInPAs are widely studied for enhancing device performance. GaAs-based compounds have much higher mobility than their silicon counterparts and are thus suitable for high-speed operations. This chapter focuses on GaN-based heterostructures (such as the traditional AlGaN/GaN) with high breakdown voltage and large carrier density with high mobility, which make them ideal for high frequency and high power applications.

Chapter 15 provides an overview of the advanced fabrication technique NEMS (nanoelectromechanical system), which is very hot topic today. Information on the scaling effect and current research on NEMS and carbon materials are discussed.

Editors

Sunipa Roy received her MTech in VLSI and microelectronics from West Bengal University of Technology in 2009 and her PhD in engineering from Jadavpur University, ETCE Dept, in 2014. She served as a senior research fellow of the Council of Scientific and Industrial Research, Government of India and junior research fellow of the Department of Science and Technology, Government of India. She is a chartered engineer of the Institution of Engineers (India) and a member of IEEE. Presently she is head in the Electronics and Telecommunication Engg. Department at Guru Nanak Institute of Technology, Kolkata. She served as an invited speaker at other Indian universities. Her research interests include nanopiezotronics, MEMS, nanocrystalline metal oxide, graphene and its application as a gas sensor. She has authored one book (CRC Press) and published more than 20 research papers in various peer-reviewed international journals and also national/international conferences. She has filed patents and is supervising PhD students under her guidance.

Dr. Chandan Kumar Ghosh is an assistant professor in the School of Materials Science and Nanotechnology, Jadavpur University. He received his PhD from the Department of Physics, Jadavpur University, in the year 2010. His main scientific interests include synthesis and study of optical properties of nanomaterials, green synthesis of noble metal nanoparticles, antibacterial activity of nanomaterials, and electronic structure calculation by density functional theory. He has published more than 45 papers in different peer-reviewed international journals.

Chandan Kumar Sarkar (SM'87) received his MSc in physics from Aligarh Muslim University, Aligarh, India, in 1975; his PhD from Calcutta University, Kolkata, India, in 1979; and his DPhil from the University of Oxford, Oxford, UK, in 1983. He was also postdoctoral research fellow of the Royal Commission for the Exhibition of 1851 at the Clarendon Laboratory, University of Oxford, from 1983 to 1985. He was a fellow of Wolfson College, Oxford. He was a tutor and lab instructor at Clarendon Laboratory. He was also a visiting fellow with the Linkoping University, Linkoping, Sweden; Max Planck Institute at Stuttgart, Germany. He joined Jadavpur University, Kolkata, in 1987 as a reader in electronics and telecommunication engineering. Subsequently, he became a professor and the head of the Department of Physics, and dean of Faculty of Science Bengal Engineering Science University (BESU) during 1996–1999. Later he once again joined Jadavpur University ETCE Department as a professor. He has served as a visiting professor in many universities such as the Tokyo Institute of Technology, Japan and Hong Kong University, Hong Kong. Since 1999, he has been a professor with the Department of Electronics and Telecommunication and has published more than 300 papers in journals of repute and in well-known international conferences. He has guided more than 20 PhD candidates. Dr. Sarkar is the chair of the IEEE Electron Devices Society (EDS), Kolkata Chapter, and former vice chair of IEEE Section. He serves as a distinguished lecturer of the IEEE, EDS and was invited to several countries. He is a fellow of IE (India) Chartered Engineer, IETE, and WBAST. Currently he is an associate of the Third World Academy of Science at National University, La Plata, Argentina. Previously he was associated with the Shanghai Institute of Metallurgy, Shanghai. He has been awarded INSA-Royal Society (UK) fellowship to visit several UK universities.

Authors

Sarosij Adak received his BSc in physics from Calcutta University, his MSc from Vidyasagar University, and his MTech in electronics and communication engineering from the Institute of Engineering and Management, Kolkata, India, in the years 2007, 2009, and 2011, respectively. Currently, he is a senior research fellow (DST) at Jadavpur University and is working toward a PhD in engineering at the Indian Institute of Engineering Science and Technology, Shibpur, Howrah, India. His field of work is in wide–band gap, compound semiconductor–based AlGaN/GaN MOSHEMT; InAlN/GaN HEMT; and MOSHEMT. He has published his research paper in many scientific journals.

Amrita Banerjee is currently an assistant professor in the Department of Electronics and Communication Engineering at Heritage Institute of Technology (Kolkata). She received her BSc (physics honors) from the University of Calcutta in 2008. She also received her BTech from the Institute of Radio Physics and Electronics (University of Calcutta) and her MTech from the same institute with a specialization in space science and communication. She received a scholarship under the "Space Science Promotional Scheme" given by ISRO while pursuing her master's degree. Presently, she is pursuing her PhD. Her research interest includes nano-generators and nanodevices.

Swapan Das received his MTech in VLSI design and microelectronics technology from Jadavpur University, Kolkata, India, in 2015. He is currently pursuing a PhD in the Department of Electronics and Telecommunication Engineering at Jadavpur University, Kolkata. His current research interests include MEMS-based gas sensor interface with CMOS circuit.

Swapnadip De graduated with a degree in radio physics and electronics from the University College of Science and Technology in 2001. He obtained his MTech in VLSI and microelectronics from Jadavpur University. Later, he was awarded a PhD in engineering from Jadavpur University. He has been working as assistant professor in the Department of ECE at the Meghnad Saha Institute of Technology since December 2002. He is a senior member of IEEE and is currently the vice chairman of IEEE EDS Kolkata Chapter. He is also the branch chapter advisor of IEEE MSIT EDS SBC. He has been an executive member of IEEE EDS Kolkata Chapter since December 2013. He is a senior member of the International Engineering and Technology Institute, Hong Kong, and also a life member of IETE. He has also been an executive committee member of IEEE SSCS Kolkata Chapter since September 2015. He has published papers in refereed international journals of reputed publishers like Elsevier, Springer, IEEE Transactions, IET, Wiley, Taylor & Francis, World Scientific and WSEAS, to name a few. He has already authored nine books in India and abroad. He is the official reviewer of many reputed international journals and conferences like *IEEE Transactions on Electron Devices*. He is the editor of journals of international publishing houses like Inderscience Publications, UK, and WASET, USA. He has organized many international conferences, workshops, and mini-colloquiums. He has also authored reports in *IEEE EDS Newsletters* and *IEEE Newsletters* of the Kolkata Section. His biography is included in the 33rd edition of *Marquis Who's Who in the World 2016*. His biography was also nominated for the 2010 edition of *Marquis Who's Who in the World.*

Arka Dutta received his BTech in electronics and communication from West Bengal University of Technology, Kolkata, India, in 2010 and completed his ME with a specialization on electron devices in 2012 in the Department of Electronics and Telecommunication Engineering at Jadavpur University, Kolkata, West Bengal, India. He started working toward his PhD in the same department in 2013 and completed his PhD from Jadavpur University as Council of Scientific and Industrial Research (CSIR) senior research fellow in 2016. His research interests include compact modeling of advanced CMOS devices, device modeling for process variation, analog/RF performance analysis of advanced CMOS devices, and reliability analysis of RF CMOS devices for circuit applications.

Dutta was a senior research fellow, CSIR, Government of India, and also served as senior laboratory engineer under the SMDP II, Government of India project at Jadavpur University from 2012 to 2013. He has also served as joint secretary in IEEE EDS, Kolkata Section, from 2014 to 2015. He is presently working as a design engineer at ARM, Bangalore.

Atanu Kundu is working as an assistant professor in the Electronics & Communication Engineering Department at Heritage Institute of Technology. He has also served as a guest lecturer in the Department of Electronics and Telecommunication Engineering at Jadavpur University. Kundu completed his PhD in 2016, MTech in 2009, and BTech in 2005. He has published 26 papers in refereed journals and international conferences and coauthored a book titled *Technology Computer Aided Design: Simulation for VLSI MOSFET* published by CRC Press, Taylor & Francis Group. His research interest includes the "study of subthreshold analog and RF performance trends of multi-gate MOSFETs in sub-nanometer regime, biosensors." At present, he is the chairman of IEEE EDS, Kolkata chapter; chapter advisor, IEEE EDS, Kalyani Govt. Engineering College Student Branch; and chapter advisor, IEEE EDS, Heritage Institute of Technology Student Branch chapter. Prior to that, he worked as the vice chairman of IEEE EDS, Kolkata chapter, and as joint secretary there and also a senior member of IEEE. He has organized several international/national conferences, workshops, and seminars.

Soumya Pandit is currently an assistant professor, Stage-II, at the Institute of Radio Physics and Electronics, University of Calcutta, India.

Dr. Pandit received his BSc (Physics, Honours), MSc (Electronic Science), and MTech (Radio Physics and Electronics) from the University of Calcutta in 1998, 2000, and 2002, respectively, and started his career as a lecturer at the Meghnad Saha Institute of Technology, under Maulana Abul Kalam Azad University of Technology. After graduating with a PhD in the domain of VLSI Design from the Indian Institute of Technology, Kharagpur, he joined the Department of Radio Physics and Electronics at the University of Calcutta as an assistant professor in 2008.

Dr. Pandit researched in developing design methodologies and associated computer-aided design tools for a high-level synthesis of CMOS analog circuits in his PhD work. He served as a research consultant at the Advanced VLSI Design Laboratory at IIT Kharagpur from 2003 to 2008. During this period, he acted as a lead project scientist in several R&D projects sponsored by semiconductor industries like National Semiconductor (Santa Clara, California) and government agencies like DeitY. He had successfully taped out several integrated circuits in 0.8 micron and 0.18 micron CMOS technology. He has published more than 35 papers in leading international journals and conferences. He has authored a book entitled *Nano-Scale CMOS Analog Circuits: Models and CAD Techniques for High-Level Design*, published by CRC Press, USA, and several book chapters in edited volumes published by CRC Press and Springer. Dr. Pandit successfully completed four R&D projects sponsored by DST, Government of India; TEQIP, Phase-II, University of Calcutta; and others. He is currently the chief investigator of the Special Manpower Development Program for Chip to System Design (SMDP-C2SD) Project at the University of Calcutta. He has developed the IC Design Laboratory at the Department of Radio Physics and Electronics, University of Calcutta, meant for postgraduate teaching and research students. His current research interest includes VLSI design, technology-aware CMOS device design, and circuit design.

Dr. Pandit is a senior member of IEEE, USA. He is currently the vice chair of SRC, Region 10, IEEE Electron Devices Society (EDS), USA. He is the founder chapter adviser of the IEEE EDS, University of Calcutta Student Chapter. He served as the chair of IEEE EDS Kolkata Chapter during 2014–2015.

Kaushik Roy completed his MTech in nanoscience and technology in 2012 from Jadavpur University, India. He is currently working as a senior research fellow in the School of Materials Science and Nanotechnology at the same university, and his field of interest includes different applications of noble metal nanoparticles.

Angsuman Sarkar is currently serving as an associate professor of electronics and communication engineering in Kalyani Government Engineering College, West Bengal. He had earlier served as lecturer in the ECE

Department, Jalpaiguri Government Engineering College, West Bengal, for 10 years. He received an MTech in VLSI and microelectronics from Jadavpur University. He completed his PhD from Jadavpur University in 2013. His current research interest spans around the study of short channel effects of sub–100 nm MOSFETs and nanodevice modeling. He is a senior member of IEEE, life member of the Indian Society for Technical Education, associate life member of Institution of Engineers (India), and executive committee member of the Electron Device Society, Kolkata Section. He has authored six books, five contributed book chapters, 48 journal papers in international refereed journals, and 24 research papers in national and international conferences.

Arghyadeep Sarkar received his MS in material science and engineering from National Chiao Tung University, Taiwan, in 2016. He is currently pursuing his PhD in an EECS International Graduate Program at National Chiao Tung University, Taiwan. He is a recipient of the Outstanding New Student scholarship award from National Chiao Tung University for his PhD studies.

Sanjit Kumar Swain received his BTech and MTech in electronics and telecommunication engineering from Biju Patnaik University of Technology, Odisha, India, in the years 2003 and 2011, respectively. He has a teaching experience of around 11 years as assistant professor in Silicon Institute of Technology, Bhubaneswar, Odisha. Currently, he is working toward a PhD in the field of nanodevices in the Department of Electronics and Telecommunication Engineering at Jadavpur University, Kolkata, India. His PhD thesis is on performance analysis of NANO-MOS devices. He has attended many national and international conferences and seminars and published his research paper in many scientific journals.

Contributors

Sarosij Adak
Nano Device Simulation Laboratory
Jadavpur University
Kolkata, India

Amrita Banerjee
Department of Electronics and
    Communication Engineering
Heritage Institute of Technology
Kolkata, India

Swapan Das
Department of Electronics and
    Telecommunication Engineering
Jadavpur University
Kolkata, India

Swapnadip De
Department of Electronics and
    Communication Engineering
Meghnad Saha Institute of Technology
Kolkata, India

Arka Dutta
Department of Electronics and
    Telecommunication Engineering
Jadavpur University
Kolkata, India

Chandan Kumar Ghosh
School of Material Science and
    Nanotechnology
Jadavpur University
Kolkata, India

Atanu Kundu
Department of Electronics and
    Communication Engineering
Heritage Institute of Technology
Kolkata, India

Soumya Pandit
Institute of Radio Physics and
    Electronics
University of Calcutta
Kolkata, India

Kaushik Roy
School of Materials Science and
    Nanotechnology
Jadavpur University
Kolkata, India

Sunipa Roy
Department of Electronics and
    Communication Engineering
Guru Nanak Institute of Technology
Kolkata, India

Angsuman Sarkar
Department of Electronics and
    Communication Engineering
Kalyani Government of Engineering
    College
Kalyani, India

Arghyadeep Sarkar
Department of Electronics
    Engineering
National Chiao Tung University
Hsinchu, Taiwan, Republic of China

Chandan Kumar Sarkar
Department of Electronics and
    Telecommunication Engineering
Jadavpur University
Kolkata, India

Sanjit Kumar Swain
Department of Electronics and
    Communication Engineering
Silicon Institute of Technology
Orissa, India

1

Introduction: Motivation for Nanotechnology

Sunipa Roy, Chandan Kumar Ghosh, and Chandan Kumar Sarkar

Nanotechnology is a completely new branch of science and engineering that hit the market a few years ago. Nanotechnology is a revolution where even a 0.1 nm variation in particle size plays a significant role. Fundamentally, nanoscale implies a range from 1 to 100 nm. Nanotechnology can be best explained as "the development, synthesis, characterization, and application of materials and devices by tailoring their shape and size at the nanoscale." Surprisingly, each permutation of shape and sizes produces a unique property with essentially new characteristics and potentiality.

The goal of this book is to provide some ideas about the optical and electrical properties of semiconducting and metallic materials at the nanoscale, and then to discuss some real-life applicational opportunities for fabricating devices. To start, we discuss basic quantum mechanics to understand the behavior of microscopic particles such as electrons and holes. We have also introduced the concept of quantum theory of radiation. It has been discussed that earlier experiments predicted the wave nature of radiation, and phenomena like interference and diffraction could be well established by Maxwell's theory. In contrast to the particle nature of matter, the wave nature of matter is discussed conceptually. We then move on to a discussion of electronic behavior in metallic or semiconducting systems where the Sommerfeld free electron theory is addressed followed by Bloch theory. We have also qualitatively discussed the band structure and density of states for the bulk system, followed by nanoparticles. This energy band theory explains the electrical transport phenomenon at nanoscale.

Nanotechnology starts with quantum dots (QDs), defined as nanoparticles exhibiting three-dimensional quantum confinements, which leads to the development of many unique optical and transport properties depending on their shape and size. QDs could be prepared either from metal or from semiconductors. The reduction in the number of atoms in QDs results in the confinement of normally delocalized energy states when the diameter of QDs approaches the de Broglie wavelength of electrons in the conduction band or hole in the valence band. The result is that the energy difference between energy bands is increased with decreasing particle size.

Here, it is worth mentioning that the effect of van der Waals force is very significant in nanomaterials. This is the force between two atoms with a closed electronic shell, as in the inert gases, when no overlap in their wave function is observed. Due to van der Waals force, the binding energy associated with individual atoms is quite small (0.1 eV per atom), though the binding energies for ionic and covalent bonds are 100 times greater than the van der Waal bond. The origin of this force is polarization (mutual polarization) mediated and is the result of temporary transient dipoles on molecules leading to localized charge fluctuations. In this context, the concept of mutual polarization arises due to the localization of the electron charge cloud at any instant of time around the nucleus and generates instantaneous fluctuation of a dipole moment even when atoms have a zero averaged dipole moment. This instantaneous dipole moment on an atom generates an electric field, which in turn induces a dipole moment on other atoms or molecules, thus polarizing any nearby neutral atom. The resulting polarization of the two nearest atoms gives rise to an instantaneous attractive force between these two atoms. van der Waal forces are always active between two atoms or molecules, which could be stretched up to 10 nm and possibly 100 nm in the case of two surfaces. It is pertinent to mention that in between two surfaces, interaction is proportional to $1/r^2$ where r is the separation between the two surfaces.

There is some confusion among newcomers about the difference between nanoscience and nanotechnology. To make it understandable, one can state that nanoscience deals with the arrangement of atoms and understanding their fundamental properties at the nanoscale, whereas nanotechnology is the controlling of matter at atomic scale while synthesizing a new material with different exotic properties.

Nanotechnology is already receiving attention across all branches of engineering as it is an interdisciplinary area of research. The general population isn't aware of its presence in daily life but it is emerging in medicine, energy and the environment, defense and security, and electronics and materials.

Research in this field mainly depends on two concepts: positional assembly and self-replication. Positional assembly is a technique to move molecular pieces into their proper places and maintain their position throughout the process. Molecular robots are one of the examples that carry out positional assembly. On the other hand, self-replication occurs by multiplying the positional arrangements in some habitual way. The applications of MEMS and nanotechnology are overlapping everywhere. An ideal example of this is the development by researchers at the Technical University of Munich of carbon nanotube–based small sensors that can be sprayed over the packet. These lilliput sensors detect the concentrations of volatile organic compound emitted by the product at very low concentrations. The output of the sensors is interfaced with a wireless device that alerts authorities to the infection of food and thus prevents damage.

This emerging technology is also a breakthrough in the domain of highly powerful computers and communication devices. According to Moore's law, there is a limit to the number of components that can be fabricated onto a silicon wafer. Conventionally, circuits have been made on the wafer by removing the unwanted portion of the material in the region. In view of the upcoming emerging trend of nanotechnology, scientists suggest that it is possible to build chips with a single atom to make the devices smaller than ever, which is not possible using traditional methods of etching. If this becomes possible, there will be no extra atoms, implying that each atom bears its own meaning and a particular purpose. Conductors like nanowire would be only one atom thick. It would be remarkable if a data bit could be represented by the presence or absence of a single electron.

Nanotechnology is the study of phenomena and fine-tuning of materials at atomic level, where a significantly different property is obtained compared to a larger scale. Very recently, individuals and groups have been working on different aspects of nanotechnology such as renewable energy harvesting and converting it into useful electrical energy. Thermal, nuclear, wind, hydrolytic, and solar energy scavenging have ushered in a new area of research, "nanopiezotronics," whose fundamental principle utilizes the coupled piezoelectric and semiconducting properties of nanowires and nanorods for fabricating electronic devices or systems such as field-effect transistors and diodes. The term nanopiezotronics was coined by Professor Zhong Lin Wang at Georgia Tech and is included in this book.

The physics of nanopiezotronics is based on the principle of a nanogenerator that converts mechanical energy into electric energy. When a piezoelectric material is twisted, electric charges collect on its surfaces. Further, bending the structures creates a charge separation, positive on one side and negative on the other. The output in the form of charge creation taken from the device, can be used to produce measurable electrical currents in a nanogenerator when an array of nanowires is bent and then released subsequently. As the structures that are responsible for the generation of electric current have a dimension at nanoscale, the term nanogenerator is most suitable.

The basic principle of a solar cell is the conversion of solar energy to chemical energy of electron-hole pairs followed by the conversion of chemical energy to electrical energy. Among all the heterojunction methods, solar cells have the greatest potential, highest efficiency, and greatest stability under light and thermal exposure due to the tunneling of electrons.

The use of graphene in solar cell technology has improved its efficiency tremendously. Graphene/Si heterojunction solar cells are a very recent area of research which have replaced dye-sensitized solar cells due to their high cost, and they have been included in this volume. Graphene absorbs only 2% of light and it is a very good conductor because it has only three covalent bonds per atom, compared to the full four in diamond. This makes it possible for electrons to move freely over a sheet of graphene to conduct electricity.

Like metals, this means it will absorb or reflect light because the free electrons can absorb the small amount of energy in the photon. Graphene/Si heterojunction solar cells can be assembled by transferring as-synthesized graphene films onto n-type Si.

With the advancement of technology, new industries are being formed, but the main concern is that they are polluting the environment and are hazardous to the health of every one of us. Here nanoparticles play a crucial role in minimizing pollution. For example, dye industries are leaving different azo dyes in the environment. Here the photocatalytic ability of nanoparticles is used to degrade these dyes into less harmful materials. In this context, it should be mentioned that the photocatalytic activity significantly depends on the shape and size of the nanoparticles since catalytic activity originates from surface atoms. It is a triumph of nanoscience and nontechnology that the fundamental relation between the properties of surface atoms and catalytic property is being examined under nanoscience where the search for new materials with superior activity is illustrated in the field of nanotechnology. Nanoparticles also exhibit a few environmental applications like antibacterial, anticancer, and antifungal activity.

This book is an amalgamation of nanoscale engineering, fundamental concepts, and novel nanodevices to prove where nano is these days, and what we can anticipate from it in the future. The chapters will highlight the fundamental ideas as well as ground-breaking applications of nano which will amaze the whole world. The authors have also provided images to make these concepts, the objectives, and the lab facilities more understandable for research students.

In the near future, nanotechnology will control the way we live, work, and communicate.

2

Introductory Quantum Mechanics for Nanoscience

Chandan Kumar Ghosh

CONTENTS

2.1 Introduction

Our universe consists of radiation and matter. Until the end of the nineteenth century, all physical phenomena associated with the classical macroscopic material world could be explained in terms of the position and momentum of the particle and with the aid of force acting on it according to Newton's law. Similarly, all the phenomena related to radiation could be well explained by the wave aspect of radiation according to Maxwell's electromagnetic theory. During the late nineteenth century and early twentieth centuries, a few experiments were carried out in the fields of radiation and matter and the outcomes could not be explained through the existing theories given by Newton and Maxwell. Thus, scientists were forced to introduce new theories or concepts after accumulating the experimental evidence. In this chapter, we will subsequently discuss some key experiments and concepts developed from these experiments: *particle nature of radiation, wave nature of particles, wave-particle duality of matter and radiation,* and *quantization of physical quantities such as energy.*

2.2 Quantum Theory of Radiation

Let us start with Young's double slit experiment that involves two slits and radiation incidenting on them (schematically shown in Figure 2.1). If one slit is kept open, other remains closed and electromagnetic radiation falls on the slits, then it would be diffracted (deviate from its initial propagation direction) from the slits and this diffraction could be detected by placing a

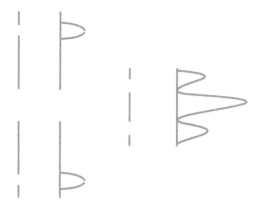

FIGURE 2.1
Young's double slit experiment.

photographic plate on the other side of the slits. When both slits are kept opened, an interference pattern is generated on the photographic plate. The phenomena of diffraction and interference are the fingerprints of the wave nature of radiation, and the interference can be explained from the interference effect of electric fields associated with two waves corresponding to radiation emerging from two slits with the help of Maxwell's theory. If the same experiment is carried out with very weak light, no interference pattern is observed on the photographic plate. Instead of an interference pattern, it produces a few spots on the photographic plate, indicating the particle nature of radiation. Therefore, these experimental results cannot be explained by the existing Maxwell's theory.

Let us consider another problem, namely, *radiation from black body* that leads to acceptance of the violation of Maxwell's law. If we have radiation incidenting on the surface of a body, a fraction of it is absorbed by the body and the rest is reflected. This fraction is generally determined by the nature of the body surface and the wavelength of the incident radiation. It has been observed that a body with a dark surface absorbs most of the fraction, while light surface reflects most. In this context, *absorption coefficient* of the body is defined at a particular wavelength to signify the fraction of the incident radiation being absorbed by the body. A *black body* is defined to be a body having absorption coefficient = 1, that is, it does not reflect any radiation. Similar to absorption, a body could also emit radiation upon heating. The spectral distribution of radiation depends on the absolute temperature (T) of the body (Stefan, 1879). In 1884, L. Boltzmann derived an expression for *total emissive power, R(T)*, which is defined as the total power emitted per unit area of a black body, and it is found to be given by $R(T)\alpha T^4$. This law is called the Stefan–Boltzmann law. Later, the spectral distribution of $R(T)$ was found to be dependent of the emitted wavelength (λ) as well, that is, emissive power is expressed as $R(T, \lambda)$. In 1899, O. Lummer first experimentally measured $R(T, \lambda)$, represented in Figure 2.2. It is evident from the figure that $R(T, \lambda)$ increases with temperature and at each temperature there exists a λ (called λ_{max}) for which $R(T, \lambda)$ exhibits maximum value. Further, an inverse relation was established by Wien between λ_{max} and T. This is known as Wien's displacement law. Instead of discussing $R(T, \lambda)$, scientists adopted the $\rho(T, \lambda)$, which is related to $R(T, \lambda)$ by the relation $\rho(T, \lambda) = 4/cR(T, \lambda)$ for further discussion and is called the *spectral distribution function* or *monochromatic energy density* [1].

Physically, these two are proportional to each other. Wien, Lord Rayleigh, and J. Jeans derived an expression for $\rho(T, \lambda)$ considering the classical law of radiation as follows.

According to classical electromagnetic theory, thermal radiation within a cavity can be considered as standing electromagnetic waves. The number of modes in the wavelength ranges between λ and $\lambda + d\lambda$ and may be shown to be $n(\lambda) = \dfrac{8\pi}{\lambda^4} d\lambda$ per unit volume of the cavity. This number is

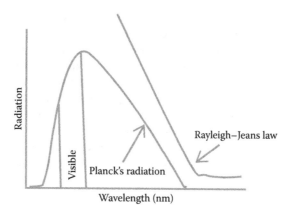

FIGURE 2.2
Planck's radiation and Rayleigh–Jeans relation.

independent of the size and shape of the cavity. If each mode of vibration has the average energy $\bar{\varepsilon}$, then $\rho(T, \lambda)$ is the product of $n(\lambda)$ and $\bar{\varepsilon}$, that is, $\rho(T, \lambda) = n(\lambda)\bar{\varepsilon}$. Since the system is at thermal equilibrium at temperature T, then according to Boltzmann's theory $\bar{\varepsilon}$ is found to be kT, where "k" represents Boltzmann's constant. Therefore, the expression of $\rho(T, \lambda)$ has been found to be $\rho(T, \lambda) = \dfrac{8\pi}{\lambda^4}kT$ by Rayleigh and Jeans. After their name, this relation is known as the Rayleigh–Jeans spectral distribution law. In the long wavelength limit ($\lambda \to \infty$), it matches experimental variation of spectral distribution, but in the short wavelength limit ($\lambda \to 0$), the Rayleigh–Jeans law diverges, and does not exhibit maxima that has been observed in the experimentally measured spectral distribution function. This behavior is known as "ultraviolet catastrophe." In 1900, Max Planck carried out a remarkable solution to this problem, postulating that the energy of the oscillating electric field, instead of having continuous frequency variation between 0 and ∞, possesses discrete frequency $n\epsilon_0$, where n is any integer and ϵ_0 represents the *quantum* of energy. The average of energy ($\bar{\varepsilon}$) is calculated to be $\bar{\varepsilon} = \dfrac{\epsilon_0}{\exp\left(\dfrac{\epsilon_0}{kT}\right) - 1}$. With this expression for $\bar{\varepsilon}$, the spectral distribution

function is found to be $\rho(T,\lambda) = \dfrac{8\pi}{\lambda^4} \dfrac{\epsilon_0}{\exp\left(\dfrac{\epsilon_0}{kT}\right) - 1}$. In this context, it may be

stated that ϵ_0 is proportional to the frequency of incident light, and hence it is inversely proportional to the wavelength of the electromagnetic radiation, that is, $\epsilon_0 = hc/\lambda$, where "h" and "c" represent Planck's constant and velocity of light, respectively. This spectral distribution, derived by Planck, is known as the *Planck spectral distribution law* for $\rho(T, \lambda)$. Expansion of the denominator of the Planck spectral distribution law reveals that in the long wavelength limit, this law reduces to the Rayleigh–Jeans distribution law. On the other hand, in the short wavelength limit ($\lambda \to 0$), $\rho(T, \lambda) \to 0$. The wavelength at which the spectral distribution exhibits maximum (λ_{max}) value can be calculated by differentiating $\rho(T, \lambda)$ with respect to λ and it gives $\lambda_{max}T = hc/4.965k$. Therefore, from the earlier discussion, it is clear that classical description of black body radiation is found to be incorrect, whereas Planck's theory predicts it correctly.

Later, the photoelectric effect that includes the ejection of electrons from a metallic surface in the presence of a high-frequency electromagnetic wave extends Planck's postulates of quantization of energy. This effect was first observed by W. Hallwachs, M. Stoletov, and P. Lenard. This experiment includes a metal plate (which acts as a cathode) in an evacuated glass tube, which liberates the electron when it is irradiated with ultraviolet light. They observed the following important features which cannot be explained on the basis of classical electromagnetic theory:

1. There exists a minimum frequency, below which no emission of electrons is observed.
2. The stopping (the anode potential at which electron flow reduced to zero) is linearly proportional to the frequency of the radiation and found to be independent of its intensity.
3. No time delay exists between irradiation and emission of electrons, and thus the instantaneous process.

In 1905, Einstein predicted a theory to explain this phenomenon. According to him, light, that is, radiation, instead of having wave nature is considered to be a collection of particles carrying quanta of energy, called *photons*. The energy of the photon (ε) is related to the frequency (ν) or λ of the radiation by the following relations: $\varepsilon = h\nu = hc/\lambda$. Photons are absorbed by an electron within the metal surface and escape from the metallic surface if ε is sufficient for electrons to cross the energy barrier, called work function W. The current resulting from photoelectric electrons, called photoelectric current, becomes proportional to the intensity of radiation. The escaping velocity (v) of electrons from the surface of metal is directly proportional to the difference of $h\nu$ and W of the metals, and thus we can write $\frac{1}{2}mv^2 = h\nu - W$, suggesting linear dependency of stopping potential on ν. If $W < h\nu$, no electrons will be ejected by the radiation, irrespective of its intensity. Absorption of photons by electrons is an instantaneous process, and thus there is no time delay.

2.3 Quantum Theory of Matter

In 1927, from the breakthrough experiment carried out by Davisson and Germer, the wave nature of material particles was first predicted. In this experiment, electrons from an electron gun were accelerated through a potential difference and then used to fall on nickel crystal. They measured the intensity of scattered electrons at various angles and noted intensity maxima near $\theta = 0°$ as well as near $\theta = 35°$. The maxima near $\theta = 0°$ is explained from particles propagating in the forward direction, whereas the maxima near $\theta = 35°$ could only be explained in terms of constructive interference of the electronic waves, scattered by the regular lattice indicating wave-like properties of material particles. Here, de Broglie postulated that the λ of the particles is related to momentum (p) by $\lambda = h/p$. Hence, it may be stated that microscopic particles may be considered as quantized bundles of energy having wave-like properties (characterized by ω and λ) and may possess particle-like properties (characterized by E and p). For example, particle-like properties of the electron are reflected by its charge and mass, whereas reflection and diffraction phenomena predict the wave-like property of the electron. Hence, it can be stated that microscopic particles exhibit *wave–particle duality*. In fact, macroscopic particles also possess a wave-like nature, but due to their higher mass, λ associated with them can't be measured, and thus the wave properties of the macroscopic particle are not detected. The other way of modeling the wave property of particles is to introduce the concept of *wavepacket*, which has properties of propagation as well as localization. For example, from a long distance, wavepacket looks like a particle moving in one direction, whereas a very close look at the wavepacket shows its spreading, that is, wave character. Therefore, both wave- and particle-like properties, that is, wave–particle duality of material particles, can be attributed to wavepacket.

But the real problem of microscopic particles starts somewhere else. It is well known that the state of any classical particle is determined by two parameters, namely, the position and momentum of the particle. So to predict the state of the particle, these two parameters must be known precisely. The future of the particles, that is, their kinematics, is determined from Newton's law $\mathbf{F} = m\dfrac{d^2 r}{dt^2}$, where $\mathbf{F}$ is the applied force (electrical or mechanical), and from a set of initial conditions. But simultaneously precise measurement of position and momentum is not possible for microscopic particles. For example, if we consider a flow of microscopic particles along x-axis and an aperture (along y-axis) normally kept at $y = 0$ with aperture width d (similar to the double slit experiment, Figure 2.1), then before entering into the aperture particles possess zero momentum along the y-axis ($p_y = 0$), that is, we have precise knowledge of their momentum, but their y-component of position is completely unknown. When the particles cross the aperture, they suffer collision at the edges of the apertures (they are diffracted), and thus their y-component of the momentum is uncertain. From the wave theory it can be calculated that the width of the diffraction pattern is given by $d \sin \theta_{max} = \lambda$, where θ and λ represent the maximum angle of diffraction and the wavelength of the incidenting electrons, that is, electrons can be detected anywhere between angle $\theta = 0$ and $\theta_{max} = \sin^{-1}\left(\dfrac{\lambda}{d}\right)$. During collision, electrons gain momentum of $p = p \sin \theta$ along the y-axis, that is, electrons detected at $\theta = 0$ have $p_y = 0$ and electrons detected at θ_{max} have $p_y = p \sin \theta$. Therefore, the y-component of the electrons is uncertain by $\Delta p_y = p \sin \theta_{max}$ and the y-component of the position is uncertain by $\Delta y = d$. Therefore, the product of $\Delta y \cdot \Delta p_y$ can be calculated to be $\Delta y \cdot \Delta p_y = dp \sin \theta_{max} = \lambda p = h$. This relation is known as Heisenberg's uncertainty relation and it indicates that the momentum and position can't be measured simultaneously without any uncertainty. Thus, it is evident from Davisson and Germer's experiment and the uncertainty relation that we can't apply Newton's law to microscopic particles and the situation demands development of new mechanics for microscopic particles consistent with the uncertainty relation and wave nature of matter [2].

2.3.1 Development of Schrödinger's Wave Mechanics

The new theoretical structure, commonly known as quantum mechanics, was processed in the years between 1925 and 1930. There are two phases in quantum mechanics formulation. The first phase was matrix mechanics that was developed in 1925 and 1926 by W. Heisenberg, M. Born, and P. Jordon. According to them, a matrix is associated with each physically observable quantity that obeys noncommutative algebra, significantly different from classical mechanics.

The second phase was wave mechanics, which was developed by E. Schrödinger, inspired by de Broglie's hypothesis about matter waves. In 1926, the equivalence between matrix mechanics and wave mechanics was established by Schrödinger for microscopic particles on the basis of the following postulates:

- For every microscopic system, there exists a state function, $\psi(x, t)$, commonly known as wave function, that contains all the information regarding the system. Instead of position and momentum, $\psi(x, t)$ is sufficient to characterize a microscopic system.
- For every physically observable quantity, there exists a linear operator that extracts observable quantity from $\psi(x, t)$ and it should be real quantity.

$\psi(x, t)$, as suggested by Max Born, manifests the probabilistic space-time behavior of a quantum system as the probability of finding quantum particles at a particular point can be determined by $|\psi(x, t)|^2$. Very briefly, it can be stated that "$P(x)$," representing the probability of finding a quantum particle between x and $x + dx$, is defined as $P(x)\, dx = |\psi(x,t)|^2\, dx$, and therefore $\int_{-\infty}^{+\infty} |\psi(x,t)|^2 \, dx$ represents the probability of finding the microscopic particle in the entire region of space, and it should be equal to one, that is, $\int_{-\infty}^{+\infty} |\psi(x,t)|^2 \, dx = 1$. This is known as the normalization process and $\psi(x,t)$ is said to be normalized wave function.

The fundamental characteristics of $\psi(x,t)$ are as follows:

- $\psi(x,t)$ must be linear so that it can follow the superposition principle, that is, if $\psi_1(x,t)$ and $\psi_2(x,t)$ are two solutions for the wave equation, then linear combination $(C_1\psi_1 + C_2\psi_2)$ will also be the solution for the wave equation.

- The wave equation comprising $\psi(x,t)$ should be consistent with de Broglie's hypothesis and Einstein's relation.

- There should be a differential equation for $\psi(x,t)$ that illustrates futuristic prediction of $\psi(x,t)$ from its initial value $\psi(x,t_0)$.

The differential equation suggested by Schrödinger for a microscopic particle moving in a potential field of $V(x,t)$ is

$$-\frac{\hbar^2}{2m}\frac{d^2\psi(x,t)}{dx^2} + V(x,t)\psi(x,t) = i\hbar\frac{d\psi(x,t)}{dt}$$

The above relation can be derived in the following manner: from de Broglie's relation, it can be written that $p = h/\lambda = 2\Pi\hbar/\lambda = \hbar k$, where $\hbar$ and k represent the reduced Planck's constant and wave number, respectively. The energy of the particle is related to the frequency (ν) of the associated wave by Einstein's relation, $E = h\nu = \hbar\omega$, where ω represents the angular frequency. For microscopic particles, moving in free space, wave function can be written in the form of plane wave, that is, $\psi(x,t) = A \exp[i(kx - \omega t)]$. Then differentiating $\psi(x,t)$ with respect to "t," it is obtained that $i\hbar\frac{\partial\psi}{\partial t} = E\psi(x,t)$, and the derivative with respect to "x" gives $-i\hbar\frac{\partial\psi}{\partial x} = p\psi(x,t)$. For free particles, it may be written that $\frac{p^2}{2m} = E$, that is, in this case it may written that $-\frac{\hbar^2}{2m}\frac{\partial^2\psi}{\partial x^2} = i\hbar\frac{\partial\psi}{\partial t}$, signifying Schrödinger's equation for free particles in one dimension.

If the particle is influenced by a force field $V(x,t)$, then total energy E, according to classical physics, becomes $E = \frac{p^2}{2m} + V(x,t)$, giving Schrödinger's equation. For three-dimensional cases the above equation becomes

$$i\hbar\frac{\partial\psi(x,t)}{\partial t} = -\frac{\hbar^2}{2m}\nabla^2\psi(x,t) + V\psi(x,t) = \left[-\frac{\hbar^2}{2m}\nabla^2 + V\right]\psi(x,t)$$

where $\nabla^2 = \dfrac{\partial^2}{\partial x^2} + \dfrac{\partial^2}{\partial y^2} + \dfrac{\partial^2}{\partial z^2}$. The operator "$-\dfrac{\hbar^2}{2m}\nabla^2 + V$" is known as the Hamiltonian operator and is denoted by H. The above equation can also be written as $i\hbar\dfrac{\partial\psi}{\partial t} = H\psi$.

2.3.2 Time-Independent Schrödinger's Equation: One-Dimensional Consideration

The time-independent Schrödinger's equation is a special case of a time-dependent wave equation, that is, when Hamiltonian operator H of any quantum system becomes independent of time. In this case, the potential energy $V(x, t)$ is also independent of time. The deduction of the time-independent from the generalized time-dependent Schrödinger's equation can be carried out as follows.

For this purpose the wave function $\psi(x, t)$ is expressed as $\psi(x, t) = \Psi(x)T(t)$. Then, putting it into the time-dependent Schrödinger's equation, we have

$$i\hbar\Psi(x)\frac{dT(t)}{dt} = -\frac{\hbar^2}{2m}T(t)\frac{d^2\Psi(x)}{dx^2} + V(x)\Psi(x)T(t)$$

or

$$\frac{1}{T(t)}i\hbar\frac{d}{dt}\big(T(t)\big) = \frac{1}{\Psi(x)}\left[-\frac{\hbar^2}{2m}\frac{d^2\Psi(x)}{dx^2} + V(x)\Psi(x)\right]$$

The left- and right-hand sides of the above equation have different variables and therefore they must be equal to some constant. Let the constant be G. Thus, we have

$$\frac{1}{T(t)}i\hbar\frac{d}{dt}\big(T(t)\big) = G$$

and

$$\frac{1}{\Psi(x)}\left[-\frac{\hbar^2}{2m}\frac{d^2\Psi(x)}{dx^2} + V(x)\Psi(x)\right] = G$$

The solution for the first equation can be written as $T(t) = \exp(-iGt)/\hbar$. From dimensionality analyses, it is obtained that G must have the dimension of energy, that is, $G = E$.

The other equation becomes

$$-\frac{\hbar^2}{2m}\frac{d^2\Psi(x)}{dx^2}+V(x)\Psi(x)=E\Psi(x)$$

and in terms of H, it turns into $H\Psi(x)=E\Psi(x)$. Therefore, the significance of "H" is that it represents the total energy operator of the microscopic system.

So, for the time-independent Schrödinger's equation, $|\psi(x,t)|^2=|\Psi(x)|^2$, that is, the wave function becomes completely independent of time, and thus the solution for the time-independent Schrödinger's equation gives stationary states.

Properties of $\Psi(x)$ are as follows:

1. $\Psi(x)$ and $\dfrac{d\Psi}{dx}$ must be a single-valued function.

2. $\Psi(x)$ and $\dfrac{d\Psi}{dx}$ should be continuous.

3. $\Psi(x)$ should be finite.

2.3.2.1 Solution of Schrödinger's Equation: One-Dimensional System—Particle in an Infinitely Deep Potential Well

In this section we will discuss the solution for the time-independent Schrödinger's equation in various one-dimensional systems like the infinitely deep quantum well. If we consider a quantum particle entrapped in an infinitely deep potential well, then the distribution of potential energy [$V(x)$] is given by (shown in Figure 2.3)

$$V(x)=0 \quad \text{for } 0\leq x\leq a=\infty \text{ elsewhere}$$

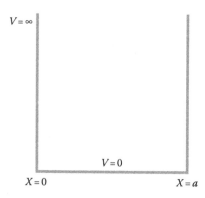

FIGURE 2.3
Schematic diagram of infinite square well potential.

Since potential energy is infinite outside the well, the probability of finding any microscopic particle outside the potential region would be zero. Therefore, a feasible solution for Schrödinger's equation is valid in the region $0 \leq x \leq a$ and it is given by

$$\frac{d^2 \Psi(x)}{dx^2} + \frac{2mE}{\hbar^2} \Psi(x) = 0$$

The general solution for the above equation is

$$\Psi(x) = A \sin kx + B \cos kx, \quad \text{where } k^2 = \frac{2mE}{\hbar^2}$$

Now using the boundary conditions that $\Psi(x)$ should equal zero at $x = 0$ and $x = a$, we have $B = 0$ and $\sin ka = 0$ gives $ka = n\pi$ (n is any integer). Therefore, it may be written that $\sqrt{\frac{2mE}{\hbar^2}} a = n\pi$, that is, $E = \frac{\pi^2 n^2 \hbar^2}{2ma^2}$, $n = 1, 2, 3$. From the expression of quantized energy, it is noted that the energy difference between two successive states is inversely proportional to the mass of the system, that is, for microscopic particles the difference is so insignificant that it can't be measured, and hence they are continuous for macroscopic particles. But for microscopic particles, the difference is significant enough to be measured due to their lighter mass. The value of "A," calculated from the normalization condition, that is, $\int_0^a |\Psi_n|^2 dx = 1$, is found to be $A = \sqrt{\frac{2}{a}}$. Thus, $\Psi_n(x) = \sqrt{\frac{2}{a}} \sin\left(\frac{n\pi}{a}\right) x$ represents the complete set of the wave function corresponding to any microscopic particle, confined within such infinite square well potential, that is, any microscopic particle must have this form of wave function or any linear combination. $\Psi_n(x)$ for different values of n is plotted in Figure 2.4 and the time-dependent form of the wave function is given by

$$\Psi_n(x, t) = \left[\sqrt{\frac{2}{a}} \sin\left(\frac{n\pi}{a}\right) x \right] \exp(-iE_n t)/\hbar.$$

2.3.2.2 Solution of Schrödinger's Equation: One-Dimensional System—Particle in One-Dimensional Quantum Well of Finite Depth

The variation of the potential corresponding to the dimensional potential well of finite depth is characterized as shown (schematically shown in Figure 2.5):

$$V(x) = 0 \quad \text{for } -\frac{a}{2} < x < \frac{a}{2}$$

$$= V_0 \quad \text{for } |x| > \frac{a}{2}$$

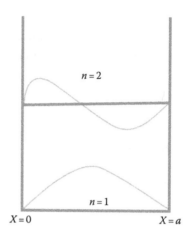

FIGURE 2.4

$\Psi_n(x)$ is plotted as a function of x for different values of n.

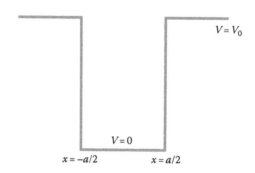

FIGURE 2.5

Potential well having finite depth of V_0.

In general, two different solutions for Schrödinger's equation are possible: (1) $E > V_0$ and (2) $E < V_0$. The first condition ($E > V_0$) gives no entrapment of the microscopic particles, and hence we will not discuss this solution and will consider the second case ($E < V_0$) to solve Schrödinger's time-independent equation. Thus, Schrödinger's equation can be written in the following form:

$$\frac{d^2\Psi(x)}{dx^2} + \frac{2mE}{\hbar^2}\Psi(x) = 0 \quad \text{for} \quad -\frac{a}{2} < x < \frac{a}{2}$$

The solution $(-a/2 < x < a/2)$ is obtained in the form

$$\Psi_1(x) = A\sin kx + B\cos kx, \quad \text{where } k^2 = \frac{2mE}{\hbar^2}$$

Schrödinger's equation in the region $|x| > \dfrac{a}{2}$ can be written in the following form:

$$\frac{d^2\Psi(x)}{dx^2} - t^2\Psi(x) = 0, \quad \text{where } t^2 = \frac{2m(V_0 - E)}{\hbar^2}$$

and its solution is given by

$$\Psi_{\mathrm{II}}(x) = Ce^{-tx} + De^{tx} \left(|x| > \frac{a}{2} \right)$$

As $x \to \infty$, $\Psi(x)$ becomes infinite, that is, giving no feasible solution, D must be set equal to 0 and the wave function can be written in the form $\Psi_{\mathrm{II}}(x) = Ce^{-tx}$. For simplicity, even wave function is considered only, that is, $\Psi_{\mathrm{I}}(x) = B\cos kx$. Applying the boundary conditions that the wave function and its first derivative should be continuous at boundaries, we have $B\cos ka/2 = Ce^{-ta/2}$ and $-Bk\sin ka/2 = -Cte^{-ta/2}$. From these two relations it may be derived that $k\tan ka/2 = t$ or it can also be written as $ka\tan ka/2 = ta$ after multiplying both sides by a. Consideration of an odd solution gives $ka\cot ka/2 = -ta$. If we consider $ka/2 = \gamma = \sqrt{\dfrac{2ma^2 E}{4\hbar^2}}$ and $\mu = \sqrt{\dfrac{2ma^2 V_0}{4\hbar^2}}$, then the above relations can be reduced to $\gamma\tan\gamma = \sqrt{\mu^2 - \gamma^2}$ and $-\gamma\cot\gamma = \sqrt{\mu^2 - \gamma^2}$, respectively. From the above relations, it can also be deduced that $\sqrt{\mu^2 - \gamma^2} = ta/2$. Here we define $\dfrac{ta}{2} = \eta$, and then in terms of η, the above relations can be further simplified into $\eta^2 + \gamma^2 = \mu^2$. The energy levels in the potential well having finite depth are determined by the points of intersection of the circle $\eta^2 + \gamma^2 = \mu^2$ having radius μ with the curves $\gamma\tan\gamma = \sqrt{\mu^2 - \gamma^2}$ (for even states) and $-\gamma\cot\gamma = \sqrt{\mu^2 - \gamma^2}$ (for odd states) as a function of γ (shown in Figure 2.6). The condition for obtaining bounded even states is $(N_E - 1)\pi < \gamma < N_E\pi$, where N_E is the number of even states. Odd states are found if $\left(N_O - \dfrac{1}{2}\right)\pi < \gamma < \left(N_O + \dfrac{1}{2}\right)\pi$, where N_O is the number of odd states. Hence, the total bound state consists of alternating even and odd states, and their number strongly depends on γ, which is termed the strength parameter. Ground state is always found to be even state.

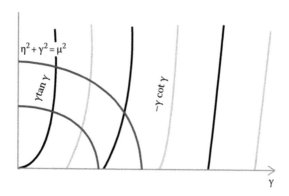

FIGURE 2.6
The variation $\gamma \tan \gamma$ (solid black line) and $-\gamma \cot \gamma$ (solid orange line) as a function of strength parameter γ. The circle $\eta^2 + \gamma^2 = \mu^2$ intersects with these lines and the points of intersections are even and odd states of the finite potential well.

2.3.3 Operators: Expectation Value

So far, we have solved Schrödinger's equation to get the possible $\psi(x, t)$. The next part of quantum mechanics is given by the second postulate of quantum mechanics as it states that we require an operator to extract information about the physically measurable quantity from $\psi(x, t)$ containing all the information about the microscopic particle. First, the probabilistic interpretation of $\psi(x, t)$ is that $|\psi(x, t)|^2$, that is, $\psi(x, t)^*\psi(x, t)$ gives the probability of finding any microscopic particle between x and $x + dx$. Therefore, the average (*expectation*) value of finding a microscopic particle can be written as

$$\langle x \rangle = \int x |\psi(x, t)|^2 \, dx = \int \psi(x, t)^* \times \psi(x, t) dx$$

The time-independent wave function of any free quantum particle can be written as $\psi(x) = e^{ikx}$. It is consistent with the uncertainty relation in the way that the particle has single wave vector k, that is, momentum $p = \hbar k$. Therefore, from the uncertainty relation its position should be completely uncertain. In fact, $|\psi(x, t)|^2$ is equal to one at every position in space, that is, the position of the particle can be measured anywhere in space with equal probability. So to extract momentum information from $\psi(x)$, we need an operator for momentum. The simplest mathematical tool to extract momentum information from $\psi(x)$ should be proportional to the differentiation with respect to x and the proportionality constant should be $-i\hbar$. Therefore, the operator for momentum (p_{op}) is assigned to be $-i\hbar \dfrac{d}{dx}$. Thus, we can write

$$p_{op}\psi(x, t) = -i\hbar \frac{de^{ikx}}{dx} = \hbar k \psi(x, t)$$

The important point to note here is that in the case of a free particle, when we have applied p_{op} on its wave function, we have obtained $\hbar k \psi(x,t)$, that is, it can be stated that without disturbing the state of the system, p_{op} extracts the momentum information from it. Such a case where the operator extracts information without disturbing the wave function is known as the *eigenfunction* of the operator and the value is known as the *eigenvalue*. It happens only when the wave function contains unique information corresponding to the measurable quantity. For any arbitrary wave function which is not the *eigenfunction* of p_{op}, the average momentum can be calculated as follows:

$$\langle p \rangle = \int \psi(x,t)^* -i\hbar \frac{d}{dx} \psi(x,t) dx = -i\hbar \int \psi(x,t)^* \frac{d\psi(x,t)}{dx} dx$$

More generally for the expectation value of any arbitrary function, $f(x, t)$ is given by

$$\langle f(x,t) \rangle = \int \psi(x,t)^* f(x,t)_{op} \psi(x,t) dx$$

Another important point to discuss here is that the operators, defined for microscopic systems, don't commute with each other, that is, if A and B are any two operators then AB is not necessarily equal to BA. The *commutators* are generally defined by their difference, that is, $AB - BA$ and are denoted by $[A, B]$. As an example, we consider position and momentum operator. For any $\psi(x)$, the commutation between position and momentum operator can be calculated in the following way:

$$[x, p_{op}]\psi(x) = (xp_{op} - p_{op}x)\psi(x) = -i\hbar \left[x\frac{d\psi(x)}{dx} - \frac{d\{x\psi(x)\}}{dx} \right] = -i\hbar\psi(x)$$

It may be generalized from the above result that any two operators corresponding to a canonically physically measurable quantity don't commute. These two quantities can't be measured simultaneously with arbitrary accuracy, and this is a consequence of Heisenberg's uncertainty relation.

2.3.4 Important Consequences of the Wave Functions

So far, we have discussed the methods to obtain wave function characterizing microscopic systems as well as the process to extract a physical measurable quantity from it. In addition, here we'll discuss a few important features of the wave functions.

2.3.4.1 Two Wave Functions, Obtained from Solution for Schrödinger's Equation, are Independent of Each Other

Let $\psi(x)_1$ and $\psi(x)_2$ be the two solutions for Schrödinger's equation, and thus we can write

$$-\frac{\hbar^2}{2m}\frac{d^2\psi(x)_1}{dx^2}+V(x)\psi(x)_1 = E\psi(x)_1$$

$$\text{and} \quad -\frac{\hbar^2}{2m}\frac{d^2\psi(x)_2}{dx^2}+V(x)\psi(x)_2 = E\psi(x)_2$$

If we multiply the first equation by ψ_2 and the second one by ψ_1, after rearrangement we may obtain $\dfrac{\dfrac{d^2\psi(x)_1}{dx^2}}{\psi(x)_1} = \dfrac{\dfrac{d^2\psi(x)_2}{dx^2}}{\psi(x)_2} = \dfrac{2m}{\hbar^2}(V-E).$

Integrating this, we have $\psi(x)_1\dfrac{d\psi(x)_2}{dx}-\psi(x)_2\dfrac{d\psi(x)_1}{dx} = \text{constant}.$

This is called the *Wronskian* of $\psi(x)_1$ and $\psi(x)_2$ and it is found to be non-zero for any wave functions; hence, these two wave functions are mutually independent.

2.3.4.2 Eigenfunctions are Mutually Orthogonal and Corresponding Eigenvalues are Real

We consider the solutions of Schrödinger's equation for two different energy values (E' and E''). Let these wave functions be $\psi_{E'}$ and $\psi_{E''}$. Thus, both of them should satisfy Schrödinger's equation, that is, $H\psi_{E'} = E'\psi_{E'}$ and $H\psi_{E''} = E''\psi_{E''}$. Taking a complex conjugate of $H\psi_{E'} = E'\psi_{E'}$, we have $\left(H\psi_{E'}\right)^* = E'^*\psi_{E'}^*$. We multiply $H\psi_{E'} = E'\psi_{E'}$, by $\psi_{E''}^*$ and $H\psi_{E''} = E''\psi_{E''}$ by $\psi_{E'}^*$. Therefore, we have $\psi_{E''}^* \cdot H\psi_{E'} = \psi_{E''}^* \cdot E'\psi_{E'} = E'^*\psi_{E''}^*\psi_{E'}$ and $\psi_{E'}^* \cdot H\psi_{E''} = \psi_{E'}^* \cdot E''\psi_{E''} = E''\psi_{E'}^*\psi_{E''}$. Now integrating the over space, we have $(E'' - E'^*)\int\psi_{E'}^*\psi_{E''}dx = \int\psi_{E''}^* \cdot H\psi_{E'}dx - \int\psi_{E'}^* \cdot H\psi_{E''}dx = 0$; therefore, in this equation if we consider $E'' \neq E'^*$, then $\int\psi_{E'}^*\psi_{E''}dx = 0$, signifying that the wave functions don't overlap, that is, they are said to be orthogonal. On the other hand, if we have $E'' = E'^*$, it signifies that energy levels exhibit real value.

2.3.5 One-Dimensional Quantum Barrier-Transmission and Reflection Coefficient

In this section we'll discuss the system of moving microscopic particles. The interpretation of $\psi(x, t)$ is that $P(x, t) = |\psi(x, t)|^2 dx$ represents the probability of finding the particle between x and $x + dx$. Now we are considering moving particles, and thus for such particles $P(x, t)$ is no longer time-independent but rather dependent. Now, differentiating $P(x, t)$ with respect to time we have

$$\frac{\partial P(x, t)}{\partial t} = \frac{\partial \psi(x, t)^*}{\partial t} \psi(x, t) + \psi(x, t)^* \frac{\partial \psi(x, t)}{\partial t}$$

Substituting $\dfrac{\partial \psi(x, t)^*}{\partial t}$ and $\dfrac{\partial \psi(x, t)}{\partial t}$ from the time-dependent Schrödinger's equation, we have

$$\frac{\partial P(x, t)}{\partial t} = \frac{i}{\hbar}\left[-\frac{\hbar^2}{2m}\frac{d^2\psi(x, t)^*}{dx^2} + V(x)\psi(x, t)^* \right]\psi(x, t)$$

$$+ \psi(x, t)^* \frac{-i}{\hbar}\left[-\frac{\hbar^2}{2m}\frac{d^2\psi(x, t)}{dx^2} + V(x)\psi(x, t) \right]$$

$$= \frac{i}{\hbar}\left[-\frac{\hbar^2}{2m}\frac{d^2\psi(x, t)}{dx^2}\psi(x, t) + V(x)\psi(x, t)^* \psi(x, t) \right.$$

$$\left. + \frac{\hbar^2}{2m}\psi(x, t)^* \frac{d^2\psi(x, t)}{dx^2} - V(x)\psi(x, t)^* \psi(x, t) \right]$$

$$= \frac{i}{\hbar}\left[-\frac{\hbar^2}{2m}\frac{d^2\psi(x, t)^*}{dx^2}\psi(x, t) + \frac{\hbar^2}{2m}\psi(x, t)^* \frac{d^2\psi(x, t)}{dx^2} \right]$$

$$= -\frac{i\hbar}{2m}\left[\frac{d}{dx}\left(\frac{d\psi(x, t)^*}{dx}\psi(x, t) \right) - \frac{d\psi(x, t)^*}{dx}\frac{d\psi(x, t)}{dx} \right.$$

$$\left. - \frac{d}{dx}\left(\psi(x, t)^* \frac{d\psi(x, t)}{dx} \right) + \frac{d\psi(x, t)^*}{dx}\frac{d\psi(x, t)}{dx} \right]$$

$$= -\frac{i\hbar}{2m}\frac{d}{dx}\left[\frac{d\psi(x, t)^*}{dx}\psi(x, t) - \psi(x, t)^* \frac{d\psi(x, t)}{dx} \right]$$

$$= -\frac{d}{dx}\left[\frac{i\hbar}{2m}\left(\frac{d\psi(x, t)^*}{dx}\psi(x, t) - \psi(x, t)^* \frac{d\psi(x, t)}{dx} \right) \right]$$

$$= -\frac{dj}{dx}, \quad \text{where } j = \frac{i\hbar}{2m}\left(\frac{d\psi(x, t)^*}{dx}\psi(x, t) - \psi(x, t)^* \frac{d\psi(x, t)}{dx} \right)$$

Therefore, it may be written as $\dfrac{\partial P(x, t)}{\partial t} + \dfrac{dj}{dx} = 0$

This equation is identical to the continuity equation of current electricity, signifying conservation of charge. Instead of flow of electron, here we have considered flow of microscopic particles. Therefore, j, defined in this differential equation, is assigned to be *probability current density* characterizing any moving quantum particles, and the differential equation signifies the *conservation rule of microscopic particles*. It should be noted here that j can be calculated from $\psi(x,t)$. In the following subsections, we'll discuss j and its relation to moving particles.

2.3.5.1 Reflection and Transmission of Quantum Particles from One-Dimensional Potential Step

We consider a one-dimensional potential step, defined as (represented in Figure 2.7)

$$V(x) = 0 \quad \text{for } x < 0 \ (\text{region I})$$

$$= V_0 \quad \text{for } x > 0 \ (\text{region II})$$

and we have flow of microscopic particles moving in the positive x-direction with energy $E > V_0$.

Then Schrödinger's equation can be written in the following form in region I:

$$\frac{d^2 \Psi(x)}{dx^2} + k^2 \Psi(x) = 0 \quad \text{where } k^2 = \frac{2mE}{\hbar^2}$$

And its solution is given by

$$\Psi(x) = Ae^{ikx} + Be^{-ikx}$$

The wave function can be interpretrated as follows: the first term represents the wave function of the microscopic particles propagating in the positive x-direction, whereas the second term is associated with quantum particles moving in the negative x-direction. The origin of this reflection is the step

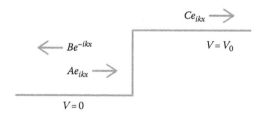

FIGURE 2.7
One-dimensional potential step having finite height V_0.

potential (V_0) at $x = 0$. Similarly, for region II, we have the wave function:

$\Psi(x) = Ce^{i\beta x} + De^{-i\beta x}$, where $\beta^2 = \dfrac{2m(E - V_0)}{\hbar^2}$.

The difference lies in the fact that there is no source of the reflected wave of quantum particles in region II, so in this region wave function is solely described by $\Psi(x) = Ce^{i\beta x}$.

At $x = 0$, wave functions and their first derivative are equal (boundary condition), that is,

$$\Psi(x = 0)_{\text{region I}} = \Psi(x = 0)_{\text{region II}} \quad \text{and} \quad \left.\frac{d\Psi(x = 0)}{dx}\right|_{\text{region I}} = \left.\frac{d\Psi(x = 0)}{dx}\right|_{\text{region II}}$$

Thus, we have $A + B = C$ and $ik(A - B) = i\beta C$. On simplifying, we obtain $B = \dfrac{k - \beta}{k + \beta} A$ and $C = \dfrac{2k}{k + \beta} A$.

As we are dealing with moving particles, they would be characterized by probability current density (j) that can be calculated using wave functions. It is calculated to be $j_{in} = \dfrac{\hbar k}{m}|A|^2$, $j_{re} = \dfrac{\hbar k}{m}|B|^2$, and $j_{tr} = \dfrac{\hbar \beta}{m}|C|^2$ for incident, reflected, and transmitted wave, respectively. In this context, transmission (T) and reflection (R) coefficients are defined by the ratio $T = \dfrac{j_{tr}}{j_{in}}$ and $R = \dfrac{j_{re}}{j_{in}}$ and using their expressions $T + R$ is found to be 1. The striking result is that although the particles have energy greater than the potential step height, not all of them are transmitted to region II, and few particles are reflected from the step, which is in contradiction with classical particles. The difference from the macroscopic system is that macroscopic particles would not be reflected from the potential step if their energy is higher than the potential step height.

A different situation will appear for incident particles having energy $E < V_0$. In this situation,

Schrödinger's equation becomes (in region I)

$$\frac{d^2\Psi(x)}{dx^2} + k^2\Psi(x) = 0, \quad \text{where } k = \sqrt{\frac{2mE}{\hbar^2}}$$

And the solution is given by $\Psi(x) = Ae^{ikx} + Be^{-ikx}$.

In region II, Schrödinger's equation is given by

$$\frac{d^2\Psi(x)}{dx^2} - \beta^2\Psi(x) = 0, \quad \text{where } \beta = i\sqrt{\frac{2m(V_0 - E)}{\hbar^2}}$$

β is now an imaginary quantity and the solution is given by $\Psi(x) = Ce^{-\beta x} + De^{\beta x}$.

As $x \rightarrow \infty$, $\Psi(x)$ would diverge due to divergence of $De^{\beta x}$ and would not give finite probability, so D is considered to be zero. Therefore, in region II, wave function is determined solely by $\Psi(x) = Ce^{-\beta x}$. Using the boundary conditions at $x = 0$, we have $A + B = C$ and $ik(A - B) = -\beta C$. After simplifying, one may obtain $A = \left(1 + \dfrac{ik}{\beta}\right)\dfrac{D}{2}$ and $B = \left(1 - \dfrac{ik}{\beta}\right)\dfrac{D}{2}$. The corresponding probability current densities are found to be $j_{in} = \dfrac{\hbar k}{m}|A|^2$, $j_{re} = \dfrac{\hbar k}{m}|B|^2$ and $j_{tr} = 0$. Using these expressions, it appears that $R = 1$ and $T = 0$. The significance of this result is that all the particles are reflected from the potential step, that is, there is no propagation of quantum particles in region II. The difference from classical mechanics lies in the fact that the classical macroscopic particles are reflected from $x = 0$ point, whereas microscopic particles can penetrate into region II and are then reflected back from any point in region II, that is, $x = 0$ is not the turning point for quantum particles.

2.3.5.2 Reflection and Transmission of Quantum Particles from Rectangular Potential Barrier

This type of quantum system has practical applicability in tunneling phenomena where there is a finite probability of tunneling through the barrier, whose potential (V_0) is sufficiently larger than the total energy (E) of a particle ($E < V_0$). A rectangular potential barrier is characterized by (potential height is schematically shown in Figure 2.8)

$$V(x) = 0 \quad x < 0$$
$$= V_0 \quad 0 \leq x \leq a$$
$$= 0 \quad x > a$$

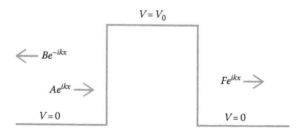

FIGURE 2.8
Rectangular quantum barrier having width "*a*" and height V_0.

Schrödinger's equation is written in the following form for regions $x < 0$ and $x > a$:

$$\frac{d^2\Psi(x)}{dx^2} + k^2\Psi(x) = 0, \quad \text{where } k = \sqrt{\frac{2mE}{\hbar^2}}$$

And the solutions are given by

$$\Psi_{\text{I}}(x) = Ae^{ikx} + Be^{-ikx} \quad \text{for } x < 0 \text{ (region I)}$$
$$\Psi_{\text{III}}(x) = Fe^{ikx} + Ge^{-ikx} \quad \text{for } x > a \text{ (region III)}$$

With the region $0 \le x \le a$ (region II), Schrödinger's equation is given by

$$\frac{d^2\Psi(x)}{dx^2} - \beta^2\Psi(x) = 0, \quad \text{where } \beta = i\sqrt{\frac{2m(V_0 - E)}{\hbar^2}} \quad \text{for region II.}$$

And the corresponding solution is

$$\Psi_{\text{II}}(x) = Ce^{\beta x} + De^{-\beta x} \quad \text{for region II.}$$

Now applying boundary conditions that the wave functions and their derivatives are continuous at $x = 0$, we have $A + B = C + D$ and $\frac{ik}{\beta}(A - B) = (C - D)$,

and after simplifying we have $C = \frac{1}{2}\left(1 + \frac{ik}{\beta}\right)A + \frac{1}{2}\left(1 - \frac{ik}{\beta}\right)B$ and

$D = \frac{1}{2}\left(1 - \frac{ik}{\beta}\right)A + \frac{1}{2}\left(1 + \frac{ik}{\beta}\right)B$. Similarly, boundary conditions at $x = a$ give $Ce^{\beta a} + De^{-\beta a} = Fe^{ika}$ and $\beta Ce^{\beta a} - \beta De^{-\beta a} = ikFe^{ika}$. Here Ge^{-ikx} is discarded as it vanishes as x tends to infinity. After substituting for C and D we have $\frac{B}{A} = \frac{(k^2 + \beta^2)(1 - e^{2\beta a})}{(k - i\beta)^2 - (k + i\beta)^2 e^{2\beta a}}$ and $\frac{F}{A} = \frac{(-i)4k\beta e^{-ika}e^{\beta a}}{(k - i\beta)^2 - (k + i\beta)^2 e^{2\beta a}}$. The reflection and transmission coefficients are calculated to be $R = \left[1 + \frac{4\delta(1 - \delta)}{\sinh^2 \beta a}\right]^{-1}$ and

$T = \left[1 + \frac{\sinh^2 \beta a}{4\delta(1 - \delta)}\right]^{-1}$, where $\delta = \frac{E}{V_0}$. Again it may be verified that $T + R = 1$

is fulfilled. In this context, we define the parameter $\beta a = \left(\frac{2m}{\hbar^2}V_0 a^2\right)^{\frac{1}{2}}(1 - \delta)^{\frac{1}{2}}$,

where $\left(\frac{2m}{\hbar^2}V_0 a^2\right)^{\frac{1}{2}}$ is called "opacity" of the barrier. If $\beta a \gg 1$, that is, opacity is very high, then the transmittance (T) vanishes according to classical mechanics, but for quantum particles $T \approx 16\delta(1-\delta)e^{-2\beta a}$, as $\sinh(\beta a) \cong 2^{-1} e^{\beta a}$. This expression implies that there is very low transmittance inside the barrier. This phenomenon has an interesting application in scanning tunneling microscopy.

The reflection (R) and transmission (T) coefficients are calculated to be

$$\left[1 + \frac{4\delta(1-\delta)}{\sin h^2 \beta a}\right]^{-1} \text{ and } \left[1 + \frac{\sin h^2 \beta a}{4\delta(1-\delta)}\right]^{-1} \text{ respectively, where } \delta = \frac{E}{V_0}. \text{ Again } T + R = 1$$

holds in this situation. The transmittance (T) is less than unity, which is in strong contradiction with classical physics where there is always significant transmission when E is much greater than V_0. $T = 1$ only for $\tau a = \pi, 2\pi, 3\pi, \ldots$ as $\sin^2 \tau a = 0$. Therefore, it may be stated that $T = 1$ would occur when barrier thickness is equal to half-integral or integral number of the de Broglie wavelength $\lambda = \frac{2\pi}{\tau}$ in region II.

2.3.6 Approximation Method for Time-Independent Hamiltonian: Perturbation Theory

Similar to classical mechanics, the situations where we can exactly apply quantum mechanics, that is, Schrödinger's equation, are limited in number. To deal with real systems, several approximation methods have been developed. The developed methods are divided into time-independent and time-dependent depending on the time dependence of the Hamiltonian of the quantum system. In this section, we'll discuss a few approximation methods, namely, the perturbation method, Wentzel-Kramers-Brillouin (WKB) approximation, and the variational method. The variational method relies on finding the wave function with the lowest energy level. Here, the other two approximation methods have been briefly described.

2.3.6.1 Rayleigh–Schrödinger Perturbation Method

The perturbation method has been developed by Rayleigh and Schrödinger, so this is also known as the Rayleigh–Schrödinger perturbation method. In this method, the Hamiltonian (H) of any real system is expressed as $H = H_0 + \chi H'$, where H_0 is the unperturbed Hamiltonian of the system for which the exact solution for Schrödinger's equation is known ($H_0 \Psi_n^{(0)} = E_n^{(0)} \Psi_n^{(0)}$, where $E_n^{(0)}$ is the unperturbed energy eigenvalues and $\Psi_n^{(0)}$ is corresponding eigenfunction) and H' is referred to as perturbation to the Hamiltonian. The parameter "χ" is a real parameter signifying various orders of the perturbation. Similar to the unperturbed Hamiltonian, the perturbed Hamiltonian can be expressed as $H\Psi_n = E_n \Psi_n$, where E_n and Ψ_n represent the energy eigenvalue and eigenfunction of the perturbed Hamiltonian. For any nondegenerate system, Ψ_n and E_n are expressed in the power series of χ, that is,

$$\Psi_n = \Psi_n^{(0)} + \chi\Psi_n^{(1)} + \chi^2\Psi_n^{(2)} + \cdots$$

$$E_n = E_n^{(0)} + \chi E_n^{(1)} + \chi^2 E_n^{(2)} + \cdots$$

Now using the expressions for eigenfunctions and eigenvalues for the perturbed Hamiltonian, we have

$$\left(H_0 + \chi H'\right)\left(\Psi_n^{(0)} + \chi\Psi_n^{(1)} + \chi^2\Psi_n^{(2)} + \cdots\right)$$

$$= \left(E_n^{(0)} + \chi E_n^{(1)} + \chi^2 E_n^{(2)} + \cdots\right)\left(\Psi_n^{(0)} + \chi\Psi_n^{(1)} + \chi^2\Psi_n^{(2)} + \cdots\right)$$

Now equating the term that is proportional to χ^0, we have $H_0\Psi_n^{(0)} = E_n^{(0)}\Psi_n^{(0)}$ representing Schrödinger's equation of the unperturbed system. Equating the terms that are proportional to χ^1, we have $H_0\Psi_n^{(1)} + H'\Psi_n^{(0)} = E_n^{(0)}\Psi_n^{(1)} + E_n^{(1)}\Psi_n^{(0)}$.

If $\Psi_n^{(0)}$ forms the complete basis of the wave function, any wave function can be expressed in terms of this basis set, and thus $\Psi_n^{(1)} = \sum_j A_j\Psi_j^{(0)}$, where A_j represents the coefficient of expansion. Substituting this form of $\Psi_n^{(1)}$ into the above equation, we have

$$H_0\sum_j A_j\Psi_j^{(0)} + H'\Psi_n^{(0)} = E_n^{(0)}\sum_j A_j\Psi_j^{(0)} + E_n^{(1)}\Psi_n^{(0)}$$

On simplification, we have $\sum_j E_j^{(0)} A_j\Psi_j^{(0)} + H^*\Psi_n^{(0)} = E_n^{(0)}\sum_j A_j\Psi_j^{(0)} + E_n^{(1)}\Psi_n^{(0)}$

In order to determine the coefficients (A_j), we would multiply both sides by $\Psi_i^{(0)*}$ and would integrate over the entire region of space; then we have

$$\sum_j E_j^{(0)} A_j\int\Psi_i^{(0)*}\Psi_j^{(0)}\,dx + \int\Psi_i^{(0)*}H'\Psi_n^{(0)}\,dx$$

$$= E_n^{(0)}\sum_j A_j\int\Psi_i^{(0)*}\Psi_j^{(0)}\,dx + E_n^{(1)}\int\Psi_i^{(0)*}\Psi_n^{(0)}\,dx$$

Since the wave functions are the eigenfunction of a Hermitian operator, they are orthogonal to each other, and thus $\int\Psi_i^{(0)*}\Psi_j^{(0)}dx = \delta_{ij}$. Substituting this condition into the above equation, we have

$$\sum_j E_j^{(0)} A_j\delta_{ij} + \int\Psi_i^{(0)*}H'\Psi_n^{(0)}\,dx = E_n^{(0)}\sum_j A_j\delta_{ij} + E_n^{(1)}\delta_{in}$$

In the summation, all the terms would be zero except $i = j$, and therefore on simplification

$$E_i^{(0)} A_i + \int \Psi_i^{(0)*} H' \Psi_n^{(0)} \, dx = E_n^{(0)} A_i + E_n^{(1)} \delta_{in}$$

When $i = n$, then we have $E_n^{(1)} = \int \Psi_n^{(0)*} H' \Psi_n^{(0)} \, dx$ signifying the first-order correction if energy is due to the presence of perturbation. From the unperturbed Hamiltonian we know $\Psi_n^{(0)}$ and the perturbation acting on the quantum particle is also known, and therefore from this equation we can calculate the first-order correction in energy.

On the other hand, for states with $i \neq n$, $A_i \left(E_n^{(0)} - E_i^{(0)} \right) = \int \Psi_i^{(0)*} H' \Psi_n^{(0)} \, dx$, that is, $A_i = \dfrac{\int \Psi_i^{(0)*} H' \Psi_n^{(0)}}{\left(E_n^{(0)} - E_i^{(0)} \right)}$, so from these relations we can calculate the A_i, which is the expansion coefficient of $\Psi_n^{(1)}$, that is, first-order correction in wave function can be calculated. In this way we can also calculate the higher-order corrections in the wave function and energy eigenvalue.

2.3.6.2 Wentzel–Kramers–Brillouin Approximation Method

This approximation method was developed by G. Wentzel, H.A. Kramers, and L. Brillouin in 1926 to calculate the wave function of any quantum system under slow varying potential with respect to position. For constant potential (V_0), the solution for Schrödinger's equation gives the wave function in the form $\Psi(x) = A \exp \left(\int \pm \dfrac{i}{\hbar} p_0 x \right)$, similar to plane electromagnetic wave, and the momentum (p_0) is given by $p_0 = [2m(E - V_0)]^{1/2}$. Now for varying potential we may write the expression for momentum $p(x) = [2m\{E - V(x)\}]^{1/2}$, and thus we may expect the wave function to be in the form $\Psi(x) = A \exp \left[\dfrac{i}{\hbar} S(x) \right]$. Now substituting this expression for $\Psi(x)$ into Schrödinger's equation, we have $-\dfrac{i\hbar}{2m} \dfrac{d^2 S(x)}{dx^2} + \dfrac{1}{2m} \left[\dfrac{dS(x)}{dx} \right]^2 + V(x) = E$. At this point, the approximation that has been adopted is that $S(x)$ can be expressed in the following way $S(x) = S_0(x) + \hbar S_1(x) + \dfrac{\hbar^2}{2} S_2(x) + \cdots$. Substituting $S(x)$ into a differential equation and equating the terms independent of $\hbar$ and proportional to $\hbar$, we have the following equations:

$$\frac{1}{2m} \left[\frac{dS_0(x)}{dx} \right]^2 + V(x) - E = 0 \quad \text{and} \quad \frac{dS_0(x)}{dx} \frac{dS_1(x)}{dx} - \frac{i}{2} \frac{d^2 S_0(x)}{dx^2} = 0.$$

From the first equation, it may be derived that $\dfrac{dS_0(x)}{dx} = \left[2m(E-V(x))\right]^{1/2}$

and the right-hand side of it represents the expression of $p(x)$. So $\dfrac{dS_0(x)}{dx}$

is simply the momentum and $S_0(x) = \int p(x)\, dx$. Substituting $\dfrac{dS_0(x)}{dx} = p(x)$

into the other equation, we have $p(x)\dfrac{dS_1(x)}{dx} - \dfrac{i}{2}\dfrac{dp(x)}{dx} = 0$. Therefore, after

integrating, we have $S_1(x) = \dfrac{i}{2}\log p(x)$. So substituting $S_0(x)$ and $S_1(x)$ into

$\Psi(x) = A\exp\left[\dfrac{i}{\hbar}S(x)\right]$, we have

$$\Psi(x) = A\exp\dfrac{i}{\hbar}\left[S_0(x) + \hbar S_1(x)\right]$$

$$= A\exp\left[\dfrac{i}{\hbar}S_0(x)\right]\exp\left[iS_1(x)\right]$$

$$= A\exp\left[\dfrac{i}{\hbar}S_0(x)\right]\exp\left[\dfrac{-1}{2}\log p(x)\right]$$

$$= A\left[p(x)\right]^{-1/2}\exp\left[\dfrac{i}{\hbar}\int p(x)\,dx\right]$$

So in this way we can calculate $\Psi(x)$ for varying potential with respect to

position. The validity of this approximation is $\left|\dfrac{\hbar m \dfrac{dV(x)}{dx}}{\left[2m(E-V(x))\right]^{3/2}}\right| \ll 1$. Cold

emission of electrons is explained well using this approximation method.

2.3.7 Time-Dependent Perturbation Theory

We consider a quantum system that is expressed by the Hamiltonian $H = H_0 + \chi H'(t)$, where H_0 is the unperturbed time-independent Hamiltonian of the system for which the solution for Schrödinger's equation is known and $\chi H'(t)$ represents time-dependent perturbation acting on the microscopic system. In the presence of $\chi H'(t)$, it is to be considered that the wave function Ψ should be time dependent and should satisfy the time-dependent Schrödinger's equation, that is, $i\hbar\dfrac{\partial\Psi(x,t)}{\partial t} = H\Psi(x,t)$. Eigenfunctions $\left(\Phi_j^{(0)}\right)$ of H_0 form a complete basis set, so $\Psi(x,t)$ may be written on this basis, that is, $\Psi(x,t) = \sum_j A_j(t)\Phi_j^{(0)}\exp\left(-iE_j^0 t/\hbar\right)$. In contrast to the time-independent system, here the wave function becomes a function of time, and hence $A_j(t)$

should be a function of time. Now substituting the expression for $\Psi(x,t)$ into Schrödinger's equation, we have

$$i\hbar \sum_j \dot{A}_j(t)\Phi_j^{(0)} \exp\left(\frac{-iE_j^0 t}{\hbar}\right) + \sum_j A_j(t) E_j^0 \Phi_j^{(0)} \exp\left(\frac{-iE_j^0 t}{\hbar}\right)$$

$$= \sum_j \left[H_0 + \chi H'(t)\right] A_j(t)\Phi_j^{(0)} \exp\left(\frac{-iE_j^0 t}{\hbar}\right)$$

On simplification, we have $i\hbar \sum_j \dot{A}_j(t)\Phi_j^{(0)} \exp\left(\dfrac{-iE_j^0 t}{\hbar}\right) = \sum_j \chi H'(t) A_j(t)\Phi_j^{(0)}$

$\exp\left(\dfrac{-iE_j^0 t}{\hbar}\right)$.

If we multiply both sides by the complex conjugate of $\Phi_b^{(0)} \exp\left(\dfrac{-iE_b^0 t}{\hbar}\right)$ and integrate, then we have

$$i\hbar \sum_j \dot{A}_j(t) \exp\left(\frac{-iE_j^0 t}{\hbar}\right) \exp\left(\frac{-iE_b^0 t}{\hbar}\right) \int \Phi_b^{(0)*} \Phi_j^{(0)} \, dx$$

$$= \sum_j \chi A_j(t) \exp\left(\frac{-iE_j^0 t}{\hbar}\right) \exp\left(\frac{-iE_b^0 t}{\hbar}\right) \int \Phi_b^{(0)*} H'(t) \Phi_j^{(0)} dx$$

Since $\Phi_j^{(0)}$ forms the orthogonal basis set, we have $\int \Phi_b^{(0)*} \Phi_j^{(0)} \, dx = \delta_{bj}$. If we write $\int \Phi_b^{(0)*} H'(t) \Phi_j^{(0)} \, dx = H'_{bj}(t, x)$ then the above expression can be simplified further, $i\hbar \dot{A}_b(t) = \sum_j \chi A_j(t) \exp \dfrac{i\left(E_b^0 - E_j^0\right)t}{\hbar} H'_{bj}(t, x)$. The approximation that has been carried out is that $A_b(t)$ could be expressed in the following way:

$$A_b(t) = A_b^0(t) + \chi A_b^1(t) + \chi^2 A_b^2(t) + \cdots$$

$$\text{i.e., } \dot{A}_b(t) = \dot{A}_b^0(t) + \chi \dot{A}_b^1(t) + \chi^2 \dot{A}_b^2(t) + \cdots$$

Now substituting this into the above equation, we have

$$i\hbar \left[\dot{A}_b(t) = \dot{A}_b^0(t) + \chi \dot{A}_b^1(t) + \chi^2 \dot{A}_b^2(t) + \cdots\right]$$

$$= \sum_j \chi A_j(t) \exp \frac{i\left(E_b^0 - E_j^0\right)t}{\hbar} H'_{bj}(t, x)$$

Equating the term that is independent of χ, we have $\dot{A}_b^0(t) = 0$, that is, $A_b^0(t) =$ constant signifying that the coefficient does not change with time, and hence it defines the initial condition of the problem. If the system is defined with n-th initial state then $A_j^0(t) = \delta_{nj}$. Equating the term that is proportional to χ, we have $i\hbar \dot{A}_b^1(t) = \exp\dfrac{i\left(E_b^0 - E_n^0\right)t}{\hbar} H'_{bn}(t, x)$. The exponential term represents the oscillation of the system between the b-th and n-th state of the system and the corresponding frequency can be written in the form $\dfrac{\left(E_b^0 - E_n^0\right)}{\hbar} = \omega_{bn}$. In terms of ω_{bn}, the above equation can be written as $\dot{A}_b^1(t) = \dfrac{1}{i\hbar}\exp\dfrac{i\omega_{bn}t}{\hbar} H'_{bn}(t, x)$, and after integration we have $A_b^1(t) = \dfrac{1}{i\hbar}\int \exp\dfrac{i\omega_{bn}t}{\hbar} H'_{bn}(t, x)dt$. So the transition probability between the two states is given by $\left|A_b^1(t)\right|^2 = \dfrac{1}{\hbar^2}\left|\int \exp\dfrac{i\omega_{bn}t}{\hbar} H'_{bn}(t, x)dt\right|^2$. Using this expression, we can calculate the transition from the initial state to the final state in the presence of time-dependent perturbation, and this has been well accepted to calculate the transition for molecules, atoms, etc.

Acknowledgment

The author thanks all of the CRC Press team for their suggestions to improve the chapter.

References

1. B.H. Bransden and C.J. Joachain, *Introduction to Quantum Mechanics*, Educational Low-Priced Books Scheme, Longman Group UK Ltd. Longman House, Burnt Mill, England, 1990.
2. R. Eisberg and R. Resnick, *Quantum Physics of Atoms, Molecules, Solids, Nuclei and Particles*, 2nd edn., John Wiley & Sons, New York, 1985.

3

Crystallography, Band Structure, and Density of States at Nanoscale

Swapnadip De

CONTENTS

3.1 Introduction

The chapter begins with the concept of crystal structures and energy bands. The density of states for 0D, 1D, and 2D systems as well as Somerfield's free electron theory are explained in detail. The Kronig–Penney model of electrons along with the concept of band structure of nanoparticles are discussed. The idea of flat band voltage is then utilized to model the subthreshold surface potential for nanodevices. This idea can be utilized by researchers to obtain mathematical expressions of subthreshold drain current, threshold voltage, transconductance, and other characteristic parameters for short-channel MOSFETs.

3.2 Concept of Crystal Structure

Matter is broadly divided into solid, liquid, and gaseous state. The molecules, atoms, or ions in solids are closely packed and hence they are rigid and incompressible. Further solids can be classified into

- Crystalline
- Amorphous

The atoms or molecules of crystalline materials are arranged in a unique order along the lattice points. The lattice point is an imaginary, mathematical concept. A group of atoms or molecules known as basis arrange themselves in a unique manner around the lattice points. A unit cell is the smallest volume of a crystal structure having the same symmetry as the whole crystal. The unit cell when repeated along all directions forms the crystal. A unit cell is known as primitive when it has only one lattice point and is known as nonprimitive when it has more than one lattice point (Figure 3.1).

The widely used extrinsic semiconductors in the electronic industry are formed by doping of crystal structures. We create free holes (in case of *p* type) and free electrons (in case of *n* type) by means of doping such that they can help in conduction. Silicon, germanium, and gallium are widely used in semiconductors. Thus, the crystal structure plays an important role in the creation of semiconductors, which are the basic building blocks of diodes and transistors.

3.3 Energy Bands

We know that the energy of the bound electron in an atom is quantized. It occupies atomic orbitals of discrete energy levels. As the atoms combine to form molecules, the atomic orbitals overlap. According to Pauli's exclusion principle, no two electrons can have the same quantum number in a molecule. In the case of crystal lattices, a large number of identical atoms combine

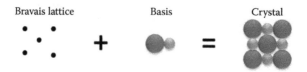

FIGURE 3.1
Crystal.

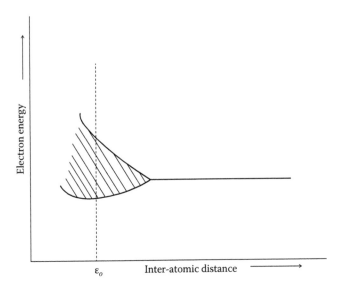

FIGURE 3.2
Splitting of energy bands.

to form a molecule. Here the atomic orbitals split into different energy levels. A large number of closely spaced energy levels are formed. These closely spaced energy levels are known as energy bands. The finite energy gap between two energy bands is known as band gap (Figure 3.2).

Band structure can be found by solving Schrödinger's equation, which gives Bloch waves as the solution, of the form (3.1)

$$\psi(x) = u(x)e^{ikx} \tag{3.1}$$

where k is the wave vector or constant of motion.

The energy E has discontinuities with forbidden gaps for the particles. The E vs k plot is shown in Figure 3.3.

3.4 Density of States and Somerfield's Free Electron Theory

Arnold Somerfield combined the Drude Model and Fermi–Dirac Statistics to develop the free electron model. This model can be effectively used to determine the Wiedemann–Franz law, the temperature dependence of the capacity, shape of electronic states, range of binding energy values, electrical conductivities and thermal and field electron emission from bulk metals. The density of state function can be derived using the free electron model.

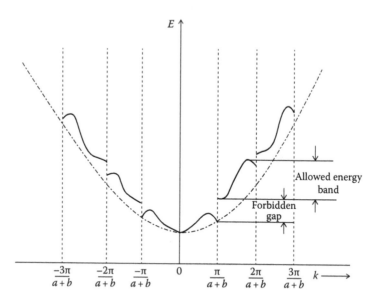

FIGURE 3.3
E vs *k* graph.

The density of states is represented by a function gE where $gEdE$ = the number of states per unit volume in the energy range $(E, E + dE)$. Let us consider a free electron confined to points k within the surface. The solution for electrons confined in a box with rigid walls can be obtained by Schrödinger's wave equation in terms of three nonzero integers, nx, ny, and nz.

According to de Broglie's wave particle duality we get (3.2)

$$k = \frac{p}{\hbar} \tag{3.2}$$

where
  k is the wave number
  $\hbar$ is the reduced Planck's constant

Now we substitute terms

$$k = \frac{p}{\hbar} \rightarrow k = \frac{mv}{\hbar} \rightarrow v = \frac{\hbar k}{m} \tag{3.3}$$

Substituting v in the equation for energy, we get (3.4)

$$E = \frac{1}{2} m \left(\frac{\hbar k}{m} \right)^2 \tag{3.4}$$

The dispersion relation for electron energy is given by (3.5)

$$E = \frac{\hbar^2 k^2}{2m^*}$$
(3.5)

where m^* is the effective mass of an electron.

The allowed value of k is found by solving Schrödinger's wave equation with certain boundary conditions.

The solution is given as (3.6)

$$u = Ae^{i(k_x x + k_y y + k_z z)}$$
(3.6)

Applying the boundary conditions (3.7) is obtained.

$$e^{i(k_x L + k_y L + k_z L)} = 1 \rightarrow (k_x, k_y, k_z) = \left(n_x \frac{2\pi}{L}, n_y \frac{2\pi}{L}, n_z \frac{2\pi}{L} \right)$$
(3.7)

The allowed values from $E = \dfrac{\hbar^2 k^2}{2m^*}$ can be pictured as a sphere near the origin with a radius k and thickness dk. The allowed states are found within the volume contained between k and $k + dk$ (Figure 3.4).

The volume of the shell with radius k and thickness dk is given by (3.8)

$$V_{shell} = 4\pi k^2 dk$$
(3.8)

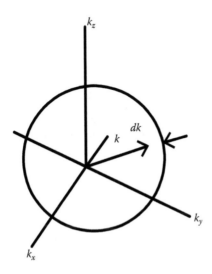

FIGURE 3.4
Coordinates of the sphere.

and number of states by

$$\left(\frac{1}{2\pi}\right)^3 4\pi k^2 dk$$

Substituting the following equation in the dispersion relation for electron energy

$$E = \frac{h^2 k^2}{2m^*} \rightarrow k = \sqrt{\frac{2m^* E}{h^2}} \tag{3.9}$$

leads to (3.10):

$$\frac{dk}{dE} = \left(\frac{2m^* E}{h^2}\right)^{-1/2} \frac{m^*}{h^2} \tag{3.10}$$

Substituting the expressions obtained for dk and k^2 in terms of E back into the expression for the number of states, the number of allowed k values within this shell gives the number of available states when divided by the shell thickness, dE. The function $g\epsilon$ is given in (3.11).

$$g(E) = \frac{1}{4\pi^2}\left(\frac{2m^*}{h^2}\right)^{3/2} E^{1/2} \tag{3.11}$$

We must now account for the fact that any k state can contain two electrons, spin-up and spin-down, so we multiply by a factor of two
    The k state can contain two electrons and is given by (3.12).

$$g(E) = \frac{1}{2\pi^2}\left(\frac{2m^*}{h^2}\right)^{3/2} E^{1/2} \tag{3.12}$$

3.4.1 Density of States of 2D, 1D, and 0D Systems

Here we consider an electron of effective mass m^* being confined in an infinite quantum well (2D) of length L. The potential energy is set to zero. On solving Schrödinger's wave equation we obtain (3.13)

$$\left(-\frac{h^2}{2m} del^2\right)\psi = E\psi$$

$$\frac{\delta^2\psi}{\delta x^2} + \frac{\delta^2\psi}{\delta y^2} + k^2\psi = 0 \tag{3.13}$$

where $k = \sqrt{\frac{2mE}{h^2}}$.

Using separation of variables we get (3.14)

$$\psi(x,y) = \psi_x(x)\psi_y(y) \tag{3.14}$$

The solution for the wave equation can be given as (3.15)

$$\psi = A\sin kx + B\cos kx \tag{3.15}$$

Now, k space volume of a single space cube is $V = \dfrac{\pi^2}{V} = \dfrac{\pi^2}{L^2}$.

K space volume of a sphere in a circle is $V_{circle} = \pi k^2$.

Number of filled states in a sphere is $\dfrac{V_{circle}}{V} \times 2 \times \dfrac{1}{2} \times \dfrac{1}{2} = \dfrac{k^2 L^2}{2\pi}$.

The density of states per unit volume, as a function of energy, is obtained by dividing by the volume of the crystal as in (3.16).

$$g(E)_{2D} = \dfrac{\dfrac{L^2 m}{h^2 \pi}}{L^2} = \dfrac{m}{\pi h^2} \tag{3.16}$$

For a 1D system, k space volume of a single space cube is $V = \dfrac{\pi}{V} = \dfrac{\pi}{L}$.

K space volume of a sphere in k space circle is $V_{Line} = \pi k$.

Number of filled states in a sphere is $\dfrac{V_{line}}{V} \times 2 \times \dfrac{1}{2} = \dfrac{kL}{\pi}$.

The density of states per unit volume, as a function of energy, is obtained by dividing by the volume of the crystal as (3.17)

$$g(E)_{1D} = \dfrac{\dfrac{(2mE)^{-1/2} mL}{h\pi}}{L} = \dfrac{m}{h\pi\sqrt{2mE}} \tag{3.17}$$

Simplifying the above equation we obtain (3.18)

$$g(E)_{1D} = \dfrac{1}{h\pi}\sqrt{\dfrac{m}{2E}} \tag{3.18}$$

As electron mass is considered as m^*, and the kinetic energy is taken as Ec, we obtain (3.19)

$$g(E)_{1D} = \dfrac{1}{h\pi}\sqrt{\dfrac{m^*}{2(E - Ec)}} \tag{3.19}$$

In a 0D structure (quantum dot), there is no free motion. Here all the k space is filled with electrons and all the states exist at discrete energy levels. The density of states for a 0D system is described by the delta function. This is given as (3.20)

$$g(E)_{0D} = 2\delta(E - Ec)$$
(3.20)

3.5 Kronig–Penney Model of Electron

The classical free electron theory based on laws of statistical mechanics suffers from certain major drawbacks such as the following:

- Maxwell–Boltzmann statistics and law of equipartition of energy fail to accurately evaluate electronic-specific heat
- It does not consider the magnetic moment of an electron due to its spin.

Some of the shortcomings of the classical theory are solved by the quantum free electron theory, such as the issue with specific heat. However, the quantum theory also could not account for the differences in conductivities of metals, semiconductors, and insulators.

Here is where the Kronig–Penney model comes in, which employs an arrangement of periodically repeating potential wells and potential barriers as shown. The size of the well should roughly correspond to the lattice spacing (Figures 3.5 and 3.6).

We need to solve Schrödinger's equation for a single period, and check whether it is continuous and smooth. The equations are (3.21 and 3.22)

$$\frac{d^2\psi}{dx^2} + \frac{2mE}{\hbar^2}\psi = 0 \quad \text{for } 0 < x < a$$
(3.21)

$$\frac{d^2\psi}{dx^2} + \frac{2m(E - V_0)}{\hbar^2}\psi = 0 \quad \text{for } -b < x < 0$$
(3.22)

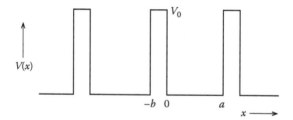

FIGURE 3.5
Potential wells.

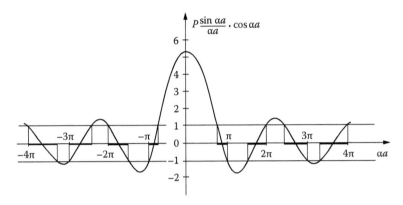

FIGURE 3.6
Energy range.

According to Bloch's theorem, the wave function solution for Schrödinger's equation, when the potential is periodic, is of the form shown in (3.23), known as Bloch's function:

$$\psi(x) = u(x)e^{ikx} \tag{3.23}$$

The solution obtained after simplification is (3.24)

$$P\frac{\sin \alpha a}{\alpha a} + \cos \alpha a = \cos Ka \tag{3.24}$$

where P is called scattering power, and is given by (3.25)

$$P = \frac{mV_0 ab}{\hbar^2} \tag{3.25}$$

And α and β are two real quantities, given by (3.26 and 3.27)

$$\alpha = \frac{(2mE)^{1/2}}{\hbar} \tag{3.26}$$

$$\beta = \frac{\left[2m(V_0 - E)\right]^{1/2}}{\hbar} \tag{3.27}$$

3.6 Band Structure of Nanoparticles

There is a clear difference between the band structure of a conductor, insulator, and semiconductor. For bulk matter, the bands are formed by merging nearby energy levels of atoms or molecules. The smaller the particle size, the lower the number of overlapping orbitals or energy levels and narrower the width of the bands. An atom or a molecule can have separate energy levels represented by single lines. The smaller the size of the particle, the larger the gap between the conduction and valence band. This is the main reason why the nanoparticles have higher band gap when compared to the corresponding bulk matter. The greater the forbidden gap, the greater the restriction on the movement of electrons. The carbon nanotube is an example of this concept. In this case, the band gap depends on the geometric shape features.

3.7 Idea of Band-to-Band Tunneling and Estimation of Subthreshold Surface Potential for Nanoscale Devices

3.7.1 Concept of Flat Band Voltage

Flat band voltage is defined as the voltage that must be applied to the gate to bring the semiconductor energy bands to a flat level. It can also be defined as the voltage applied to the gate such that there is no band bending in the semiconductor. For development of VLSI technology, the channel length of MOSFET is decreased so that the electric field is increased and various effects emerge for short-channel devices such as hot-electron effect and DIBL. It has been noted that with the reduction of channel length, the threshold voltage reduces with scaling. Also, as the drain-to-source voltage increases, the off-state current increases as in Figure 3.7. Such a phenomenon is termed DIBL.

When a drain voltage is applied, the DIBL effect is caused as a result of decrease of the barrier potential at the source end as in Figure 3.8.

A measure of DIBL is (3.28)

$$\text{DIBL} = \frac{V_{t,lin} - V_{t,sat}}{V_{dd} - V_{d,lin}} \tag{3.28}$$

where
  V_{dd} is the supply voltage
  $V_{d,lin}$ is the linear drain voltage
  $V_{t,lin}$ and $V_{t,sat}$ are the threshold voltages in the linear and the saturated operations, respectively

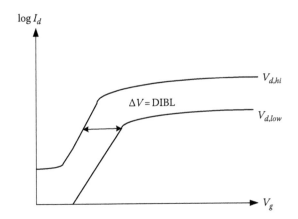

FIGURE 3.7
Drain current versus gate voltage for different drain biases, showing DIBL.

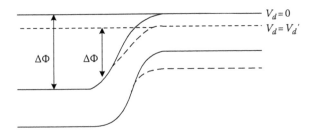

FIGURE 3.8
The energy diagram at the source end with and without drain voltage.

Hot carrier refers to high-energy electrons or holes which are accelerated due to the high horizontal and vertical electric fields due to the gate and drain bias [1,2]. In the case of constant voltage scaling, as the channel length is reduced, the oxide thickness also decreases, keeping the terminal voltages constant. So the vertical electric field increases due to scaling (Figure 3.9). The electrons are attracted to the oxide–silicon interface. Some of these high-energy electrons overcome the oxide–silicon potential barrier and get trapped in the oxide. These electrons are called hot electrons and they degrade the performance of oxide (Figures 3.10 and 3.11).

To overcome the hot-electron effect, the channel engineering, gate engineering, and their combinations are used.

The DGDMDH-MOS transistor shown in Figure 3.12 uses two laterally contacted materials of different work functions as the gate. The structure of the n-channel DGDMDH-MOS transistor used two different materials, M1 and M2, with lengths L1 and L2, and work functions $\Phi 1$ and $\Phi 2$, respectively, contacted laterally are used as the gate (Figure 3.13) [3–5].

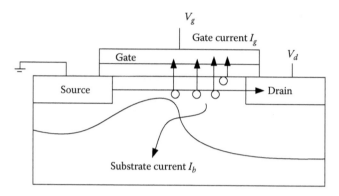

FIGURE 3.9
Hot-carrier effect.

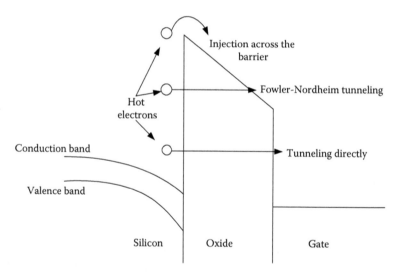

FIGURE 3.10
Three different types of carrier injection into the gate resulting in hot-carrier effects.

The band bending for the strong inversion case is shown in Figure 3.14. In the case of strong inversion, Figure 3.14 shows the band bending of the semiconductor [6]. Here $\Phi(x)$ denotes the location of the intrinsic Fermi level. The bending of bands at the oxide–semiconductor interface is denoted as Φ_S. $\Phi_S = 0$ denotes the flat band condition of MOSFET. $\Phi_S < 0$ gives the condition for accumulation of holes and the corresponding band bending. $\Phi_S > 0$ denotes the condition of MOSFET to be in depletion mode, and for $\Phi_S > 0$ or greater than Φ_F, E_i is below E_F, which causes inversion.

When the MOSFET is in the weak inversion regime, $\Phi_F \leq \Phi_S \leq 2\Phi_F$ denotes the condition for the MOSFET to be in the weak inversion regime. In the case

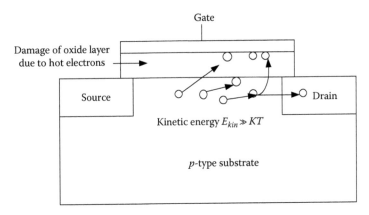

FIGURE 3.11
Damage of oxide due to hot carriers in short-channel MOSFET.

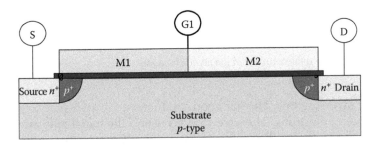

FIGURE 3.12
Double-halo single-gate double-metals MOSFET.

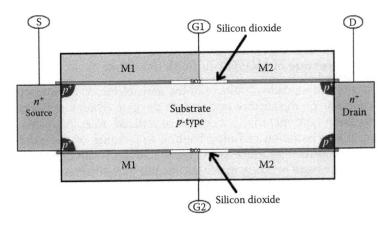

FIGURE 3.13
Double-gate dual-material double-halo MOSFET.

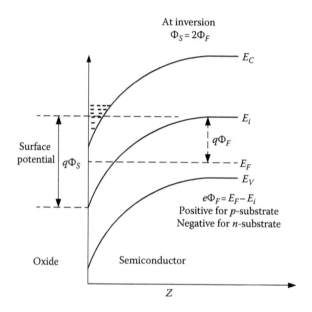

FIGURE 3.14
Band bending of the semiconductor in the inversion mode.

of strong inversion $\Phi_S \geq 2\Phi_F$. For good conductors, the inversion layer has an abundance of electrons. The inversion layer and the metal gate are the two conducting electrodes. The capacitance of the MOS structure in accumulation mode is C_{ox}.

The electrons are the majority carriers for MOSFET in the inversion regime. As the gate bias is increased, the electron concentration in the inversion layer increases, and so inversion charge Q_{inv} is increased. However, the inversion layer is not very thick (Figure 3.15).

3.7.2 Band Structure of MOSFET in Depletion Mode

On application of a positive voltage in the gate, electrons are attracted and they accumulate on the surface underneath the gate. Therefore the negative charge in the p-type material is because of ionized acceptors. This causes lowering of concentration of holes, leading to bending of the band down near the semiconductor surface as in Figure 3.16 [6]. When the positive bias is increased, the band bending increases, resulting in inversion.

The potential drop across the oxide. V_{gb} is the gate bias voltage and V_{fb} is the flat band voltage under the gate (Figure 3.17).

The flux lines, terminating on the interface charge per unit area Q_0, will not contribute to the same since it is modeled as the effective interface charge which resides on the oxide side of the interface [7–9]. Thus, only flux lines

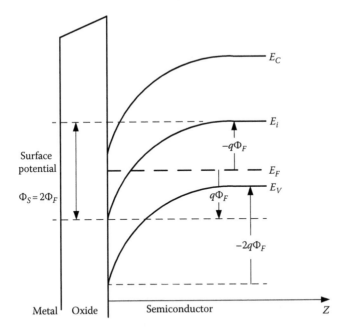

FIGURE 3.15
Band bending under strong inversion at the surface.

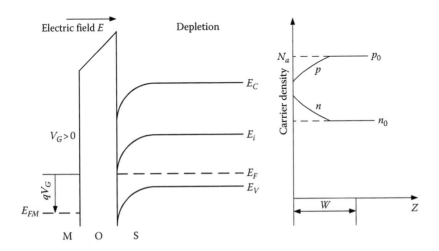

FIGURE 3.16
Effect of applied electric field on the interface charge density in the ideal MOS capacitor: positive gate voltage (V_G) creates depletion region.

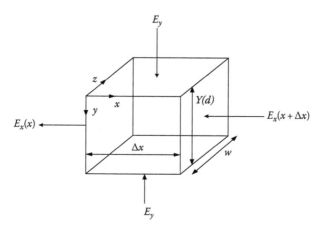

FIGURE 3.17
Gaussian surface.

terminating on Q_c will contribute to the same, and for an oxide thickness of t_{ox}, the corresponding vertical field E_{y1} on the top along the y-direction can be obtained as (3.29)

$$E_{y1} = \frac{V_{gb} - V_{fb} - \psi_s}{t_{ox}} \qquad (3.29)$$

Application of Gauss's law to the said surface leads to (3.30)

$$\varepsilon \int_{surface} E \cdot ds = -qN_a y_d \Delta x W \qquad (3.30)$$

A pseudo-2D Poisson's equation can be obtained as in [7–11]. The variation of depletion layer depth is shown in [9]. The unknown potentials V_2, V_3, V_4, V_5, and V_6 are found by solving the following simultaneous equations obtained by applying the continuity of the derivative of potential at the boundaries between two regions. Substituting the values of V_2, V_3, V_4, V_5, and V_6, Ψ_s in all the six regions may easily be computed. The SGDMDH structure uses a relatively lower doped substrate than the halo region in the channel [7–9]. The subthreshold surface potential variation is estimated as a function of channel length in [7–11]. From the expression of subthreshold surface potential, the minimum value is computed. Using the minimum value of surface potential, threshold voltage, drain current, and transconductance can be computed for nanodevices.

References

1. S. De, A. Sarkar, and C.K. Sarkar, Modelling of characteristic parameters for asymmetric DHDMG MOSFET, *WSEAS Transactions on Circuits and Systems*, 11(11), November 2012, 371–380.
2. S. De, A. Sarkar, and C.K. Sarkar, Fringing capacitance based surface potential model for pocket DMG n-MOSFETs, *Journal of Electron Devices*, 12, 2012, 704–712.
3. S. Ninomiya et al., Vth control by halo implantation using the SEN's MIND system, *Proceedings of Ninth International Workshop on Junction Technology*, 2009, pp. 100–103, Kyoto, Japan.
4. S. Baishya, A. Mallik, and C.K. Sarkar, A subthreshold surface potential and drain current model for lateral asymmetric channel(LAC) MOSFETs, *IETE Journal of Research*, 52, September–October 2006, 379–390.
5. S. Baishya, A. Mallik, and C.K. Sarkar, A subthreshold surface potential model for short-channel MOSFET taking into account the varying depth of channel depletion layer due to source and drain junctions, *IEEE Transactions on Electron Devices*, 53, March 2006, 507–514.
6. S. De, Basic semiconductor and MOS physics, in *Technology Computer Aided Design: Simulation for VLSI MOSFET*, C.K. Sarkar, Ed., CRC Press, Taylor & Francis Group, LLC, Boca Raton, FL, 2013, pp. 45–144, ISBN: 978-1-4665-1265-8.
7. S. De, A. Kumari, P. Dutta, I. Gupta, and M. Bhattacharya, A review of subthreshold surface potential for single gate dual material double halo MOSFET, *International Journal of VLSI Design and Technology*, 2(1), 2016, 1–20.
8. S. De and A. Sarkar, Modeling of characteristic parameter for submicron MOSFET, *The IUP Journal of Electrical & Electronics Engineering*, 5(3), July 2012, 67–76.
9. D. Das, S. De, M. Chanda, and C.K. Sarkar, Modelling of sub threshold surface potential for short channel double gate dual material double halo MOSFET, *The IUP Journal of Electrical & Electronics Engineering*, 7(4), October 2014, 19–42.
10. S. De, A. Sarkar, and C.K. Sarkar, Modelling of parameters for asymmetric halo and symmetric DHDMG n-MOSFETs, *International Journal of Electronics*, 98(10), October 2011, 1365–1381.
11. S. De, M. Chanda, and A. Sarkar, Analytical sub-threshold surface potential and drain current model for linearly doped double halo DMG MOSFET, *Proceedings of Second National Conference on Advanced Communication Systems and Design Techniques 2012*, Vol. 1, pp. 67–70, September 29–30, 2012, Haldia Institute of Technology, West Bengal, India.

4

Growth Techniques and Characterization Tools of Nanomaterials

Chandan Kumar Ghosh and Arka Dutta

CONTENTS

4.1 Introduction

Nanotechnology encompasses the process of engineering and fabricating objects with a size smaller than 100 nm. They often exhibit exotic properties completely different from bulk material, and it has been noted that a large fraction of surface atoms generates these properties. This fraction of surface atoms and related physical–chemical properties are found to be highly dependent on the growth processes of the nanoparticles, which is inherent to the synthesis protocols. Hence, to incorporate nanoparticles in a wide range of technological aspects, various synthesis processes of nanomaterials have evolved over time. However, they can be broadly classified under two major approaches depending upon the adopted methodology, namely, the top-down approach and the bottom-up approach. Both approaches play a crucial role in the development of nanoscience and nanotechnology. The fundamentals behind the top-down and bottom-up approaches lie in whether the starting material is bulk material or an atom. In the top-down approach, a bulk material is sheared down by various etching or milling techniques to reduce the dimension of the bulk system into nanomaterials and nanostructures, and in the bottom-up approach, atomic species are gradually coalesced to create the same. The following sections pertaining to the top-down and bottom-up manufacturing approaches present a thorough discussion on various nanomaterial synthesis techniques.

4.1.1 Bottom-Up Approaches to Nonmaterial and Nanostructure Synthesis

Bottom-up approaches to nanostructure and nanomaterial fabrication are capable of countering some of these shortcomings as they do not involve breaking down steps. In the following subsections, some of the common bottom-up approaches to nanostructure and nanomaterial synthesis are described in detail.

4.1.1.1 Sol-Gel Processing

The sol-gel process was developed almost 40 years ago during preparation of glass and ceramic materials at considerably low temperature. In the initial stage, it started with a solution subjected to polymerization followed by polycondensation, leading to the gradual formation of a solid-phase network that gives the final product. It is a wet chemical synthesis method for oxides, composites, and organic–inorganic hybrids from a liquid precursor. Its basic principle involves the formation of inorganic networks from a liquid phase, which is a colloidal suspension, described as sol, into a semisolid phase where a network-enclosing liquid phase, described as gel, is formed. There are various definitions of gel in the literature. In order to avoid confusion, gel, as defined by Flory, exhibits the following characteristics [1]: (1) well-ordered lamellar structures; (2) completely disordered covalently bonded polymeric structure; (3) physical aggregation of disordered polymeric network; and (4) disordered structure of particular materials. Gel exhibits a diphasic system in both the liquid and solid phase, in which the morphology of the polymer changes continuously. Often, the volume fraction of the polymer is found to be very low, that is, then it becomes necessary to increase the volume fraction. This is carried out either by allowing time for sedimentation or by a drying process. Removal of water introduces porosity within the gel. Precursor materials, used in the sol-gel process, are taken in the form of metal acetates, chloride, alkoxides, aloxysilanes, etc. As an example, tetramethoxysilane (TMOS) and tetraethoxysilanes (TEOS) are frequently used as gels during the preparation of SiO_2. In this context, it may be stated that synthesis of SiO_2 is accompanied by the formation of sol from inorganic metal salt or from dispersible oxides like organic metal alkoxides, produced by hydrolysis followed by polymerization. Hydrolysis appears after the addition of water that causes replacement of the OR group of TEOS or TMOS by the OH⁻ group, followed by attacking of Si atoms within silica gel by oxygen-giving siloxane bonds (Si–O–Si). The hydrolysis step can be expressed as follows:

$$Si(OH)_4 + H_2O \rightarrow HO-Si(OR)_3 + R-OH$$

Two partially hydrolyzed molecules react together to form the Si–O–Si bond. The next step is the polymerization by producing either water or alcohol that results in the formation of monomer, dimer, tetramer, etc., and it can be written as

$$2HOSi(OR)_3 \rightarrow (OR)_3SiOSi(OR)_3 + H_2O$$

or

$$2ORSi(OR)_3 + 2OHSi(OR)_3 \rightarrow (OR)_3Si-O-Si(OR)_3 + R-OH$$

Therefore, it may be stated that polymerization is followed by the formation of a one-, two-, or three-dimensional network of Si–O–Si bonds and H_2O and R–OH productions. Sol with a dense, porous, or polymeric substructure results from aggregation of subcolloidal amorphous or crystalline chemical. Sometimes, HCl or NH_3 is used to accelerate the reaction. In this context, hydrolysis is tuned by pH, regent concentration, aging, drying, etc. The appropriate choice of these parameters results in different structures of the derived inorganic networks. Particles grow due to a difference in solubility between smaller and larger particles, and it has been shown that when this difference becomes indistinguishable, growth stops. It is the general tendency of sols that they aggregate due to attractive van der Waals force that minimizes the energy of the sol. It has been examined that aggregation results in greater particle size, but aggregation is prevented by encapsulating the colloid particles by charged ions to induce repulsive forces among particles, or by surrounding them with reactive ligands to introduce steric hindrance. Rapid evaporation of solvent increases the viscosity of the sols until they form immobile gels. The process of gelation sometimes requires agglomeration-boosting additive to overcome repulsive forces in stabilized sols. It has been shown that sol to gel transition rate controls the size of the particle formed in this process. When pore fluid of the gel is removed in the gaseous phase, then in a hypercritical condition, the network remains unchanged and very low density aerogel is formed. Aerogel may have a pore volume greater than 95% and low density ~80 kg/m^3. When pore fluid is being removed at ambient pressure by thermal evaporation, then the network shrinks further into dense gel, called xerogels. Often, drying stress destroys monolithic gel bodies to form powders. The sol-gel process is generally adopted to obtain various oxides by calcination of the gels. The primary advantages of the sol-gel process are its versatility and flexible rheology that allows easy shaping and embedding. This method provides a significant homogeneity for multicomponent systems, particularly in the synthesis of laminating materials. The process offers unique prospects for accessing advanced organic–inorganic materials.

Another technique that has been adapted to synthesize monoliths using the sol–gel method is gelation of a solution in the form of colloidal powders, defined as the suspension where the dispersed phase with dimensions ranging from 1 to 100 nm is so short that gravitational force acting on them is negligible. Here, sol is redefined as the dispersed colloidal particles within liquid. When sol particles grow and collide with each other, condensation leads to the formation of nanoparticles. In this case, sol becomes gel when it is supported by stress elastically. This process does not require any activation energy and does not possess any critical point where the sol transforms into an elastic gel, and this change is associated with increasingly interconnected particles [2]. A large variety of oxide materials including ZnO [3], TiO$_2$ [4], MgO [5], CuAlO$_2$ [6], etc., have been synthesized by this method. For the synthesis of TiO$_2$ nanoparticles, the following protocols have been proposed [7]:

$$Ti(OR)_3 + H_2O \rightarrow Ti(OH)_3 + R_2O$$

$$Ti(OH)_3 \rightarrow TiO_2 + H_2O$$

The sol-gel process has also been adopted to prepare thin film of metal oxides. As an example, CuCo$_2$O$_4$ thin film on glass substrate has been synthesized by this technique using copper acetate and cobalt acetate as a precursor material in alcoholic medium. Here, triethanolamine is added as a chelating agent [8]. In addition to the preparation of virgin material, this technique has also been used to prepare several doped thin films. As an example, Mn-doped ZnO thin film was prepared by Stefan et al. starting from zinc acetate and manganese acetate in ethanolic solution [9]. In this reaction monoethanolamine is used as a stabilizer. Recently, a new type of material, namely, layered double hydroxide (LDH), has been developed with potential application in the fields of supercapacitors, catalysts, drug delivery carriers, etc. It is also synthesized by the sol-gel method. As an example, Ni-Al-LDH has been prepared in alcoholic medium using nickel acetate, aluminum acetate, and ammonium hydroxide [10]. Ferrite-based magnetic material such as NiFe$_2$O$_4$ has been synthesized in this process starting from iron nitrate, nickel nitrate, citric acid, and water [11]. In order to investigate the effect of temperature on the magnetic properties of $NiFe_2O_4$, the same research group sintered the product at different temperatures and observed that the magnetic coercive field is reduced with sintering temperature. Ferroelectric material like $SrTiO_3$ was synthesized by Visuttipitukul et al. [12]. The sol-gel technique is also adopted to prepare nanocomposites. In this context, Grigorie et al. prepared $ZnO–SiO_2$ nanocomposites using zinc acetate and TEOS as precursor materials for ZnO and SiO_2, respectively [13]. Noble metals incorporating oxide thin films have been synthesized by Epifani et al. [14].

4.1.1.2 Chemical Vapor Deposition

Chemical vapor deposition (CVD) is a technique for depositing high-quality thin films, specific to the semiconductor industry. It has been developed based on the chemical reaction of a chemical component in the vapor phase taking place close to or on a hot substrate. When the vapor phase precursor is brought in contact with a hot surface, under a critical condition, nucleation of the particles in the vapor phase occurs rather than deposition of a film on the substrate. The first step of this process is the volatilization of a solid or liquid feed containing the deposit material. In the next step, the gaseous component is transferred onto the hot activated substrate with the help of a carrier gas (nitrogen or argon). Generally, the substrate is activated by heating, radiation, or plasma. Here, formation of thin film takes place either by powder formation or by heterogeneous chemical reactions in the vapor phase near hot substrate. The general reaction mechanism during the CVD process can be written as follows:

$$AB_{2(\text{solid or liquid})} \rightarrow AB_{2(\text{gas})} \rightarrow A_{(\text{solid})} + B_{2(\text{gas})}$$

The feasibility of the CVD reaction could also be analyzed by considering transfer of free energy associated with this reaction, expressed as follows:

$$\Delta G_r^0 = \Sigma \Delta G_{f,\text{product}}^0 - \Sigma \Delta G_{f,\text{reactant}}^0$$

If ΔG_r^0 is negative then the reaction becomes feasible. Reaction kinetics depend on the quantity of gas in the low-pressure and low-temperature region, whereas in the high-temperature and high-pressure region, kinetics are determined by the diffusion rate of the reactant gases, that is, the reaction is dominated by mass transport that could be expressed by $\exp(-E/kT)$, where "E" represents the activation energy, and "k" and "T" are the Boltzmann's constant and absolute temperature, respectively. Materials with different properties are generated by appropriate choice of substrate temperature, the composition of the reaction gas mixture, pressure, etc. It can be stated that when the reactant and deposit are of the same material, volatilization is associated with the reaction between the reactant and gaseous species. A completely opposite phenomenon occurs during solid deposit. Transport of the gas is found to be dependent on the equilibrium constant, determined by temperature and the reactant source. In order to avoid deposition at unnecessary places, all parts of the system are kept as hot as the vapor pressure. The substrate is generally maintained at a temperature higher than the vapor supply. One important point to mention here is that the deposit is to be carried out at low vapor pressure to stop volatilization. Chemical by-products of the reaction are exhaust and unreacted precursor gases. As an example, SiC is synthesized by CVD according to the following reaction:

$$CH_3SiCl_{3(gas)} \rightarrow SiC_{(solid)} + 3HCl_{(gas)}$$

Solid silicon could also be synthesized from $SiCl_4$ according to the following reaction:

$$SiCl_{4(gas)} + 2H_2 \rightarrow Si_{(solid)} + 4HCl$$

Metallic particles could also be prepared from the following CVD reaction:

$$TiCl_{4(gas)} + 2Mg_{(solid)} \rightarrow Ti_{(solid)} + 2MgCl_{2(gas)}$$

Oxide particles like SnO_2 nanoparticles have been synthesized by the CVD process. The following reaction is assigned:

$$SnCl_{4(gas)} + 2H_2O_{(gas)} \rightarrow SnO_{2(solid)} + 4HCl_{(gas)}$$

In CVD, chloride salts are always preferred due to low vaporization temperature and low cost. Multicomponent materials are also synthesized by CVD. As an example, Eu-doped Y_2O_3 has been prepared from an organometallic precursor of yttrium and europium. Synthesis of silver nanoparticles by this technique seems to be somewhat difficult due to lack of a stable, volatile silver complex. It has been found that organo-silver compounds are very light and moisture sensitive, which leads to low thermal stability. On the other hand, inorganic complexes of silver are not volatile [15]. CVD is classified into various types depending on the type of technique used for activation and the type of precursor gas used. When substrate activation in the CVD process is carried out at high temperatures (~200°C–1600°C) at subtorr total pressures, it is described as thermal CVD. The most frequently used activation process is by plasma, called plasma-enhanced chemical vapor deposition (PECVD). PECVD works at relatively low temperatures and has recently been widely used for carbon nanotube synthesis. Pure silver film has been deposited using thermal and PECVD using $AgC–(CF_3)–CF(CF)_3$ as precursor material [16]. The main difficulty here is the handling of fluoroalkyl compound. ZnO thin film has been grown by metal-organic chemical vapor deposition (MOCVD) [17].

Not only thin films but also a few nanostructures are also synthesized using this process. For example, TiO_2 nanorods have been synthesized on WC-Co substrate. In this context, the same research group has observed that there exists an optimum temperature for the formation of nanorods [18]. Other activation processes such as ions, photons, lasers, hot filaments, or combustion reactions are used successively in CVD. Depending on the type of precursor gas, CVD can be classified into MOCVD or organo-metallic chemical vapor deposition. Generally, metal atoms with a number of alkyl radicals like methyl, ethyl, and isopropyl are used to prepare thin film using MOCVD. The main disadvantage of this technique is that the metalorganic compounds are highly reactive in the presence of oxygen, and thus they often get contaminated. To avoid contamination, H_2 is used to flow through the reaction chamber [19]. The III–V group of semiconductors are synthesized by the MOCVD technique using AsH_3, PH_3, and NH_3 as precursor materials. The reaction mechanism that has been proposed here is as follows:

$$\left(CH_3\right)_3 Ga + AsH_3 \rightarrow GaAs + 3CH_4$$

The main advantage of MOCVD is the fact that the reaction chamber includes a large temperature gradient creating convective loops and high flow of gas, leading to turbulence in gas flow, etc. CVD is described as metal-organic vapor phase epitaxy or organometallic vapor phase epitaxy if the process is used for thin epitaxial growth rather than polycrystalline or amorphous film deposition. Epitaxial ferroelectric thin film of $BaTiO_3$ was deposited by Kaiser et al. at 600°C [20]. CVD possesses several advantages such as conformal growth of layers representing equal and uniform thickness of the film. The films and layers deposited by CVD are of high quality and are free from impurities due to its self-cleaning nature in the form of exhaust gases. However, CVD is not free from disadvantages, such as the need for precursors that are volatile at room temperature, which are very expensive. In addition, by-products of the process such as $Ni(CO)_4$, explosive B_2H_6, corrosive $SiCl_4$, hazardous CO, H_2, or HF are found to be highly toxic and require proper handling systems, which adds costs. In CVD, due to high temperature, uneven thermal expansion of the substrate introduces unwanted stress that causes mechanical instabilities in the grown layers. The latter issue, however, is minimized to some extent by using low temperature activation in PECVD. The schematic diagram for the PECVD setup consisting of vacuum chamber, substrate, substrate heater, and RF generator is shown in Figure 4.1.

4.1.1.3 Flame Spraying Synthesis

The process of plasma or flame spraying synthesis is one of the most suitable techniques to synthesize nanopowders, noble metal nanoparticles on substrate, thin films, and multilayer systems. It requires a vertically or horizontally oriented thermally insulated chamber containing setup to produce

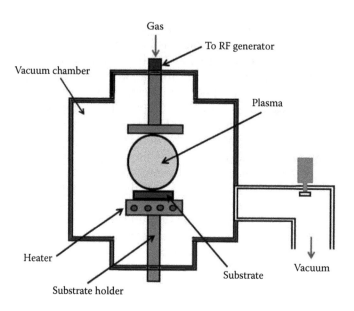

FIGURE 4.1
Schematic diagram of PECVD setup.

high-energy flames, such as thermal plasma, an electric heater, a microwave, and a laser. In the thermal plasma technique, extremely high temperatures ranging from 1,000 to 20,000 K are produced by a direct current arc or by inductively coupled radio frequency discharge. A stream of reactants either in gaseous state or in liquid state in the form of aerosols, or a mixture of both, is thrust or sprayed through a small nozzle into the flame to form thermal plasmas as illustrated in Figure 4.2. In this respect, there exists another process, namely flame aerosol, close to the flame spraying where droplet-to-particle and gas-to-particle are two mechanisms of conversion of

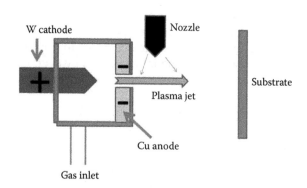

FIGURE 4.2
Schematic diagram of flame spraying technique.

the metal precursor. The former uses gaseous precursors, and is called the vapor-fed flame process, while the latter utilizes a liquid-fed flame. The difference between them is that in the liquid-fed process, if the supporting fuel gets mixed with metal then it is called the flame spray synthesis process. If the precursor is sprayed separately then it is called the flame-assisted spray synthesis process. In the flame spray technique, precursor droplets containing fuel are evaporated and burned out. Particle formation is driven by this burning process. Plasma initiates in situ chemical reactions where the precursor materials are instantly decomposed into various nanoparticles of oxide and nitride in the presence of radicalized oxygen and nitrogen. Using high-energy flame, microscale solid precursors such as metals are instantly vaporized. The subsequent vapors, if cooled rapidly, form nanoparticles by homogeneous nucleation. The size of the synthesized particles depends on the diameter of the nozzle tip, the surface tension of the starting solution, the viscosity, and the pressure difference between the inside and outside of the nozzle. The main advantage of this technique is that the high voltage applied between the sprayer and the substrate helps in uniform deposition of the thin film. In addition, as this is a high-temperature method, no further heat treatment is required for the high-temperature crystallite phase.

This process is also used for synthesizing nanoparticles and powders of various metallic alloys. For example, after spraying nitrates of Mn, Zn, Fe, etc., with an atomizer it is possible to synthesize ferrite nanoparticles [21]. Pt nanoparticles have been successively synthesized on Al_2O_3 substrate [22]. Sometimes the chamber is maintained in atmospheric conditions, and in this case thermal decomposition is accompanied by oxidation of metallic species, giving rise to metal oxide nanopowders. Instead of a nozzle, an ultrasonic nebulizer, attached to the horizontal tubular chamber, is sometimes used, whereby it is possible to synthesize submicron-sized particles with uniform distribution. It has been shown that larger particles are formed due to drying and collapse. Densification of droplets results from insufficient enthalpy during the conversion of liquid solvent into vapor. Often, the frequency of the ultrasonic nebulizer is used to tune the size of the synthesized particles, and the empirical relation between particle size (D) and nebulizer frequency (f in MHz) has been proposed in this context, given by the following expression:

$$D = 0.34 \left(\frac{8\pi\gamma}{\rho f^2} \right)^{1/3}$$

where γ and ρ represent the liquid's surface tension (in N/m) and density (in kg/m^3), respectively. For example, metallic silver nanoparticles have been synthesized with an ultrasonically atomized spray technique from aqueous silver nitrate solution at a temperature above 650°C within a chamber filled with argon [23]. Often, the gaseous composition is modified to incorporate doping into the final product. As an example, partial fluorination of Ti precursor introduced F doping into TiO_2 [24]. It is often found that hollow particles are formed due to precipitation of the precursor surrounding the droplet prior to the evaporation of the solvent.

4.1.1.4 Self-Assembly Synthesis

Self-assembly is one of the most promising bottom-up chemical approaches for the synthesis of nanostructures. Wide ranges of nanoparticles have been synthesized by this method, ranging from organic, biological compounds to inorganic oxides, metals, and semiconductors. It starts with the spontaneous formation of an organized structure from elementary particles or molecules [25]. Briefly, it can be stated that the process of self-assembly is the selective attachment of a particle or molecule on a surface or another molecule by natural or catalyzed interactions forming ordered patterns, guided by thermodynamic equilibrium with its surroundings. The interaction, acting at molecular level, could be of several types, such as weak covalent interactions, van der Waals interactions, electrostatic interaction, hydrophobic interactions, or interfacial hydrogen bonding. The type of interaction determines the structure formed in this process. These techniques can be classified into chemical, physical, and colloidal self-assembly. In chemical self-assembly, molecules from a solution assemble by noncovalent interaction to form new crystal or chained structures such as self-assembled monolayers. The mobility of the molecules plays a significant role during the self-assembly process. In physically assembled molecules, they align directly on a substrate surface to form ordered layers. In colloidal self-assembly, molecules assemble inside a liquid suspension to form nanostructures. Four strategies have been developed to prepare assemblies: (1) controlled formation of covalent bond, (2) polymerization utilizing covalence, (3) self-organization, and (4) self-assembly of molecules. The first process has been developed on the basis of assembling molecules depending on the sequential formation of covalent bonds, which is used to generate arrays of covalently bonded atoms with well-defined compositions, suitable to prepare nanostructures, far from thermodynamic equilibrium. The second method includes covalent polymerization of small molecules into larger molecules that indirectly guide stable nanostructure. The basic strategy behind it is the abandoned covalent bonds establishing connections between atoms to organize them into a structure. Molecular crystals, liquid crystals, emulsions, and colloids are examples of such structures, synthesized by this technique. The fourth technique is the spontaneous assembly of molecules into a structured, noncovalently joined aggregate. A variety of self-assembled noble metal nanoparticles have been synthesized by this method for different applicational aspects. For example, layer-by-layer self-assembly of silver nanoparticles was synthesized by Zhao et al. [26]. In order to limit their size, they are capped by mercaptosulfonic acid. Layer-by-layer structure has been investigated through cyclic voltammetric measurement, while the synthesized layer structure was characterized for surface-enhanced Raman spectroscopy (SERS) activity. Self-assembled silver aggregates have been synthesized on a large scale under alkaline conditions using resorcinol, silver nitrate, and sodium hydroxide as precursor materials [27]. In this synthesis process, 10 mL of resorcinol solution, taken in a beaker, was mixed with 0.5 mL of 0.1 M $AgNO_3$ and 10 mL of 0.5 M NaOH solutions. Self-assembly of silver nanoparticles occurs at a temperature of 80°C. The SERS activity of the synthesized assembled structure has been investigated using crystal violet and cresyl fast violet as probe molecules. It was also noted that SERS activity corresponding to crystal violet is enhanced with increasing size of the silver aggregate. Self-assembly of a silver nanosheet into a nanoflower occurs in a typical reaction between $HAuCl_4$ and poly(vinyl pyrrolidone) in ethanolic solution [28]. The SERS activity of this self-assembled structure is also examined using 4-aminothiophenol as a probe molecule. Self-assembly of gold nanorods has been successfully synthesized by Nikoobakht et al. [29]. Not only single-compound self-assembled structure but also self-assembly of a two-component system has been produced. For example, self-assembly of FePt nanoparticles has been synthesized by Sun [30]. Synthesized assembly is found to have potential application in data storage and biomedicine. Several self-assembly nanostructures of oxides such as ZnO have been synthesized by this method [31].

4.1.1.5 Atomic or Molecular Gas Condensation Process

The atomic or molecular gas condensation process was primarily used for production of metallic nanoparticles. It starts with the production of a stream of vaporized matter in a vacuum environment. Vaporization is achieved by thermal evaporation, obtained by Joule heating of refractory crucibles, electron beam evaporation devices, or sputtering sources. Vapor is then channeled into a reaction chamber containing either an inert or reactive gas atmosphere for rapid cooling. Generally, vapor is cooled rapidly due to collision with the gas molecules, resulting in a condensed liquid phase of the nanoparticles. Nanoparticles, in the liquid phase, are then allowed to coalesce and solidify in a controlled environment to form uniformly shaped nanoparticles. There are also other physical processes like expansion of the chamber, transferring heat into the surroundings to cool vapor. Figure 4.3 illustrates the schematic diagram of the system, used for atomic or molecular gas condensation. Depending upon heating and cooling processes, it is classified into several groups. For example, a furnace flow reactor has been developed to produce saturated vapor of the material at an intermediate temperature. Here, a crucible containing the precursor material is kept in

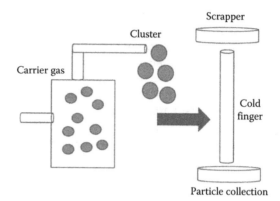

FIGURE 4.3
Schematic diagram of the system for atomic or molecular gas condensation method.

the flow of the inert carrier gas and the nanoparticles thus formed by a natural cooling process. This method is limited by the type of crucible. It has often been found that the crucible introduces impurities into the synthesized material. In the laser reactor, flowing reactant gas is heated with an IR laser and the decomposition of the gas phase into nanoparticles occurs when they collide with the carrier gas. As an example, pyrolysis of SiH_4 results in Si nanoparticle formation.

In the flame reactor, a heating-mediated flame is used to cause chemical reactions during condensable moments. This process has been used to synthesize nanoparticles of TiO_2, SiO_2, and SnO_2 (diameter ~ 10–100 nm) starting from $TiCl_4$, $SiCl_4$, and $SnCl_4$. In a plasma reactor, plasma energy is utilized to initiate a chemical reaction that starts with decomposition of precursor material into ions, atoms, and nanoparticles formed during cooling. Different types of plasma like dc plasma, ds arc plasma, RF induction plasma, and inductively coupled plasma are adopted here. In the spark method, a high-current spark between two electrodes is used to evaporate electrode material, which leads to the formation of nanoparticles. For example, SiC nanoparticles have been synthesized using this method, where discharge occurred between two Si electrodes in a CH_4 atmosphere. Al_2O_3 nanoparticle are formed from a spark between Al electrodes in the presence of an O_2 atmosphere. In the inert gas condensation technique, inter gas like He or Ar is allowed to flow over the evaporation source that carries the nanoparticles formed above the evaporative source through the process of thermophoresis toward the substrate. It has been shown that increased vapor pressure or higher molecular weight increases the mean diameter of the synthesized particles. Sometimes, O_2 is used to flow to prepare oxide nanoparticles. Iron- and cobalt-based magnetic fluid has been synthesized by this technique for targeted drug delivery and hyperthermia activity [32]. TiO_2 nanoparticles have been synthesized by Wu et al. in a supercritical CO_2 condition [33]. In this typical synthesis, titanium

(IV) isopropoxide, dissolved in isopropanol, was stirred in a high-pressure cell and 1 mL of ZONYL FSJ, which acts as a surfactant, was added to the stirred solution. Then liquid CO_2 was pumped into the high-pressure cell at a temperature of 60°C until the pressure reached 30 MPa. TiO_2 nanopowder formed after 4 hours of reaction.

4.1.1.6 Supercritical Fluid Synthesis

Supercritical fluid is defined as a state of material that neither condenses nor evaporates to form a liquid or gas above a critical temperature and pressure, that is, at supercritical temperature, no phase boundary exists between the gas and liquid phases. As a consequence, supercritical fluids exhibit a hybrid property between the liquid and gas phase in such a way that they could be continuously tunable with small differences in pressure and temperature. Supercritical fluid is created by subjecting a gas or a medium to a very high temperature and pressure depending upon the type of medium used to reach a state above its critical point. Since the 1990s, supercritical fluids have been adopted by researchers to synthesize nanomaterials, particularly functional and hybrid nanomaterials. The physicochemical properties of the synthesized nanomaterials are found to be highly dependent on the nature of the materials, their composition, the crystallinity rate of the formed nanopowders, etc. Common media used for supercritical fluid formation are CO_2 and H_2O because of their nontoxic, nonflammable, inexpensive, and environmentally safe nature. The supercritical fluids thus formed, such as supercritical carbon dioxide ($scCO_2$) and water (scH_2O), are then used as solvents to dissolve the desired compatible substance. The nanoparticles are formed through precipitation of the solvent by either rapid expansion of supercritical solutions (RESS) or by using as supercritical antisolvent (SAS) (Figure 4.4). The RESS process involves depressurizing the supercritical solution of the substance through a nozzle that gives extremely fast precipitation of nanoparticles, and the solvent is evaporated. SAS has been developed to achieve nanosized hydrophobic materials that can't be processed by REES due to their poor solubility. This method relies on the fact that when a solution is expanded by a gas, its liquid phase seems to no longer be a good solvent for the solute, and nucleation begins [34].

Various nanoparticles have been formed by this technique and the method is found to have several advantages. For example, the formation of metallic Cu nanoparticles is found to occur at relatively lower temperatures in the presence of a mixture of NH_3 and CO_2 rather than only CO_2. The reactive nature of NH_3 is taken into consideration for such reduction in formation temperature. In nanostructure formation of oxide materials by the hydrothermal process, the supercritical condition plays a significant role. A multication-based system is also synthesized using this process. $Ba_xSr_{1-x}TiO_3$ has been synthesized by this method starting from TiO_2, $BaCO_3$ at a temperature of around 150°C–380°C at

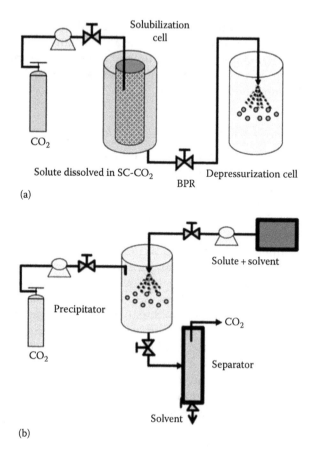

FIGURE 4.4
Schematic diagram of (a) REES and (b) SAS process.

26 MPa of pressure. The crystallinity of the synthesized nanomaterials is also tuned by supercritical fluids. Amorphous and boehmite alumina have been synthesized using CO_2/ethanol and H_2O/ethanol mixtures, respectively. During synthesis of $BaTiO_3$ using this method, the water to ethanol ratio plays a significant role in the crystallinity of $BaTiO_3$. Many organic/inorganic hybrid structures are also found to be prepared by this method. Pd and Ag nanoparticles, hybridized with polymers like perfluoroalkyl, polysiloxane, and oligoethylene glycol, have been synthesized. In this context, supercritical CO_2 swells into polymer matrices, opening routes to prepare nanostructure formation into polymer matrices. Using this process, polypropylene/silica composite has been synthesized by impregnating polypropylene with TEOS. In this reaction, supercritical CO_2 swells into matrix, followed by hydrolysis/condensation of TEOS in the polymer matrix. Supercritical H_2O concentration also plays a role in preparing nanoparticles. Alkane-thiol-stabilized Cu

nanoparticles have been synthesized under the supercritical H_2O condition. This technique overcomes the drawbacks of many major synthesis techniques such as excessive use of solvent, thermal and chemical solute degradation, structural changes, high residual solvent concentration, lack of control over particle size, nonuniform distribution of particle size, etc.

4.1.1.7 Electrochemical Deposition/Electrodeposition

Electrochemical deposition or the electrodeposition technique for the synthesis of thin film of semiconducting or metallic systems on conducting substrate has attracted interest due to its simplicity and cost-effectiveness and the possibility of scaling up. Its basic mechanism involves the reduction of cations on a conducting substrate using electrical current (Figure 4.5). For example, in the aqueous solution of $CuSO_4$, copper and sulfur are present in ionic form Cu^{+2} and SO_4^{-2}. In the presence of applied potential, Cu^{+2} ions migrate toward a cathode where they lose their charge by accepting two electrons and are deposited on the cathode. This step may be expressed as:

$$Cu^{+2} + 2e \rightarrow Cu \text{ (metallic)}$$

If the anode material is made of metallic copper, then they dissolve into the solution by releasing two electrons to balance charge neutrality. On the other hand, if the anode is made of a noble metal like Ag or Pt, water is oxidized at the anode surface. Then the reaction may be written in the following form:

$$2H_2O \rightarrow 4H^+ + O_2 + 4e$$

When Cu is used as the anode, the Cu^{+2} ion concentration remains unchanged, whereas the use of a noble metal would reduce Cu^{+2} ion concentration followed by the enhancement of H^+ ions. In the presence of an external potential

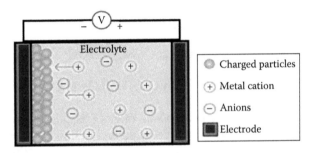

FIGURE 4.5
Schematic diagram of electrodeposition process.

there would be rearrangement of the ions at the surface of the electrode, and an electrical double layer, called the Helmholtz double layer, would form followed by diffusion of the layer. These two layers are called the Gouy-Chapman layer. Deposition of thin film can be described in the following steps: (1) hydrated metal ion migrates toward the cathode under the influence of current by a method of diffusion and convection; (2) hydrated metal ions diffuse into the double layer where water molecules are aligned. Then metal ions enter into the Helmholtz double layer where they lose their hydrated part; (3) hydrated ions are neutralized and adsorbed on the cathode surface; (4) adsorbed ions, on the cathode surface, migrate toward the growth point. This technique is found to be very useful for the preparation of thin film with a large area suitable for solar cell application. Film thickness, its morphology, and optical properties are controlled by various operational parameters like current density, applied potential, deposition time, electrolyte concentration, etc. Normally, nitrate or chloride salt is taken as a precursor material, and the deposition steps involved here significantly depend on the nature of the salt. For example, during deposition of ZnO thin film using $ZnCl_2$ as the precursor, ZnO is formed by electro-generation followed by electroprecipitation. The reaction steps can be summarized as

$$\tfrac{1}{2}O_2 + H_2O + 2e^- \rightarrow 2OH^-$$

$$Zn^{+2} + 2OH^- \rightarrow Zn(OH)_2$$

It has been shown that if the bath is maintained below 40°C, then $Zn(OH)_2$ is formed. On the other hand, ZnO forms when the bath temperature is kept above 40°C following the reaction:

$$Zn(OH)_2 \rightarrow ZnO + H_2O$$

If the starting material is $ZnNO_3$ instead of $ZnCl_2$, then two anionic groups, NO_2^- and OH^-, are formed and the formation of ZnO is the result of the following reactions:

$$Zn(NO_3)_2 \rightarrow Zn^{+2} + 2NO_3^-$$

$$2NO_3^- + H_2O + 2e^- \rightarrow NO_2^- + 2OH^-$$

$$Zn^{+2} + 2OH^- \rightarrow Zn(OH)_2 \rightarrow ZnO + H_2O$$

In addition to pure ZnO, nanocomposite of ZnO has also been synthesized by this method. For example, ZnO/polypyrole thin film was synthesized by Moghaddam et al. [35]. Si thin film is also deposited by this method. As silicon exhibits high reduction potential and a high level of surface reactivity, it can't be produced in aqueous medium. Mainly silica or fluororthosilicate is used as the solute. Amorphous Si has also been synthesized using this technique, as suggested by Agrawal and Austin, starting from $SiHCl_3$ as silicon source and propylene and 0.1 M tetrabutyl ammonium chloride as supporting electrolyte [36]. Similar to Si, the synthesis of Ni nanodots thin film has been carried out by electrodeposition on an anodic aluminum oxide template at constant current of 4 mA and taking Pt as counterelectrode [37]. This typical electrodeposition was carried out using 0.084 M of $NiCl_2$, 1.6 M of nickel sulfamate tetrahydrate, and 0.3 M of boric acid as precursor. Gold nanoparticles are also formed by electrodeposition on glassy carbon substrate [38].

TiO$_2$ nanowire has also been synthesized on anodic aluminum membrane by electrodeposition. In this typical process, electrodeposition was carried out with 0.2 M $TiCl_3$ solution at pH_2. Here TiO_2 is deposited into the pore of the membrane. After heating the membrane at 500°C for 4 hours, the template is removed and the pure anatase phase of TiO_2 is obtained [39]. $Ni(OH)_2$ nanoparticles are deposited on B-doped diamond electrodes by the process of electrodeposition [40]. In addition to the synthesization of nanoparticles, efforts have also been made to synthesize template-free nanostructures of various metallic or oxide materials using this technique. For example, Au nanorods have been synthesized by Chang et al. [41]. The deposition is carried out at a current density of 3 mA/cm^2 in the presence of hexadecyltrimethyl-ammoniumbromide and tetradodecylammonium bromide. Hexadecyltrimethyl-ammoniumbromide, in this reaction, acts as a stabilizing agent that prevents nanorods from agglomeration. The length of the nanorods is found to be dependent on gold ion concentration and their release rate. Lu et al. has synthesized gold nanowire using nanoporous polycarbonate [42].

4.1.2 Top-Down Approach

Top-down approaches play a crucial role in the synthesis process, specifically manufacturing in large quantities; however, there are a few shortcomings associated with each of the techniques. The major problems associated with these approaches can be broadly described as surface structure imperfection, crystallographic damage, and contamination due to impurities. These issues significantly impact the physical properties and surface chemistry of nanostructures and nanomaterial. The effect of these imperfections such as reduced conductivity, unwanted catalysis, etc., become pronounced for nanostructures due to the large surface over volume ratio, thereby reducing the reliability of the process. In the following sections, we'll discuss a few top-down approaches to synthesize nanoparticles.

4.1.2.1 Ball Milling

Among the latest top-down methods, the ball milling and rod milling techniques are simpler, cost-effective, and productive ways to synthesize nanomaterials and their alloys by transferring mechanical energy. The basic

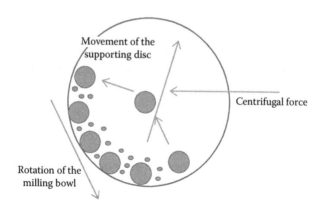

FIGURE 4.6
Schematic diagram of ball milling.

principle behind the milling process is impingement and grinding (schematically shown in Figure 4.6). The process relies on the grinding of a pow-
dered material inside a mill chamber encapsulating a drum with rotating
balls made of steel, silicon carbide, or tungsten. During the milling process,
balls rotate inside the drum and undergo high-energy collision with the
powder of the material due to centrifugal force and crush the solid particles
of the powder, generating nanoparticles. The nanoparticles are produced in
this process not by cluster assemble, but by structural decomposition due to
plastic deformation. Nanoparticles with a size between 2 and 20 nm are synthesized by this process depending on two primary factors: speed of rotation and duration of the milling process. In the process of ball milling, the
ratio of ball and powder material (generally termed RBP) is typically maintained at 2:1 for achieving substantially uniform attrition but sometimes is
varied between 1:1 and 220:1. An interesting feature of this process is that
it is carried out at room temperature. With the advancement of technology,
several milling processes such as high-energy ball milling, planetary ball
milling, attrition ball milling, centrifugal ball milling, vibratory ball milling,
low-energy tumbling mill, etc., have been developed depending upon methods. In planetary ball milling, rotation of the supporting disc and turning
of the vial introduces centrifugal force that causes charging of the powder
alternatively on the inner wall and throws off at high speed. In vibration ball
milling, powder and milling tools are agitated in a perpendicular direction
at a very high speed (1200 rpm). This method is generally used to prepare
amorphous materials. Low-energy tumbling milling is used to synthesize
mechanically alloyed powder. The attrition ball milling process relies on
the stirring action of an agitator having a vertical rotator central shaft with
horizontal arms. The main problem with this process lies in the fact that
mainly irregular-shaped nanoparticles are generated. But this problem is
overcome by increasing the energy of collision, and the milling system that

has been developed on this mechanism is high-energy ball milling. Recently, the energy has been further increased by either using magnetic force placed outside the drum or by using a planetary system. Commonly, the milling energy used here is 1000 times higher than the energy of conventional system and the process is classified into two types: (1) dry milling and (2) wet milling. Parameters such as milling time and temperature are used to tune the nanoparticle's shape and size. Additionally, in recent milling systems, atmospheric pressure, the nature of the atmosphere, etc., are also used to synthesize nanoparticles with different compositions. In the dry milling process, the particles generally tend to be agglomerated; this is overcome in the wet milling process, which involves stability of chemicals, density of materials, and vapor pressure. The advantage of the wet milling process is that it prevents reagglomeration. It has been noted that in the wet milling process, small balls favor new surface formation, whereas larger balls favor amorphization in the dry milling process. Recent studies have shown that HEBM can be used efficiently to form nanostructures such as carbon nanotubes, boron nitride nanotubes, and Zn nanowire by using a combination of ball milling and annealing. Ball milling boosts nucleation and new surface formation for growth, whereas annealing plays a significant role in nanostructure formation, which can be controlled by adjusting the mill temperatures. It has been shown earlier that temperature changes play a significant role in the diffusion mechanism involved in nanostructure formation [43]. Although the HEBM technique is lucrative for low-cost synthesis of nanoparticles and nanostructures, it requires extra measures to restrict contamination from the milling system environment.

4.1.2.2 Nanolithography Process

Nanolithography is the process of transferring a pattern with at least one dimension less than 100 nm onto a thin film or substrate by engraving the pattern on a reactive polymer film, termed photoresist followed by selective etching. This field emerged with the development of semiconductor-integrated circuits for microelectromechanical systems (MEMS). Methods like optical photolithography, suited for MEMS, were developed, but due to diffraction phenomena of light waves, this technique places limits on the preparation of suboptical-sized particles or patterns for nanoelectromechanical systems. The fabrication processes developed here are electron-beam lithography (EBL), X-ray lithography (XRL), etc. Although EBL exhibits a resolution of ~1–2 nm, its commercial usage is limited by its inherent serial processing. In contrast to EBL, parallel processing techniques enable XRL to fabricate a large number of nanostructures, but its disadvantages lie in the fact that its resolution is limited to 20–50 nm due to photoelectron range and diffraction limit. Therefore, technology demands a nanofabrication technique combining the resolution of EBL and the throughput of XRL. In this context, nanolithography based on scanning tunneling microscope (STM) or dip pen lithography using atomic force microscope has gained the attention of researchers due to the fact that it can be used for imaging as well as manipulating matter on an atomic scale. The limitation of serial processing for STM-based nanolithography has been overcome by exploring a few parallel nanolithography processes, including diffusion-controlled aggregation at surfaces, lased-focused atom deposition, nanoscale template formation from 2D crystalline protein monolayers, reactive-ion etching, etc. Over the past decade, a few techniques like evanescent near-field photolithography [44,45], surface plasmon-based lithography [46–48], and the thermal dip pen lithography method have been build up in the field of nanolithography. In Near field nanolithography, a nanometer scale circular or square aperture as a mask is used and a deep UV or Extreme UV light with extremely low wavelength is transmitted through it. Its basic principle remains same as traditional optical lithography, but cost and complexity optical system increase dramatically. It has recently been shown that optical transmission and nanoscale spatial resolution increase significantly with the use of C- and H-shaped ridge apertures. Two ridges of this shape help in confining waveguide propagation mode in the gap between ridges. In the following sections, a few of the lithography techniques are briefly discussed.

4.1.2.2.1 Electron-Beam Lithography and Focused Ion-Beam Lithography

With miniaturization of electronic devices, a method that has grown rapidly is the fabrication technology for these devices. Among the different technologies, those that have emerged steadily in this regard are EBL and the Focused Ion Beam (FIB) method, and these have gained the most attention from researchers. Since their discovery, these have proved to be powerful tools for integrated circuit debugging, device modification, mask repairing, etc. The EBL process involves exposure by a highly focused stream of electrons to modify the solubility of a resist material followed by a subsequent development step. Initially, this technique involved a scanning electron microscope, attached to a pattern generator and beam blanker for viewing the field of the exposed area. At present, EBL is a fully dedicated patterning system with a high-density electron source for faster throughput and mechanical stages. Resolution of the mechanical stage should be very high in order to expose large substrates in the presence of a narrow field of electron beam. Recently, in semiconductor industries EBL has been routinely used to fabricate master masks and reticles from computer-aided design files. Generally, EBL is being operated under two different schemes: projection printing and direct writing. In the projection printing scheme, relatively large electron beams are allowed to pass through a mask using a high-precision lens system onto a resist-coated substrate. In the direct writing method, a small spot of electrons is used to write directly on a substrate. The main advantage of the direct writing method is high resolution and the capacity for arbitrary pattern generating without masking. But its main disadvantage lies in the fact that it takes a long time for pattern generation, but this has been overcome by developing methods utilizing parallel beams of electrons. The basic objective of the EBL technique is to fabricate an arbitrary pattern in the resist with high resolution, sensitivity, density, and reliability. The process starts with a stable, high-brightness electron source focused in a small spot (size approximately a few nanometers) that should be free of astigmatism. This is achieved by electron optics and degree of focus where electrons are allowed to flow through a high-vacuum column. Mutual repulsion between electrons causes divergence in the electron stream, but this is overcome by using lower current and electrons having higher kinetic energy. As the electrons enter into the resist, they suffer from a series of low-energy elastic collisions and are deflected either in the forward or backward direction. Backscattering broadens the spot size of the incident electrons, whereas electrons scattered in the forward direction pass through the resist and deeply penetrate into the substrate. In addition to elastic scattering, electrons entered into the resist also suffer from inelastic collision, resulting in ionization (secondary electron generation into the resist) followed by physicochemical changes in the resist. Two types of resist are available. Positive tone resist possesses conversion from low to high solubility molecules upon exposure to electrons. Polymethyl methacrylate (PMMA) is the classic example of this type of resist. Its long polymer chain is broken into smaller and more soluble fragments in the presence of electrons. Hydrogen silsequioxane (HSQ) is the example of negative resist that turns into a less soluble loner chain polymer, caused by a cross-linking reaction in the presence of electrons. After being exposed to electrons, resists are immersed in a liquid developer to remove fragments in the case of positive resist or to remove non-cross-linked molecules in the case of negative resist (Figure 4.7). Temperature and duration are found to be important factors in removing the resist. During the development step, the solvent penetrates into the resist and starts to surround the fragments. A gel is formed when the molecule starts to react with the fragment. After being completely surrounded, the fragments detach from the polymer and diffuse into the solvent. In this context, powerful solvents remove longer fragments, but this reduces resolution. Removal of the fragment is found to be a diffusion-controlled process where the mobility of the molecules plays a significant role. It has been shown that diffusivity (D) at temperature T can be written as $D \sim n^{-\alpha} \exp(-U/kT)$, where $n^{-\alpha}$ represents the mobility of the fragments with size n in a medium characterized by α and U signifies the activation energy. A is found to be ~1 in dilute solution of smaller molecules and ~2 for longer chain molecules in denser medium. Depending on the kinetics of this diffusion control process, development duration is fixed. It has been observed that nanoscale resolution increases with decreasing developing temperature. Several structures such as single electron transistors which are one of the major electronic components of recent time, metal–insulator–metal tunneling junctions and, resistive nanowires have been fabricated by this method [49].

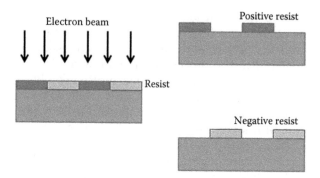

FIGURE 4.7
Schematic diagram of lithographic process.

The basic operational principle of FIB is the same as for EBL. The difference lies in the fact that in FIB, accelerated ion beams are used to incident on the substrate. Similar to EBL, it also involves two working methods: ion beam direct write and ion beam projection. The advantage of FIS over EBL lies in the fact that FIB could easily punch metallic film directly due to its heavier mass. Using this technique, now it becomes possible to deposit tungsten, platinum, gallium, carbon, etc. When precursor gas (tungsten hexacarbonyl in the case of tungsten) is used to flow into the chamber, gas molecules collide with the ion beam, leading to the decomposition of gas into the nonvolatile component (tungsten here) on the surface [50]. The direct writing process, also known as the FIB milling process, relies on transferring patterns by direct impingement of the ion beam on the substrate on which the pattern is to be generated. Similar to EBL, it also requires a process to remove materials from the substrate, and a large number of 3D patterns have been fabricated using a wide range of materials. Ion beam projection possesses a collimated beam of ions through a stencil mask, and the reduced image of the mask is projected onto the substrate underneath. Similar to EBL, pattern generation by FIB is found to be dependent on the ion-induced changes in the resist. Ion-triggered reaction, ion-induced etching, and ion-induced deposition are the generally accepted mechanisms for the changes in resist. When ions enter into the solid resist, they suffer elastic and inelastic collisions. Elastic collision causes beam broadening, amorphization of resist, and ion implantation into the resist. Inelastic collision involves transfer of energy either to electrons of the substrate or to other nuclei or atoms of the substrate. It may also give rise to emission of secondary electrons after ionizing atoms. These secondary electrons with energy between 1 and 50 eV break molecular bonds. FIB has several advantages over EBL such as its higher resolution due to the absence of proximity effects and higher resist sensitivity. Since there are no backscattered electrons, the pixel size of FIB is found to equal spot size. In addition, FIB generates almost

200 secondary electrons by single ion irradiation, and thus it works at a much faster rate than EBL. In this context, Horiuchi et al. have obtained a 200 nm line width in PMMA using He^+ ion beam [51], whereas a sub-20 nm line width on PMMA has been achieved by Kubenal et al. on the basis of a Ga^+ ion beam [52]. Recently, van Kan et al. fabricated a 22 nm line width in HSQ using a 2 MeV H_2^+ ion beam [53].

4.1.2.2.2 Dip Pen Lithography

This is the direct-writing method utilizing the scanning probe tip of AFM to draw nanostructures and it works under ambient conditions. In this pattern writing process, the tip of the probe is coated with liquid ink, which flows on the surface when it makes contact with the surface of the matter on which the pattern is to be generated. The driving force behind this writing process is chemisorption, which moves the molecules of the ink from the probe tip to the substrate via a water-filled capillary as the tip is moved across the surface of the matter. The resolution of the pattern in this process is governed by several factors, such as the grain size of the substrate, tip–substrate contact time (scan time), rate of chemisorption and self-assembly of the molecules, and relative humidity, which influences water meniscus. Dip pen lithography is classified into different groups such as classical dip pen lithography, solvent-assisted dip pen lithography, matrix-assisted dip pen lithography, and solvent- and matrix-assisted dip pen lithography (Figure 4.8).

4.1.2.2.3 Thermal Dip Pen Lithography

The basic principle of thermal dip pen lithography is the same as a soldering iron, and this has been developed on the basis of customized heatable AFM [54]. The AFM tip is coated with a material having the property that coating material remains solid at room temperature, whereas it melts at higher temperatures. When the AFM tip is heated, the coating material melts and it exits the tip and is deposited, generating a pattern on the surface. The main advantage of this technique lies in the fact that the fluidity of the coating material depends on the tip temperature that could be used to

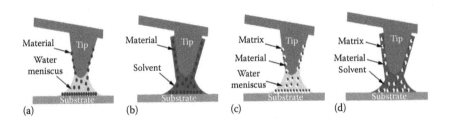

FIGURE 4.8
(a) Classical dip pen lithography, (b) solvent-assisted dip pen lithography, (c) matrix-assisted dip pen lithography, and (d) solvent- and matrix-assisted dip pen lithography.

turn on or off the pattern generation. Secondly, three-dimensional patterns can be fabricated by putting new layers on top of a previously deposited layer. Although this technique works under vacuum condition, a large variety of material such as metal, polymers, and self-assembly structure could be fabricated by this method. This technique is particularly helpful in generating a pattern of polymer material including insulating and conducting polymer, since it only requires melting of polymer before it is decomposed. Another important aspect of this method is that depositing material with high thermal energy is organized into a well-structured monolayer before solidifying. Using this technique, deposition of poly(3-dodecylthiophene), a conducting polymer, commonly known as PDDT is possible on the SiO_2 substrate. Fabrication of this particular polymer is useful for organic electronics. The smallest pattern that has been generated by this method is 75 nm in size. Now, with the advancement of technology, it is possible to heat the AFM tip up to 1000°C, allowing different material to be fabricated by this technique. In this context, using this technique, it is possible to generate a pattern of indium metal (size < 80 nm) on glass and silicon substrate [55]. Full commercial utilization of this technique has been realized by developing parallel lithography. IBM recently developed an array of 4096 controlled thermal cantilevers in AFM. They are designed in such a manner that the cantilever material exhibits rapid heating (1–20 μs) and rapid cooling (1–50 μs) times that lead to parallel patterning and fabrication of the integrated circuit based on this technique. Another important advantage of this method lies in the fact that it requires very little ink material, making it cost-effective, and it does not generate any toxic etchant waste, making it eco-friendly.

4.2 Characterization of Nanoparticles

It should be noted from the previous section that the synthesis of nanoparticles can be achieved by various methods and each method has its own characteristic advantages and disadvantages. By tuning the synthesis parameters, it is possible to prepare nanoparticles with different sizes and shapes. Nanoparticles possess a large fraction of atoms on their surface, and it has been shown that these surface atoms significantly modulate their properties. For example, the color and melting point of gold nanoparticles strongly depend on their size. It has been observed that the melting point of gold decreases from 1337 K in its bulk state to 600 K for the nanoparticle state with a diameter of 2 nm. In order to characterize the change in properties, behavior, and structural dimensions of nanoparticles, highly sensitive instrumentations are required which can measure with high accuracy and atomic-level resolution. There are two primary modes of nanoparticle and nanomaterial characterization. The microscopy method of characterization is based on the principle of imaging by using radiation or a particle beam, whereas the spectroscopy method of characterization is based on the principle of measuring shifts in the wavelength of radiation as a result of its interaction with matter. This section presents discussions on some of the common microscopy and spectroscopy methods used in the characterization of nanomaterials and nanostructures.

4.2.1 Characterization X-Ray Diffraction

X-ray is used to investigate proper phase formation of the synthesized material. We know that crystalline material is characterized by regular arrangement of atoms, and therefore when x-rays are incidenting on such material, they are diffracted. The phenomenon is similar to the diffraction of ordinary light from gratings. As amorphous materials don't exhibit any crystallinity, they would not produce any diffraction, and thus x-ray can be used to differentiate between crystalline and amorphous material. It is also well established that each material possesses its own crystal structure with definite lattice parameters, and hence x-ray can also be used to determine these parameters. The basic operational principle of x-ray diffraction lies in the fact that when x-rays are incidenting on a lattice plane (defined as the plane through atomic positions), they are reflected. The path difference between two successive reflected beams would be 2d sin θ, where the angle θ represents the angle between the incident x-ray and the sample surface, as depicted in Figure 4.9. Now if we vary the angle θ, then the corresponding path difference should also vary. Therefore, in certain conditions this path difference could become an integral multiple

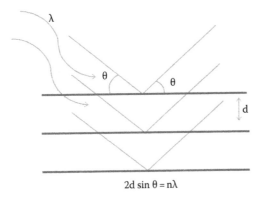

$$2d \sin \theta = n\lambda$$

FIGURE 4.9
Schematic diagram of x-ray diffraction.

of the wavelength (λ) of the incidenting x-ray, that is, it may be written as $2d \sin \theta = n\lambda$. If this condition is satisfied, then reflected x-rays are in the same phase, that is, they would interfere constructively, giving intensity maxima for the reflected rays. This relation is known as Bragg's law of reflection. At other angles, they would interfere destructively, producing zero intensity. Therefore, by noting the angle at which we are attaining maxima, it is possible to determine the value of "d" which is the characteristic of a particular crystalline material [56]. Depending on the variable parameter to satisfy Bragg's law, diffraction techniques are classified into three categories: (1) the Laue method where θ is kept fixed, whereas λ is varied. Generally, single-crystal material is investigated using the Laue method. (2) Powder diffraction, where λ is kept fixed and θ is varied. The powder diffraction technique is generally used to examine polycrystalline samples. (3) The rotating crystal method, where θ is partly varied and is used for polycrystalline samples. The powder x-ray diffractometer consists of a sample holder, a source of x-ray, and an x-ray detector. Generally, Cu, Co, and Fe are used as a target for x-ray production and a scintillation counter is used for detection of x-ray. Experimentally, satisfaction of Bragg's law is achieved by either moving the x-ray source and detector by an angle θ (called θ–θ geometry) or by moving the x-ray sample holder by an angle θ and the x-ray detector by an angle 2θ (called θ–2θ geometry). It is well established that a crystalline material may exhibit more than one lattice plane depending on the three-dimensional arrangement of atoms. But they don't possess the same intensity. The intensities of the diffraction peaks are found to be dependent on various factors:

1. *Scattering of x-ray by an electron*: The intensity of the x-ray scattered by an electron (charge e and mass m) and detected at a distance "r" was calculated according to J.J. Thomson and is given by $I = I_0 \dfrac{e^4}{r^2 m^2 c^4} \left(\dfrac{1+\cos^2 2\theta}{2} \right)$. This is also known as polarization factor.

2. *Scattering of x-ray by an atom*: The intensity of the x-ray scattered by an atom is illustrated by atomic scattering factor (f), which is defined by the ratio of amplitude of the wave scattered by an atom and amplitude of the wave scattered by an atom.

3. *Scattering of x-ray by a unit cell*: The intensity of the x-ray, scattered by a unit cell (the smallest repetitive pattern that generates the periodic system), is determined from the structure factor (F) of the material. It is defined by the ratio of the amplitudes of the wave scattered by a unit cell and the wave scattered by an electron. Its mathematical expression is given by

$$F_{hkl} = \sum_{i=0}^{N} f_i e^{2\pi i (hu_i + kv_i + lw_i)}$$

where (hkl), (u_i, v_i, w_i), and f_i are the Miller indices of the reflecting plane, fractional coordinates, and atomic scattering factor of the i-th atom, respectively. Here, summation includes all the atoms within the unit cell. Real calculation would show that F_{hkl} is nonzero only for a few combinations of Miller indices, that is, any arbitrary plane would not produce a diffraction peak.

4. *Multiplicity of the planes*: It has been shown that the intensity of the diffraction peak depends on the number of planes giving rise to diffraction maxima at that particular angle.

In addition to these, there are several factors like Lorentz factors, the x-ray absorption coefficient of the material, and temperature affecting the intensity of the diffraction peak. In addition to phase identification, diffraction technique is also used to determine the particle size of the material under investigation. This is done in the following way. Let us consider a diffraction peak at 2θ with intensity I_{max}. Let $2\theta_1$ $(<2\theta)$ and $2\theta_2$ $(>2\theta)$ be the two angles beyond which the diffraction pattern corresponds to zero intensity, and these two arise from planes separated by distance "t." If we apply Bragg's law to these two sets of planes, then we may write

$$2t \sin\theta_1 = (m-1)\lambda \quad \text{and} \quad 2t \sin\theta_2 = (m+1)\lambda$$

where "m" represents any integer number. Now from these two equations, we may write $t(\sin\theta_2 - \sin\theta_1) = \lambda$. After simplifying, we have

$$2t \cos\left(\frac{\theta_2 + \theta_1}{2}\right)\sin\left(\frac{\theta_2 - \theta_1}{2}\right) = \lambda$$

If we consider $\frac{\theta_2 + \theta_1}{2} = \theta$ and $\theta_2 - \theta_1 \beta$ that represent the full width at half maxima of the diffraction pattern, then the above equation may be simplified into $t \cos\theta \sin\beta = \lambda$. Therefore, from this relation it is possible to calculate particle size (t). More accurate calculation gives $t = \frac{0.9\lambda}{\beta \cos\theta}$. This expression, known as Debye–Scherrer's relation, has been derived on the basis of the diffraction peak broadening due to the finite size of the particle. Nonuniformity in the lattice parameter (strain effect) also introduces broadening of the lattice parameter. This can be investigated using the following method. Differentiating Bragg's law with respect to θ, we have $\frac{\Delta d}{d} + \frac{\cos\theta}{\sin\theta}\Delta\theta = 0$.

It may be written in the following form: $\Delta\theta = \frac{-\sin\theta}{\cos\theta}\frac{\Delta d}{d}$. $\frac{\Delta d}{d}$ is defined as the fractional change in the lattice parameter, that is, strain (ε). Therefore, considering these two contributions (neglecting sign), we have $\beta = \frac{\lambda}{t\cos\theta} + \frac{\sin\theta}{\cos\theta}\varepsilon$.

This expression may be simplified into $\beta \cos\theta = \frac{\lambda}{t} + \varepsilon \sin\theta$. If we plot the $\beta \cos\theta$ vs. $\sin\theta$ curve, then it would be a straight line with slope ε and intercept on $\beta \cos\theta$ axis equal to $\frac{\lambda}{t}$. Therefore, from this analysis it is possible to calculate strain as well as particle size. This plot is known as the Williamson–Hall plot.

4.2.2 Scanning Electron Microscopy

Scanning electron microscope (SEM) is one of the most commonly used techniques for characterization of the shape and size of nanostructures. It uses electrons rather than light for imaging. A typical setup of SEM is shown in Figure 4.10. Its basic component starts with an electron gun, attached to the top of the microscope, and it liberates electrons either by thermionic emission or by the electric field emission mechanism. In the case of thermionic emission, current is allowed to flow through a W-shaped tungsten wire and electrons are generated due to thermal energy. In the case of field emission, LaB_6 is used and electrons are liberated by means of tunneling in the presence of an electric field. The emitted electrons are then formed into a beam by electromagnetic lenses and are allowed to pass through a column where they accelerate. When the beam hits the sample surface, it knocks out a few electrons from the sample. The primary principle behind SEM operation is

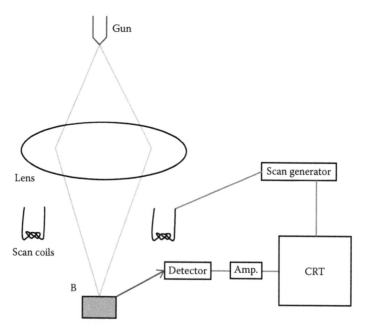

FIGURE 4.10
Schematic diagram of scanning electron microscope.

the elastic and inelastic collision of the incidenting electrons with electrons inside the sample [57]. In each type of collision, incidenting electrons after penetrating into the sample transfer their energy to electrons inside the sample surface, resulting in the ejection of secondary electrons from the sample. Since the penetration depth is finite, the depth of ejection out of the secondary electrons is limited within the sample, that is, secondary electrons emerge only from finite depths. Thus, it may be stated that secondary electrons carry information from the sample surface only. Therefore, imaging of these secondary electrons reveals topographical information, orientation of the sample materials, crystalline structure, and also the chemical composition of the sample. The magnification of the SEM is defined as the ratio of the electron beam sweep distance on the sample and sweep distance on the monitor. The magnification of the SEM images ranges from 10 to over 300,000 with a resolution of a few nm. The first step in the process of SEM imaging is sample preparation, which involves removing all water to avoid vaporization in the vacuum environment used for SEM. Also, SEM requires the sample to be conductive, and hence nonconductive samples such as nonmetals are coated with conductive materials such as gold by "sputter coater." Once the sample is prepared, an electron beam from a source is focused using magnetic lenses into a very fine spot size of ~5 nm, which is scanned over the sample by deflection coils. These electrons with energy typically ranging between 100 eV and 50 keV penetrate the sample and scatter. The scattered electrons are captured using a cathode ray tube for imaging.

The electrons may undergo elastic scattering, resulting in ejection of electrons from the sample with the same energy as incident electrons. These electrons are termed backscattered electrons and provide information regarding distribution of materials with different atomic numbers of the constituting elements. Electrons primarily also undergo inelastic scattering, which results in ejection of electrons from the sample with energy less than incident electrons, typically less than 50 eV. These are termed secondary electrons and they form the SEM images with topological information. Finally, some electrons excite the core electrons, which after ejection return to their ground state by emitting either a characteristic x-ray photon or an Auger electron, which provide information on the chemical composition of the sample. Although characterization by SEM is comparatively easy to carry out, it is limited to use in solid dry material samples only.

4.2.3 Transmission Electron Microscopy

Transmission electron microscope (TEM) is one of the most sophisticated instruments in recent times, and has been developed utilizing the wave nature of electrons. A typical TEM setup is described in Figure 4.11a. Unlike SEM, which involves scattered electrons from the sample for imaging, TEM relies on transmission or penetration of the electrons through thin samples, typically less than 200 nm [58]. Also, the energy of the accelerated electrons used

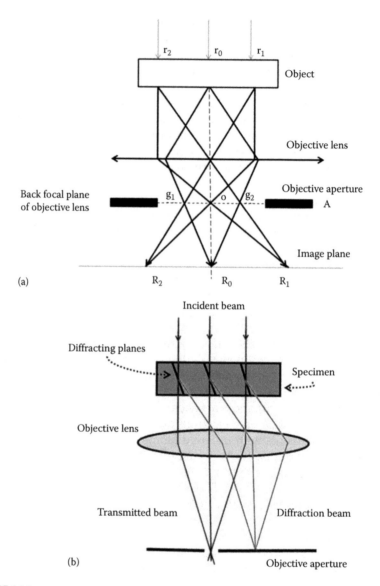

FIGURE 4.11
Schematic diagram of (a) transmission electron microscope and (b) bright field image.

in TEM is of the order of a few 100 keV, which is significantly higher than that used in SEM. The working principle of TEM generally involves diffraction of electrons from the sample, and thus they must have a wavelength of the order of lattice spacing. Thus, such high is necessary to achieve this wavelength. The characterization is carried out based on the diffraction pattern imaging of the transmitted electrons, which are either deflected or undeflected depending upon the properties of the sample, as shown in Figure 4.11b. The diffraction patterns of the electrons depend upon the scattering events during the transmission, which can either be elastic, resulting in high-intensity diffraction patters, or inelastic, depending upon heterogeneities in the sample such as grain boundaries, second-phase particles, defects, dislocations, density variations, etc., causing spatial variation of intensity in the patterns.

The TEM imaging can be of two types. Firstly, there is diffraction-contrast imaging where only a small fraction of transmitted electrons are used for the final image. Most of the scattered electrons are filtered by using a small objective aperture. This mode enables the determination of the orientation of the sample by studying the contrast of the image obtained. The second mode is the high-resolution mode in which no objective aperture is used and all the transmitted electrons post diffraction are used to obtain the diffraction pattern. This mode is defined as phase-contrast imaging mode of operation and is used to characterize structural properties of the sample including measurement of atomic patterns. The TEM technique offers a very high magnification ranging from 50 to 10^6 due to the extremely small wavelength associated with the electron beams, which is in the order of 1 Å.

4.2.4 Scanning Probe Microscopy

Scanning probe microscope (SPM) has been developed on the basis of scanning a surface using a sharp tip with a radius of curvature of ~3–10 nm. The tip, used to scan the surface, is mounted on a flexible cantilever to allow the tip to follow the surface profile. In contrast to imaging by TEM, it requires the sample to be sliced down into thin layers; sample damage doesn't arise here. The main advantage of SPM over other microscopies like SEM or HRTEM is that SPM provides 3D imaging of the sample surface. It also provides atomic-level resolution. In addition, vacuum environment conditions are not needed for SPM in contrast to TEM and SEM where a vacuum is mandatory. The SPM techniques are categorized into two primary types depending upon the type of measured parameter and the sample surface, and are described as follows.

Scanning Tunneling Microscopy: STM is based on the principle of measuring the *tunneling current* with a metallic tip, which is in very close proximity to a conducting substrate but not actually in physical contact [59]. Here, under bias voltage applied between the tip and the substrate, current flows between them depending on the distance between the tip and the substrate (Figure 4.12a). During scanning as the distance changes, tunneling current appears to be changed. This variation of current is taken as a parameter to measure height, that is, undulation of the film surface. This mode of operation is known as constant height mode. There is also another mode of operation where tunneling current is used to keep constant, while tip–substrate distance gets varied (Figure 4.12b).

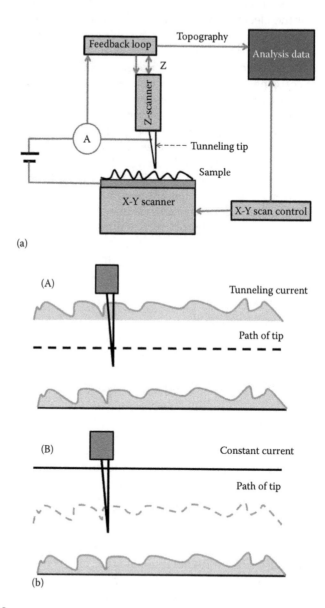

FIGURE 4.12
(a) Schematic diagram of STM and (b) variation of tunneling current upon tip–surface distance.

Atomic Force Microscopy: AFM is based on the principle of measuring the van der Waals force between the tip and the substrate surface, which may be the short-range repulsive force in contact mode or the longer-range attractive force in noncontact mode [60]. The changes in force of attraction or repulsion as per the demography of the sample are measured from the bending of a cantilever on which the probe is mounted. A mirror is attached to the back of the cantilever, on which the laser incidents, which after deflection is incidenting on a position-sensitive detector. During the interaction between the tip and sample surface, the deflection of the tip changes and this deflection is measured with a position-sensitive detector.

References

1. P.J. Flory, *Principles of Polymer Chemistry*, Cornell University Press, Ithaca, NY (1953).
2. C.J. Brinker and G.W. Scherer, *J. Non Cryst. Solids* 70 (1985), 301.
3. J.N. Hasnidawani, H.N. Azlina, H. Norita, N.N. Bonnia, S. Ratim, and E.S. Ali, *Procedia Chem.* 19 (2014), 211–216.
4. P.K. Singh, J. Roy, S. Mukharjee, C. Ghosh, and S. Maitra, *Adv. Ceram. Sci. Eng. (ACSE)* 3 (2014), 46–51.
5. P. Tamilselvi, A. Yelilarasi, M. Hema, and R. Anbarasan, *Nano Bull.*, 2 (2013), 130106.
6. C.K. Ghosh, S.R. Popuri, T.U. Mahesh, and K.K. Chattopadhyay, *J. Sol-Gel Sci. Technol.* 52 (2009), 75–81.
7. F. Kamil, K.A. Hubiter, T.K. Abed, and A.A. Al-Amiery, *J. Nanosci. Nanotechnol.* 2(1) (2016), 37–39.
8. H.Y. Chen and J.H. Chen, *Mater. Lett.* 188 (2017), 63–65.
9. M. Stefan, D. Ghica, S.V. Nistor, A.V. Maraloiu, and R. Plugaru, *Appl. Surf. Sci.* 396 (2017), 1880–1889.
10. P. Saikia, N.B. Allou, A. Borah, and R.L. Goswamee, *Mater. Chem. Phys.* 186 (2017), 52–60.
11. L. Sun, R. Zhang, Z. Wang, L. Ju, E. Cao, and Y. Zhang, *J. Magn. Magn. Mater.* 421 (2017), 65–70.
12. P. Visuttipitukul, P. Sooksaen, and N. Yongvanich, *Ferroelectrics* 457 (2013), 82–88.
13. A.C. Grigorie, C. Muntean, T. Vlase, C. Locovei, and M. Stefanescu, *Mater. Chem. Phys.* 186 (2017), 399–406.
14. M. Epifani, C. Giannini, L. Tapfer, and L. Vasanelli, *J. Am. Ceram. Soc.* 83(10) (2000), 2385–2393.
15. Z. Yuan, N.H. Dryden, J.J. Vittal, and R.J. Puddephatt, *Chem. Mater.* 7 (1995), 1696–1702.

16. C. Oehr and H. Suhr, *Appl. Phys. A* 49 (1989), 691.
17. M. Pan et al., *J. Cryst. Growth* 287 (2006), 688–693.
18. S.K. Pradhan, P.J. Reucroft, F. Yang, and A. Dozier, *J. Cryst. Growth* 256 (2003), 83–88.
19. A.G. Thomson, *Mater. Lett.* 30 (1997), 255–263.
20. D.L. Kaiser, M.D. Vaudin, L.D. Rotter, Z.L. Wang, J.P. Cline, C.S. Hwang, R.B. Marinenko, and J.G. Gillen, *Appl. Phys. Lett.* 66(21) (1995), 2801.
21. S. Lorenzou, C.C. Agrafiotis, and A.G. Konstantinopoulos, *Granul. Matter.* 10 (2008), 113–122.
22. R. Strobel, W.J. Stark, L. Mädler, S.E. Pratsinis, and A. Baiker, *J. Catal.* 213 (2003), 296–304.
23. K.C. Pingali, D.A. Rockstraw, and S. Deng, *Aerosol Sci. Technol.* 39 (2005), 1010–1014.
24. D. Haneda, H. Li, S. Hishita, N. Ohashi, and N.K. Labhsetwar, *J. Fluor. Chem.* 126 (2005), 69–77.
25. G.M. Whitesides, J.P. Mathias, and C.T. Seto, *Science* 254 (1991), 1312–1319.
26. S. Zhao, K. Zhang, J. An, Y. Sun, and C. Sun, *Mater. Lett.* 60 (2006), 1215–1218.
27. S. Panigrahi, S. Praharaj, S. Basu, S.K. Ghosh, S. Jana, S. Pande, T.V. Dinh, H. Jiang, and T. Pal, *J. Phys. Chem. B* 110 (2006), 13436–13444.
28. L. Wang, G. Wei, C. Guo, L. Sun, Y. Sun, Y. Song, T. Yang, and Z. Li, *Colloids Surf. A: Physicochem. Eng. Asp.* 312 (2–3) (2008), 148–153.
29. B. Nikoobakht, Z.L. Wang, and M.A. El-Sayed, *J. Phys. Chem. B* 104 (2000), 8635–8640.
30. S. Sun, *Adv. Mater.* 18 (2006), 393–403.
31. B. Zhang, F. Wang, C. Zhu, Q. Li, J. Song, M. Zheng, L. Ma, and W. Shen, *Nano-Micro Lett.* 8 (2016), 137–142.
32. N.H. Hai, R. Lemoine, S. Remboldt, M. Strand, J.E. Shield, D. Schmitter, R.H. Kraus, M. Espy Jr., and D.L. Leslie-Pelecky, *J. Magn. Magn. Mater.* 293 (2005), 75–79.
33. C. Wu, J.W. Huang, Y.L. Wen, S.B. Wen, Y.H. Shen, and M.Y. Yeh, *Mater. Lett.* 62 (2008), 1923–1926.
34. K. Byrappa, S. Ohara, and T. Adschiri, *Adv. Drug Deliv. Rev.* 60 (2008), 299–327.
35. A.B. Moghaddam, T. Nazari, J. Badraghi, and M. Kazemzad, *Int. J. Electrochem. Sci.* 4 (2009), 247–257.
36. K. Agrawal and A.E. Austin, *J. Electrochem. Soc.* 128 (1981), 2292.
37. J.S. Jung, E.M. Kim, W.S. Chae, L.M. Malkinski, J.H. Lim, C. O'Connor, and J.H. Jun, *Bull. Kor. Chem. Soc.* 29 (2008), 2169.
38. T. Hezard, K. Fajerwerg, D. Evrard, V. Colliere, P. Behra, and P. Gros, *Electrochim. Acta* 73 (2012), 15–22.
39. Y. Lei, L.D. Zhang, and J.C. Fan, *Chem. Phys. Lett.* 338 (2001), 231.
40. L.A. Hutton, M. Vidotti, A.N. Patel, M.E. Newton, P.R. Unwin, and J.V. Macpherson, *J. Phys. Chem. C* 115 (2011), 1649–1658.
41. S.S. Chang, C.W. Shih, C.D. Chen, W.C. Lai, and C.R.C. Wang, *Langmuir* 15 (1999), 701.
42. Y. Lu, M. Yang, F. Qu, G. Shen, and R. Yu, *Bioelectrochemistry* 71 (2007), 211.
43. C. Suryanarayana, *Prog. Mater. Sci.* 46(1–2) (January 2001), 1–184.
44. J. Aizenberg, R.J. Rogers, K.E. Paul, and G.M. Whitesides, *Appl. Phys. Lett.* 71 (1997), 3773–3775.
45. M.M. Alkaisi, R.J. Blaikie, S.J. McNab, R. Cheung, and D.R.S. Cumming *Appl. Phys. Lett.* 75 (1999), 3560–3562.

46. W. Srituravanich, N. Fang, C. Sun, Q. Luo, and X. Zhang, *Nano Lett.* 4 (2004), 1085–1088.
47. X.G. Luo and T. Ishihara, *Appl. Phys. Lett.* 84 (2004), 4780–4782.
48. Z. Liu, Q. Wei, and X. Zhang, *Nano Lett.* 5 (2005), 957–961.
49. Y. Chen, *Microelectron. Eng.* 135 (2015), 57–72.
50. J. Orloff, M. Utlaut, and L. Swanson, *High Resolution Focused Ion Beams: FIB and Its Applications*, Springer, New York (2003).
51. K. Horiuchi, T. Itakura, and H. Ishikawa, *J. Vac. Sci. Technol. B* 6(1) (1988), 241.
52. R.L. Kubena et al., *J. Vac. Sci. Technol. B* 7(6) (1989), 1798.
53. A.J. van Kan, A. Andrew, and F. Frank, *Nano Lett.* 6(3) (2006), 579–582.
54. P.E. Sheehan, L.J. Whitman, W.P. King, and B.A. Nelson, *Appl. Phys. Lett.* 85 (2004), 1589.
55. B.A. Nelson, W.P. King, A.R. Laracuente, P.E. Sheehan, and L.J. Whitman, *Appl. Phys. Lett.* 88 (2006), 033104.
56. B.D. Cullity, *Elements of X-Ray Diffraction*, Addison-Wesley, Reading, MA, Vol. 2, (1978).
57. C.W. Oatley, W.C. Nixon, and R.F.W. Pease, *Adv. Electron. Electron Phys.*, 21 (1966), 181–247.
58. D.B. Williams and C.B. Carter, The transmission electron microscope, *Transmission Electron Microscopy: A Textbook for Materials Science*, Springer, New York (1996), pp. 3–17.
59. G. Binnig and H. Rohrer, *Surf. Sci.* 126(1–3) (1983), 236–244.
60. G. Binnig, F. Calvinand, and C. Gerber, *Phys. Rev. Lett.* 56(9) (1986), 930.

5

Electrical Transport in Nanostructures

Angsuman Sarkar

CONTENTS

5.1 Introduction

Electrical transport in low-dimensional structures such as 2D quantum wells (one-dimensional confinement), 1D quantum wires (two-dimensional confinement), and 0D quantum dots (three-dimensional confinement) is mostly dominated by quantum effects. In contrast, in 3D structures, where carriers are free to move in any of three directions, classical transport equations like Boltzmann's transport are used and quantum effects are not included. The transport mechanism in low-dimensional structures has emerged as a dynamic field of research due to the advancement of fabrication and synthesis of materials with possibilities of synthesis in nanodimensional size. In this chapter, an overview of and highlights of the transport mechanism in low-dimensional nanostructures is provided. The concept of current flow in a mesoscopic conductor explained with the "transmission function" is provided. Moreover, the recent applications based on mesoscopic transport have also been provided. From this concept, Landauer expression and its applicability on different mesoscopic objects has been examined.

5.2 Quantum Effects in Low-Dimensional Systems

In a low-dimensional system, quantum effect will dominate if de Broglie wavelength given by

$$\lambda_{\text{de Broglie}} = \frac{\hbar}{\sqrt{2m^*E}} \tag{5.1}$$

becomes larger than the characteristic length of the quantum structure [1]. For a tunneling device/structure, de Broglie wavelength exceeds the length of the tunneling barrier [2]. Here $\hbar$ denotes the reduced Plank's constant, m^* denotes the effective mass of the carrier, and E is the carrier's kinetic energy. It is worth mentioning that for a low-dimensional structure with a size less than $\lambda_{\text{de Broglie}}$, wave properties of electrons are expected to be prominent.

Moreover, the thermal energy given by K_BT needs to be less than the bandgap separation energy ΔE in order to observe the quantum effects, that is, $K_BT \ll \Delta E$. Since quantum effects depend on the phase coherence of carriers, in order to increase quantum effects, scattering needs to be minimized [3]. The mean free time between two successive collisions of carriers needs to be less than the size or dimension of the low-dimensional quantum structures [4].

The limit beyond which quantum effects become dominant has been given the name mesoscopic physics, where the carrier behaves like particles as well as waves. Moreover, in the case of ballistic transport, a carrier does not suffer any collision for transmitting charge or energy [5].

A mesoscopic object refers to an object whose size is between that of microscopic and macroscopic objects. A typical mesoscopic conductor is much larger than microscopic objects like atoms but not as large as a macroscopic object, so that the mesoscopic object demonstrates ohmic behavior [6]. It is worth mentioning that a conductor will exhibit ohmic behavior if it satisfies the following:

1. Dimension is larger than characteristic length of de Broglie wavelength, associated with kinetic energy of electrons

2. Dimension is larger than mean free path which is the average distance that an electron travels before its initial momentum is destroyed by elastic scattering.

3. Dimension is larger than phase-relaxation length, interrelated to the distance it moves before its initial phase is obliterated.

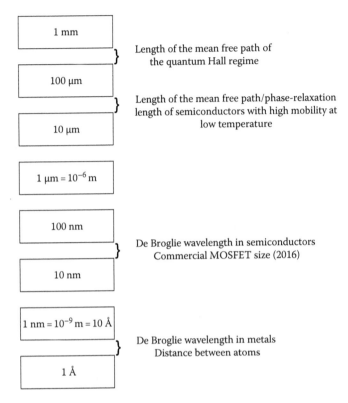

FIGURE 5.1
Length scales relevant to mesoscopic objects.

It is worth mentioning that the three characteristic lengths described earlier are material dependent, temperature dependent, and magnetic field dependent [7]. This is the reason why the size of the mesoscopic objects varies from a few nm to 100s of μm, as shown in Figure 5.1.

The low dimension, essential for observation of quantum in quantum wells, quantum wires, and quantum dots, can be fabricated or synthesized directly. For example, to realize a quantum well, z-directions are very thin for a thin-film quantum well structure, but the X–Y plane where the carriers will move are of macroscopic dimension [8]. In an alternate approach, under the gate region of a Field Effect Transistor of low dimension, by the application of potential in the gate terminal, 2DEG (two-dimensional electron gas) is able to form as shown in Figure 5.2. The inversion layer of a MOSFET contains carriers that are confined within the potential well very close to the silicon surface [9]. The well is created and shaped by the oxide barrier, which is effectively infinite except for tunneling calculations. The silicon conduction band bends downward where the amount of bending is proportional to the applied gate bias. The electrons

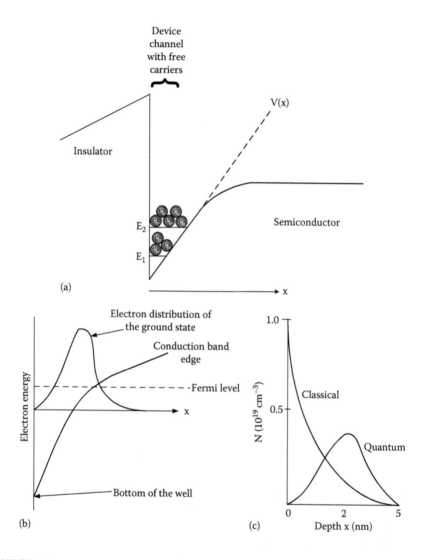

FIGURE 5.2
(a) Schematic of triangular potential well of an inverted MOSFET where the free electrons are confined. (b) An example of quantum-mechanically calculated band bending and energy levels of inversion layer electrons near the surface of the MOS device. (c) Classical and quantum-mechanical electron density versus depth for a <100> silicon inversion layer.

present in the inversion layer cannot move in the direction normal to the oxide layer. Therefore, it is convenient to treat those confined electrons as quantum-mechanically as 2D electron gas, especially at high normal fields [10]. As a result of this quantum confinement, the energy levels of those electrons are arranged in discrete subbands, where each subband corresponds to some quantized level for movement in the normal direction with a continuum for motion in the plane parallel to the surface [11].

In a similar manner, 2DEG can also be formed if a low-energy bandgap material is inserted between two high-energy bandgap materials as shown in Figure 5.3. This process of obtaining high-mobility electrons is known as modulation doping [12], shown schematically in Figure 5.4.

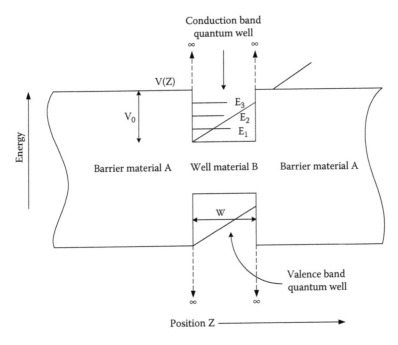

FIGURE 5.3
Band diagram of a potential well created within a heterostructure.

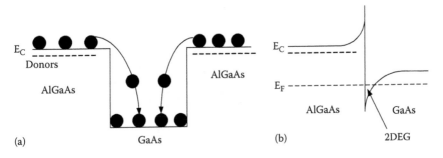

FIGURE 5.4
Modulation doping: (a) AlGaAs electrons are donors that drop into a GaAs potential well and become trapped. (b) Use of a simple AlGaAs/GaAs heterojunction to trap electrons in the undoped small-bandgap GaAs. The 2DEG resulting from the thin shift of charge due to free electrons at the interface.

5.3 Mobility

At room temperature, without an external electric field, electrons present in the conduction band of a semiconductor move at random with their net drift equals zero constituting no current in any direction. However, under the influence of an external electric field, the electron acquires a drift velocity, thus constituting a net current opposite to the direction of the drifted electrons [13] as shown in Figure 5.5.

In the steady state, the rate of collection of momentum from the applied external electric field is equal to the rate of loss of momentum [14] due to scattering, that is,

$$\frac{dp}{dt}\bigg|_{scattering} = \frac{dp}{dt}\bigg|_{Field} \tag{5.2}$$

$$\frac{mv_d}{\tau_m} = eE \tag{5.3}$$

$$v_d = \frac{e\tau_m}{m}E \tag{5.4}$$

τ_m is the momentum relaxation time, which increases with decrease in temperature due to decrease in phonon scattering. Therefore, the mobility is defined as drift velocity to the electric field ratio and is given by

$$\mu = \left|\frac{v_d}{E}\right| = \frac{|e|\tau_m}{m} \tag{5.5}$$

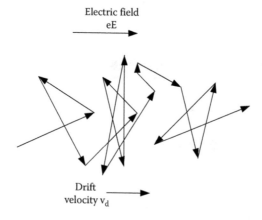

Electric field
eE

Drift
velocity v_d

FIGURE 5.5
Drift velocity superimposed on random motion of electrons in the presence of an electric field.

In the bulk semiconductors, as we reduce temperature, a corresponding rise in mobility is expected due to the decrease in scattering. However, the increase in mobility is limited within some range as ionized impurity scattering starts dominating over phonon scattering as temperature decreases [15]. The highest mobility obtained is 10^4 cm²/V-s for a doping concentration of 10^{17}/cc. If doping concentration is further reduced, higher mobility can be achieved. However, reduction of doping concentration is not fruitful as few carriers will be available for conduction.

However, in a 2DEG, the situation is something different. In this case, a large value of mobility is expected as ionized atoms are physically separated from the mobile carriers due to modulation doping, resulting in a reduction in ionized impurity scattering [16]. Mobility in the range of 10^6 cm²/V-s can be achieved due to reduction in scattering.

5.4 Density of States in Low-Dimensional Systems

Within a quantum well structure, since the $E_{n,z}$ ($n = 0, 1, 2,...$) energies are discrete, the band of energy for every n is called a subband and its total energy is obtained by superimposing electron energy on quantum well energy and their corresponding periodic solution obtained from the 2D periodic potential is given by

$$E_n\left(k_x, k_y\right) = E_{n,z} + \frac{\hbar^2}{2m_{xx}}k_x^2 + \frac{\hbar^2}{2m_{yy}}k_y^2 \tag{5.6}$$

Figure 5.6 plots the subbands and the energy levels associated with the quantum well.

Assuming a 2D circle, the number of electrons per unit area is given by

$$N_{2D} = \frac{2}{\left(2\pi\right)^2}\pi k_\perp^2 \tag{5.7}$$

where

$$k_\perp^2 = k_x^2 + k_y^2 \tag{5.8}$$

and

$$E_\perp = \frac{\hbar^2 k_\perp^2}{2m^*} \tag{5.9}$$

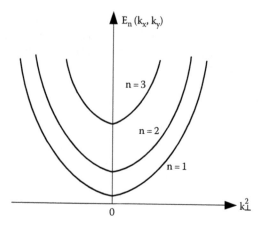

FIGURE 5.6
2DEG and the associated subbands and energy levels.

Therefore, we have density of states, which is constant for every 2D subband and is given by

$$\frac{\partial N_{2D}}{\partial E} = g_{2D}(E) = \frac{m^*}{\pi \hbar^2} \tag{5.10}$$

Figure 5.7 plots the staircase density of states for the rectangular quantum well structure. Figure 5.8 plots the g(E) for a modulation-doped GaAs–AlGaAs interface with the lowest subband filled.

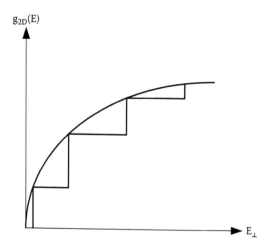

FIGURE 5.7
Staircase density of states $g_{2D}(E)$ for rectangular quantum well structure.

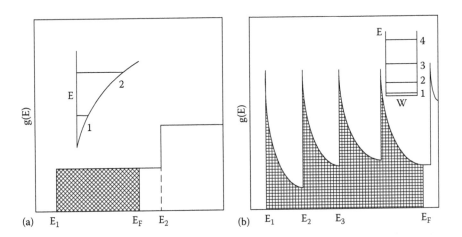

FIGURE 5.8

(a) Variation of quasi-2D density of states with energy; shaded region indicates that only the lowest subband is occupied. (Inset) Discrete energy levels associated with the bottom of the first and second 2D subbands. (b) Variation of quasi-1D density of states with energy; shaded region indicates that only the lowest four 1D subband is occupied. [Inset] Discrete energy levels of square potential well indicating the 1D subband.

In a similar manner, the density of states for a 1D electron gas is given by

$$g_{1D}(E) = \frac{\partial N_{1D}}{\partial E} \tag{5.11}$$

Now N_{1D} is expressed by

$$N_{1D} = \frac{1}{\pi}(k) \tag{5.12}$$

Considering a parabolic band structure (neglecting nonparabolicity) where $E = E_n + \hbar^2 k^2/(2m^*)$

$$N_{1D} = \frac{1}{\pi}(k) = \frac{1}{\pi}\left(\frac{2m^*(E - E_n)}{\hbar^2}\right)^{0.5} \tag{5.13}$$

thus yielding g_{1D} as

$$g_{1D}(E) = \frac{1}{2\pi}\left(\frac{2m^*}{\hbar^2}\right)^{0.5}(E - E_n)^{-0.5} \tag{5.14}$$

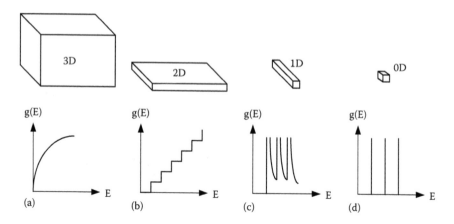

FIGURE 5.9
Density of states as a function energy for (a) 3D bulk, (b) 2D quantum well, (c) 1D quantum wire, and (d) 0D quantum dot.

This indicates that for every bound state-level E_n, there is a singularity in the density of states as shown in Figure 5.8b, with the lowest four states being occupied. For a quantum dot that is a 3D-confined and 0D system, the density of states is a series of delta functions positioned at the energy of a localized state [17]. Figure 5.9 shows the comparison between density of states as a function of energy for 3D to 0D semiconductors.

5.5 From Diffusive Transport to Ballistic Transport

According to the Drude formula of classical transport theory [18], current density J is given by

$$\sigma = \frac{ne^2\tau}{m^*} \tag{5.15}$$

This expression is valid for carrier transport through a solid where scattering dominates as shown in Figure 5.10a. But for low-dimensional structures, in the path of an electron, scattering becomes less, and other kinds of transport such as quasiballistic and ballistic transport dominate, as shown in Figure 5.10b, and c. Figure 5.10b shows that ionized impurity scattering and boundary scattering are of equal importance for quasiballistic transport [19]. However, for ballistic transport, nonzero resistance appears due to the backscattering at the

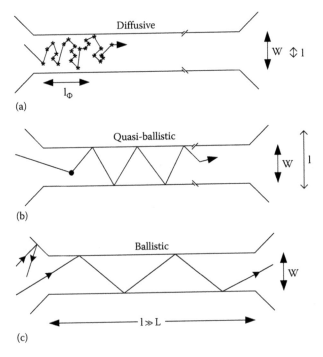

FIGURE 5.10
Comparison of trajectories of electron within a channel of width W and length L for (a) diffusive (l < W, L), (b) quasiballistic (L < l > W), and (c) ballistic transport (l > W, L), considering boundary scattering and ionized impurity scattering (asterisks).

junction between the narrow-width channel and the wide 2DEG region [20]. When collision becomes less important, it is convenient to describe transport using either the Einstein relation or Landauer formula [21].

From the expression of the continuity equation, it is possible to write that

$$j = eD\nabla n \qquad (5.16)$$

where
  ∇n is the gradient of carrier density
  D denotes the diffusion coefficient

Under the equilibrium condition, electrochemical force $\nabla\mu$ is zero due to balance of electrical force, and change in the Fermi energy is expressed as [22]

$$\nabla\mu = 0 = -eE + \nabla n \frac{dE_F}{dn} = -eE + \frac{\nabla n}{g(E_F)} \qquad (5.17)$$

Substituting (5.17) in (5.16), we obtain the Einstein relation [23]:

$$j = eDg(E_F)eE = \sigma E \tag{5.18}$$

$$\sigma = e^2 Dg(E_F) \tag{5.19}$$

The Einstein relation is considered a general expression for any low-dimensional structure. The Landauer approach is a more useful approach to define mesoscopic transport, where current passing through a conductor is defined in terms of the probability of an electron passing through it [18]. The Landauer formula is based upon the conductance G given by

$$I = GV \tag{5.20}$$

It is known that for a the conductance for rectangular 2D conductor increases with increase in width W and decrease in length L, therefore it is possible to write

$$G = \left(\frac{W}{L}\right)\sigma \tag{5.21}$$

where is the conductivity, which is a dimensionless material property. Now the question that has intrigued scientists is how much it is possible to reduce W and L before ohmic properties break down. If W and L are large compared to mean free path l, the transport method is considered diffusive [24]. However, if l is larger than W and L, the transport mechanism is ballistic, expressed by the Landauer formula given by

$$G = \frac{2e^2}{h} \sum_{\alpha,\beta}^{N} |t_{\alpha,\beta}|^2 \tag{5.22}$$

where N is the number of quantum modes and the quantum-mechanical transition probability of coupling from one channel to another $|t_{\alpha,\beta}|^2$ is $\pi L/(2LN)$, which is valid in the ballistic limit.

In general, the theoretical work by Landauer shows that transport through a nanodimensional channel can be described by summing up all the conductance for every possible transmission mode, described by a transmission coefficient t_{mn}. Therefore, it is quite reasonable to consider that quantum quantization appears in the channel conductance owing to ballistic transport [25].

Therefore, the conductance of a 1D channel is given by

$$G = \frac{e^2}{\pi\hbar} \sum_{n,m=1}^{N_c} |t_{m,n}|^2 \tag{5.23}$$

where N_C denotes the number of occupied subbands. It is worth mentioning that under perfect quantization $|t_{mn}|^2 = 0 = \delta_{mn}$.

This is the condition for no scattering and no back reflections or modes mixing in the channel, resulting in purely ballistic transport.

A more explicit expression for Landauer expression for a 1D channel can be derived by expressing the current given by

$$I_j = \int_{E_i}^{E_F} eg_j(E)v_z(E)T_j(E)dE \tag{5.24}$$

where electron velocity is expressed as

$$m^*v_z = \hbar k_z \tag{5.25}$$

$$E = E_j + \frac{\hbar^2 k_z^2}{2m^*} \tag{5.26}$$

Density of states is expressed as

$$g_j(E) = \frac{(2m^*)^{0.5}}{h(E-E_g)^{0.5}} \tag{5.27}$$

$T_j(E)$ is the probability that an electron having energy E injected into sub-band j will pass through the channel in a ballistic manner. Substituting (5.25) through (5.27) in (5.24) we obtain

$$I_j = \frac{2e}{h}\int_{E_i}^{E_j} T_j(E)dE = \frac{2e}{h}T_j\Delta V \tag{5.28}$$

where ΔV denotes the potential energy difference between initial and final energy and is expressed as

$$e\Delta V = E_f - E_i \tag{5.29}$$

Taking summation over all occupied states, we obtain the Landauer formula

$$G = \frac{2e^2}{h}\sum_j T_j \tag{5.30}$$

It is worth mentioning that there are several restrictions to be considered for ballistic transport and a subsequent quantization of the channel conductance [26].

1. Mean free path must be larger than channel length
2. Adiabatic transition at the input and output to minimize reflection
3. Fermi wavelength must be smaller than L to introduce a large number of carriers in the channel
4. Thermal energy k_BT needs to be smaller than subsequent energy level difference between two consecutive subbands. In order to satisfy this, quantum conductance needs to be measured at very low temperature.

5.6 Quantization of Electronic Charge

In this section, the role of coulomb interaction in a mesoscopic system is described. It can be considered as a type of fluctuation in mesoscopic systems which is not affected by the quantization phenomenon and the overlap of the different wave functions. At low temperature, these effects are visible in ultrathin tunnel junctions [27] or coulomb islands [28].

The energy required for localizing electrons in a system is

$$U = \frac{q^2}{C} \tag{5.31}$$

where the charge of an electron is denoted by q and the capacitance of the system is symbolized by C.

Normally, C is in the order of 10^{-18} F for small coulomb islands. This indicates that the energy for charging a single electron is noteworthy. Moreover, it also indicates that single-electron behavior dominates the flow of the carriers that are crossing a tunnel junction or coulomb islands at low temperatures (<10 K). Figure 5.11a shows the parabolic curve between charging energy and the charge Q, which indicates that when $|Q| > e/2$, discrete transfer of electrons takes place. Instability for electron transfer $|Q| > e/2$ appears owing to the discrete nature of the electron charge. Therefore, if a current source is placed, periodical charging and recharging will occur, resulting in oscillation in voltage and charge close to the junction [29]. Therefore, coulomb blockade range is described as $-e/2 < |Q| < e/2$. In the presence of the surface charge Q_0 across the tunnel junction, the charging energy turns out to be

$$U = \frac{(q - q_0)^2}{C} \tag{5.32}$$

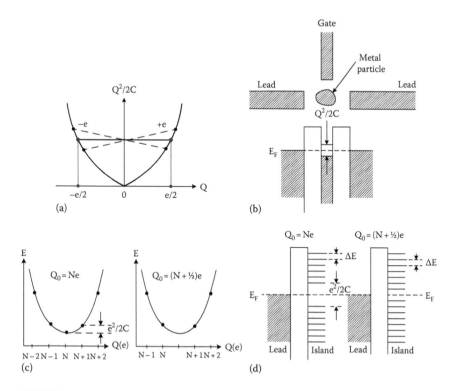

FIGURE 5.11
(a) Parabolic variation of charging energy with polarization charge Q; (b) coulomb island with an energy gap equal $e^2/2C$ and its associated energy levels; (c) charging energy versus polarization charge showing coulomb blockade, that is, when $Q_0 = Ne$, the activation energy is $e^2/2C$ for current to flow; (d) effect of energy quantization at the coulomb island.

This single-electron phenomenon has a wide application in the field of nano-electronic devices [30,31]. Figure 5.11b shows the symbolical diagram of a single-electron transistor where the region between two barriers is the quantum dot or coulomb island. The idea is to control the number of charges that is present within the coulomb island by the polarization of the applied gate bias [32]. Assuming the charges present within the coulomb island are of a quantized nature, polarized charges in the gate cause charge transfer in the coulomb island, and the change in the energy of the systems is equal to

$$U = \frac{(Q - Q_0)^2}{2C} = -qV_G + \frac{Q^2}{2C} + \text{const} \tag{5.33}$$

where $Q_0 = CV_G$ is the polarization charge that is induced by the gate bias. The first term, $-QV_G$, indicates an attractive force of opposite charges in the gate electrode and the other electrode is isolated in nature, whereas $Q^2/2C$

presents the repulsive force between electrons located within the isolated regions, as shown in Figure 5.11c and d. Considering the number of electrons that is transferred equals N, that is, Q = eN, it is possible to write

$$Q_0 = \left(N + \frac{1}{2}\right)e \qquad (5.34)$$

U turns out to be a degenerate for Q = Ne and Q = (N + 1)e. In other words, at a particular gate bias, the difference between the Nth and (N + 1)th electron vanishes. Therefore, only at that value of gate voltage is resonant transfer possible, indicating that coulomb blockade is the activation energy equal to $q^2/2C$ for flow of current.

5.7 Application of Mesoscopic Transport

5.7.1 Single-Electron Sources

The controlled production of a single and several particle excitations in electronics offers the possibility to probe the microscopic physics which underlies quantum transport for certain unitary transformation u(t) for single-source ancila preparation for transportation using the insecure communication channel shown in Figure 5.12 and promises a variety of applications ranging from quantum computation to quantum metrology. The required level of control was recently achieved experimentally for single-electron sources provided by a mesoscopic capacitor [33] and Lorentzian voltage pulses [34]. As a theory group, our goal is to reach new insights into quantum transport by characterizing setups involving such sources and to propose feasible experiments which make their underlying principles detectable. Recent developments include the analysis of the mesoscopic capacitor in a 2D topological insulator [35], where it can be used to produce entanglement as well

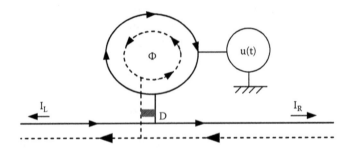

FIGURE 5.12
Unitary transformation for single-source transport using insecure channel.

as spin-polarized currents [36], and as a source that allows for the control of the amplitude and phase of a single-electron wavefunction using microwave excitations [37].

5.7.2 Noise and Fluctuations

In contrast to the electronic properties of most macroscopic objects, where measuring the current as a function of the applied voltage is in most cases sufficient for a complete characterization, the quantum regime of transport through nanostructures shows a richer phenomenology. Here, the investigation of not only the current but also its fluctuations at different frequencies (noise) reveals important insights into the physics. Moreover, one can study the full probability distribution of charge transfers within a given time interval to obtain all cumulants of the current, also known as "full counting statistics" [38]. In a complementary approach, we are interested in the probability distribution of times passing between two or more successive electrons traversing the conductor, called the "waiting time distribution" [39]. It is most suitable to investigate transport properties on short to intermediate time scales and becomes especially important for the characterization of periodically driven quantum systems such as the single-electron sources described here. Moreover, it reveals surprising insights into correlations between electrons in a phase-coherent conductor. We use scattering theory [40] as well as generalized master equations [41] to investigate noise and fluctuations in mesoscopic systems.

5.7.3 Quantum Heat Engines

Thermoelectric energy harvesting can help to recover waste heat back into useful electricity, for example, in small electronic circuits. Unfortunately, thermoelectric materials typically offer a lower efficiency of heat to work conversion. Mesoscopic physics can help overcome these limitations by providing highly efficient nanoscale heat engines. A particular realization of such nanoheat engines is based on double quantum dots in a three-terminal geometry. The two dots are capacitively coupled so they can exchange only energy but no particles. This allows for a crossed flow of heat and charge without direct electrical contact between the hot and the cold bath. For quantum dots in the coulomb blockade regime, it was found that they can reach Carnot efficiency when operated adiabatically, while they still reach half the Carnot efficiency when operated at maximum power [42]. We showed that heat engines based on chaotic cavities reach much larger currents [43]. However, their output power and efficiency drops as the number of open transport channels connecting the dots to their reservoirs is increased. We could demonstrate that the optimal energy harvester is based on resonant tunneling through either quantum dot [44]. It can yield high output power while reaching decent efficiencies of about 20% of Carnot efficiency.

In addition, the quantum dot heat engine can be scaled up to macroscopic dimension by strong parallelization in the form of a Swiss cheese sandwich heat engine. A similar proposal based on quantum wells is twice as powerful but only about half as efficient [45]. We also considered other types of heat engines that involve bosonic degrees of freedom such as magnons (spin waves) [46] and microwave cavity photons [47], thereby establishing connections to spin caloritonics and circuit QED.

5.8 Conclusion

The developments in electronic fabrication in recent years have enabled the possibility of exploring nanodimensional structures containing a thin layer of highly mobile electrons. In this chapter, the conceptual framework for describing current flow in conductors with dimensions less than the mean free length of the carrier has been attempted. The issues discussed here are based on general physics of transport, and the chapter provides very little about device applications. This chapter mainly reviews mesoscopic transport along with the application of nanodimensional transport with a theoretical framework to understand its behavior. To understand the mesoscopic system, the quantum confined system, the density of states, transmission in nanostructures, and the quantum dot and single-electron phenomena have been discussed in order to investigate the transport in such short dimensions.

References

1. Cibert, J., P. M. Petroff, G. J. Dolan, S. J. Pearton, A. C. Gossard, and J. H. English. Optically detected carrier confinement to one and zero dimension in GaAs quantum well wires and boxes. *Applied Physics Letters* 49(19) (1986): 1275–1277.
2. Bell, R. P. *The Tunnel Effect in Chemistry.* New York: Springer Science & Business Media, 2013.
3. Yu, C., J. Li, K. Gao, T. Lin, Q. Liu, S. Dun, Z. He, S. Cai, and Z. Feng. Observation of quantum Hall effect and weak localization in p-type bilayer epitaxial graphene on SiC (0001). *Solid State Communications* 175 (2013): 119–122.
4. Fischetti, M. V. and W. G. Vandenberghe, eds. Generalities about scattering in semiconductors. In *Advanced Physics of Electron Transport in Semiconductors and Nanostructures*, pp. 255–268. Cham, Switzerland: Springer International Publishing, 2016.
5. Iñiguez, B. and T. A. Fjeldly, eds. *Frontiers in Electronics: Advanced Modeling of Nanoscale Electron Devices.* Singapore: World Scientific, 2014.

6. Sohn, L. L., L. P. Kouwenhoven, and G. Schön, eds. *Mesoscopic Electron Transport*, Vol. 345. Dordrecht, the Netherlands: Springer Science & Business Media, 2013.

7. Kirk, W., ed. *Nanostructures and Mesoscopic Systems*. San Diego, CA: Academic Press, 2012; Metalidis, G. Electronic transport in mesoscopic systems. PhD dissertation, Martin-Luther-Universität Halle-Wittenberg, Saxony-Anhalt, Germany, 2007.

8. Einspruch, N. G. and W. R. Frensley, eds. *Heterostructures and Quantum Devices*, Vol. 24. Saint Louis, MO: Elsevier, 2014.

9. Park, B.-g., S. W. Hwang, and Y. J. Park. *Nanoelectronic Devices*, Vol. 1. Boca Raton, FL: CRC Press, 2012.

10. Chakraborty, T. and P. Pietiläinen. *The Quantum Hall Effects: Integral and Fractional*, Vol. 85. Springer Science & Business Media, 2013.

11. Grundmann, M. *The Physics of Semiconductors: An Introduction Including Nanophysics and Applications*. Heidelberg, Germany: Springer-Verlag, 2016.

12. Weisbuch, C. and B. Vinter. *Quantum Semiconductor Structures: Fundamentals and Applications*. San Diego, CA: Academic Press, 2014.

13. Blicher, A. *Field-Effect and Bipolar Power Transistor Physics*. New York: Academic Press, Elsevier, 2012.

14. Datta, S. *Electronic Transport in Mesoscopic Systems*. Cambridge, U.K.: Cambridge University Press, 1997.

15. Bandyopadhyay, S. *Physics of Nanostructured Solid State Devices*. New York: Springer Science & Business Media, 2012.

16. Jacoboni, C. and P. Lugli. *The Monte Carlo Method for Semiconductor Device Simulation*. Wien, Austria: Springer-Verlag, 2012.

17. Bimberg, D., M. Grundmann, and N. N. Ledentsov. *Quantum Dot Heterostructures*. West Sussex, England: John Wiley & Sons, 1999.

18. Di Ventra, M. *Electrical Transport in Nanoscale Systems*, Vol. 14. Cambridge, U.K.: Cambridge University Press, 2008.

19. Heinrich, H., G. Bauer, and F. Kuchar. *Physics and Technology of Submicron Structures: Proceedings of the Fifth International Winter School, Mauterndorf, Austria, February 22–26, 1988*, Vol. 83. Berlin, Germany: Springer Verlag, 1988.

20. Natori, K. Ballistic/quasi-ballistic transport in nanoscale transistor. *Applied Surface Science* 254(19) (2008): 6194–6198.

21. Lundstrom, M. and C. Jeong. *Fundamentals and Applications*. Chichester, U.K.: World Scientific Publishing Company Incorporated, 2013.

22. Datta, S. *Quantum Transport: Atom to Transistor*. Cambridge, U.K.: Cambridge University Press, 2005.

23. Roichman, Y. and N. Tessler. Generalized Einstein relation for disordered semi-conductors—Implications for device performance. *Applied Physics Letters* 80(11) (2002): 1948–1950.

24. Ferry, D. and S. M. Goodnick. *Transport in Nanostructures*, Vol. 6. Cambridge University Press, 1997.

25. Imry, Y. and R. Landauer. Conductance viewed as transmission. In Bederson, B., ed., *More Things in Heaven and Earth*, pp. 515–525. New York: Springer, 1999.

26. Chu, C. S. and R. S. Sorbello. Effect of impurities on the quantized conductance of narrow channels. *Physical Review B* 40(9) (1989): 5941.

27. Yoshihira, M., S. Moriyama, H. Guerin, Y. Ochi, H. Kura, T. Ogawa, T. Sato, and H. Maki. Single electron transistors with ultra-thin Au nanowires as a single Coulomb island. *Applied Physics Letters* 102(20) (2013): 203117.

28. Shin, S. J., C. S. Jung, B. J. Park, T. K. Yoon, J. J. Lee, S. J. Kim, J. B. Choi, Y. Takahashi, and D. G. Hasko. Si-based ultrasmall multiswitching single-electron transistor operating at room-temperature. *Applied Physics Letters* 97(10) (2010): 103101.
29. Grabert, H. and M. H. Devoret, eds. *Single Charge Tunneling: Coulomb Blockade Phenomena in Nanostructures*, Vol. 294. Springer Science & Business Media, 2013.
30. Kastner, M. A. The single-electron transistor. *Reviews of Modern Physics* 64(3) (1992): 849, AIP Publishing, Melville, NY.
31. Van Houten, H., C. W. J. Beenakker, and A. A. M. Staring. Coulomb-blockade oscillations in semiconductor nanostructures. In *Single Charge Tunneling*, pp. 167–216. Springer US, 1992.
32. Bird, J. P., ed. *Electron Transport in Quantum Dots*. New York: Springer Science & Business Media, 2013.
33. Fève, G., A. Mahe, J.-M. Berroir, T. Kontos, B. Placais, D. C. Glattli, A. Cavanna, B. Etienne, and Y. Jin. An on-demand coherent single-electron source. *Science* 316(5828) (2007): 1169–1172, American Association for the Advancement of Science (AAAS), Washington, DC.
34. Dubois, J., T. Jullien, F. Portier, P. Roche, A. Cavanna, Y. Jin, W. Wegscheider, P. Roulleau, and D. C. Glattli. Minimal-excitation states for electron quantum optics using levitons. *Nature* 502(7473) (2013): 659–663.
35. Hofer, P. P. and M. Büttiker. Emission of time-bin entangled particles into helical edge states. *Physical Review B* 88(24) (2013): 241308.
36. Hofer, P. P., H. Aramberri, C. Schenke, and P. A. L. Delplace. Proposal for an ac spin current source. *EPL (Europhysics Letters)* 107(2) (2014): 27003.
37. Tancredi, G., G. Ithier, and P. J. Meeson. Bifurcation, mode coupling and noise in a nonlinear multimode superconducting microwave resonator. *Applied Physics Letters* 103(6) (2013): 063504.
38. Maisi, V. F., D. Kambly, C. Flindt, and J. P. Pekola. Full counting statistics of Andreev tunneling. *Physical Review Letters* 112(3) (2014): 036801.
39. Albert, M., G. Haack, C. Flindt, and M. Büttiker. Electron waiting times in mesoscopic conductors. *Physical Review Letters* 108(18) (2012): 186806.
40. Dasenbrook, D., C. Flindt, and M. Büttiker. Floquet theory of electron waiting times in quantum-coherent conductors. *Physical Review Letters* 112(14) (2014): 146801.
41. Thomas, K. H. and C. Flindt. Electron waiting times in non-Markovian quantum transport. *Physical Review B* 87(12) (2013): 121405.
42. Sánchez, R. and M. Büttiker. Optimal energy quanta to current conversion. *Physical Review B* 83(8) (2011): 085428.
43. Sothmann, B., R. Sánchez, A. N. Jordan, and M. Büttiker. Rectification of thermal fluctuations in a chaotic cavity heat engine. *Physical Review B* 85(20) (2012): 205301.
44. Jordan, A. N., B. Sothmann, R. Sánchez, and M. Büttiker. Powerful and efficient energy harvester with resonant-tunneling quantum dots. *Physical Review B* 87(7) (2013): 075312.
45. Sothmann, B., R. Sánchez, A. N. Jordan, and M. Büttiker. Powerful energy harvester based on resonant-tunneling quantum wells. *New Journal of Physics* 15(9) (2013): 095021.
46. Sothmann, B. and M. Büttiker. Magnon-driven quantum-dot heat engine. *EPL (Europhysics Letters)* 99(2) (2012): 27001.
47. Bergenfeldt, C., P. Samuelsson, B. Sothmann, C. Flindt, and M. Büttiker. Hybrid microwave-cavity heat engine. *Physical Review Letters* 112(7) (2014): 076803.

6

Synthesis of Noble Metal Nanoparticles: Chemical and Physical Routes

Chandan Kumar Ghosh

CONTENTS

6.1 Introduction

Particles having a size <100 nm are termed nanoparticles, and their utility in various fields has attracted researchers. A closely related term, "cluster," is often adopted to indicate particles with much smaller size (few nm) containing <10^4 atoms or less. In this context, nanoparticles or clusters often exhibit exotic electronic, magnetic, chemical, optical, thermal, mechanical, etc., properties due to the large fraction of the surface atoms. It is interesting to note that the size of the nanoparticles or clusters lies between atom and bulk systems, and thus they often possess properties between atom and bulk material that remain unexplained to date. In order to vary the properties of these nanoparticles or clusters, they are sometimes dispersed into a continuous matrix of ceramic or polymer to introduce compositional heterogeneity. Among different nanoparticles or clusters such as metal oxide semiconducting nanoparticles, magnetic nanoparticles, ferroelectric nanoparticles, etc., noble metal (Ag, Au, Pt) nanoparticles or clusters have been established to be a potential material for dye industries due to their inherent photocatalytic activity, sensors due to surface-enhanced Raman scattering (SERS), electrodes in the case of multilayer ceramic capacitors, and food storage industries due to antibacterial, antifungi activity, etc. Zijlstra et al. have developed a DVD with a capacity of ~10 terabytes utilizing the optical properties of Au nanorods [1]. Surface electrons are often used to convert radiation into heat energy, thus enabling them to be used in photothermal applications. These noble metal nanoparticles exhibit high concentrations of free electrons at their surface oscillating at a characteristic frequency called plasmon frequency. If these nanoparticles are excited at plasmon frequency, surface electrons oscillate with large amplitude even in the presence of a very weak incident electric field, and this phenomenon can be adopted in the field of imaging technology. A scattering and absorption cross section of the electromagnetic radiation is enhanced at plasmon frequency and has led to the use of noble metal nanoparticles or clusters in the fields of biolabeling, filters, chemosensors, etc. [2,3]. In this context, the size and shape of the nanoparticles or clusters play a crucial role in terms of the properties of the surface electrons, and thus tuning of the size and shape of the nanoparticles or clusters seems to be very important. At present, the methods to prepare noble metal nanoparticles or clusters can be classified into physical, chemical, and biological methods. Each method has its own advantages and disadvantages with common issues such as costs, scalability, etc. The bottom-up chemical approach relies on the atom-by-atom formation of the nanoparticles in the liquid, solid, or gas phase. The physical method requires photochemical reactions at high temperature, under vacuum condition, and thus expensive high-end equipment is necessary. Chemical methods have advantages in terms of lower cost and simpler equipment. The major advantage of the latter process lies in the fact that the colloidal dispersion of nanoparticles or clusters in water or organic solvent is easily achieved. In this chapter, we'll mainly discuss the different physical and chemical methods to synthesize noble metal nanoparticles or clusters, followed by their reaction mechanism and factors affecting their shape and size. Biological methods will be discussed briefly in a separate chapter.

6.2 Chemical Reduction Method to Synthesize Noble Metal Nanoparticles

Many chemical methods have largely been adopted to prepare noble metal nanoparticles, as monodisperse nanoparticles or clusters with tunable size, shape, crystallinity, and composition are easily synthesized using chemical reaction–based bottom-up approaches. Among the various chemical methods, chemical reduction (e.g., reduction of Ag^+ ions into Ag^0 nanoparticles) is mostly used to synthesize noble metal nanoparticles where borohydride, citrate, elemental hydrogen, etc. are used as reducing agents. The main difficulty of this technique is that the synthesized nanoparticles are easily agglomerated. To avoid agglomeration, the technique has been modified slightly, including reduction followed by stabilization. Different stabilizing agents have been examined for this purpose [4]. For example, silver nanoparticles are synthesized after reducing silver nitrate ($AgNO_3$) in polyvinylpyrrolidone (PVP) aqueous solution that acts as a stabilizing agent and glucose. In this reaction, glucose serves as a reducing agent. Often sodium hydroxide (NaOH) is added into the reaction to accelerate it, which can be written in the following way:

$$Ag^+ + PVP \rightarrow Ag[PVP]^+$$

$$CH_2OH-(CHOH)_4-CHO+Ag[PVP]^+ +2OH^- \rightarrow CH_2OH-(CHOH)_4 \\ -COOH+2Ag[PVP]+H_2O$$

$$2Ag^+ +2OH^- \rightarrow Ag_2O+H_2O$$

$$Ag_2O+CH_2OH-(CHOH)_4-CHO+2PVP \rightarrow CH_2OH-(CHOH)_4 \\ -COOH+2Ag[PVP]$$

First, the pathways involve an initial reaction between Ag^+ and PVP to form the $Ag[PVP]^+$ compound, and secondly hydroxyl ions give gluconate ions that undergo a nucleophiic addition reaction to reduce Ag^+ into Ag^0. In the second reaction, Ag^+ reacts with hydroxyl ions and produces Ag_2O, which is further reduced by gluconate, and silver nanoparticles are formed. This protocol involves homogeneous nucleation by minimizing the free energy of the system and the growth of the nuclei in favorable reaction conditions depending on pH and glucose, $AgNO_3$, and NaOH concentrations [5,6]. Oleylamine is also used to obtain Ag nanoparticles from Ag^+. Homogenous nucleation of Ag instantly occurs when aqueous solution of $AgNO_3$ is added into the oleylamine–diphenyl ether mixture, preheated at 160°C–240°C. It has been shown that irrespective of the synthesis condition, namely, temperature, oleylamine gives rise to spherical or nearly spherical Ag nanoparticles [7]. Controlled growth of facets and ripening at a lower temperature provides better control of the growth of individual crystallites [8]. It has been demonstrated that Ag nanoparticles are formed by stacking Ag atoms when they are saturated within the solution. The solubility of the nuclei is found to be an important factor in determining the stage at which nucleation stops and aggregation takes place. The effect of the concentration of reducing agent on the nanostructures has been investigated and the outcome shows that lower oleylamine concentration is not able to minimize the agglomeration of silver nanoparticles, and hence larger silver nanoparticles are produced. At higher oleylamine concentration, nanoparticles are coated with oleylamine that increases electrostatic repulsion, and hence agglomeration of nanoparticles stops [7]. Monodisperse Ag nanoparticles are synthesized by reducing $AgNO_3$ using ethylene glycol in the presence of PVP. This process is known as "polyol-process." Ethylene glycol serves as a reducing agent as well as a solvent. The general mechanism of reduction of Ag^+ by ethylene glycol can be represented by the following reactions:

$$CH_2OH-CH_2OH \rightarrow CH_3CHO+H_2O$$

$$2CH_3CHO+2Ag^+ \rightarrow 2Ag+2H^+ + CH_3COCOCH_3$$

PVP containing the polar group in the form of nitrogen and oxygen exhibits a strong coordinative field that leads to the formation of a coordinative complex between silver ions and PVP. The polar group of ethylene glycol containing high electron concentration in the ligand of –N, and C=O gives up electrons to Ag^+ ion, and hence Ag^+ ions are reduced into Ag^0 [9]. When the concentration of Ag^0 reaches a critical concentration within the solution, nucleation of some Ag atoms starts and the Ag^0 nuclei grow to form Ag nanoparticles [10]. The molar ratio of PVP and $AgNO_3$ plays an important role in determining the shape and size of the nanoparticles. The synthesis of Ag nanoparticles from $AgNO_3$ using sodium borohydride as a reducing agent in the presence stabilizers including polyvinyl alcohol (PVA), PVP, bovine serum albumin (BSA), citrate, and cellulose is well accepted [11,12]. The reduction by $NaBH_4$ is expressed by:

$$AgNO_3+NaBH_4 \rightarrow Ag+\tfrac{1}{2}H_2+\tfrac{1}{2}B_2H_6+NaNO_3$$

Recently, trisodium citrate has also been found to be a reducing agent that reduces Ag^+ very slowly in alcoholic medium. Very low dispersion in the particle size (5–10 nm) is observed in this well-controlled synthesis process. Double stabilizing agents are sometimes used to synthesize monodispersed Ag nanoparticles. For example, hydrazine hydrate is used as a main reducing agent along with sodium dodecyl sulfate as primary stabilizing agent and citrate of sodium solution as secondary stabilizing agent. *N,N*-dimethylformamide (DMF), a standard polar organic compound which is generally used as a solvent in chemistry, is also adopted to reduce Ag^+ ions into Ag^0, and in this reaction DMF is oxidized into carbamic acid [13]. The chemical reaction can be written as follows:

$$HCONMe_2+2Ag^++H_2O \rightarrow 2Ag^0+Me_2NCOOH+2H^+$$

Along with DMF, water-soluble polymer PVP is frequently used to stabilize Ag nanoparticles. It has been investigated by many researchers that $AgNO_3$:PVP ratio, reaction time play crucial role on the size of the Ag nanoparticles [14]. The chemical reduction method is not limited to synthesis of silver nanoparticles; it is also useful in synthesizing gold (Au) nanoparticles. The first reduction method was employed in 1957 where gold chloride was reduced by phosphorous in aqueous medium. Later, citrate-stabilized Au nanoparticles (~20 nm) were synthesized from gold tetra-chloroauric acid using trisodium citrate as reducing agent. The synthesis reaction of Au nanoparticles using citrate is written in the following way:

$$2HAuCl_4 + 3C_6H_8O_7 \rightarrow 2Au + 3C_5H_6O_5 + 8HCl + 3CO_2$$

In this context, the size of the Au nanoparticles is varied by changing the synthesis temperature, reagent concentration, etc. [15]. Later, a method known as the Brust–Schiffrin method was developed to synthesize Au nanoparticles using $NaBH_4$ as reducing agent followed by adding alkane thiol for long-term stability. Here, relative amounts of Au precursor and the rate of adsorption of the stabilizer are found to play a crucial role in controlling the size of the nanoparticles. It has been observed that fast reactions make the particle size more sensitive to the concentrations of precursors, rate of addition of precursors, state of mixing in the reactor, and other variables, and introduce greater dispersion in the particle size [16]. Au^{+4} ions from an aqueous solution are extracted to a hydrocarbon phase using tetraoctylammonium bromide as the phase transfer agent, and after transfer it can be subsequently reduced in the organic solution using a strong reductant $NaBH_4$ in the presence of an alkanethiol, giving Au nanoparticles with an average diameter of ~2.5 nm. This reaction mechanism is summarized as follows:

$$AuCl_4 \text{ (aq)} + N(C_8H_{17})_4 + (C_6H_5Me) \rightarrow N(C_8H_{17})_4 + AuCl_4^-(C_6H_5Me)$$

$$mAuCl_4^-(C_6H_5Me) + nC_{12}H_{25}SH(C_6H_5Me) + 3me^- \rightarrow 4mCl^- \text{ (aq)} \\ + (Au_m)(C_{12}H_{25}SH)_n(C_6H_5Me)$$

Here, $NaBH_4$ acts as the electron source [17], and the nucleation and growth of the Au nanoparticles as well as the attachment of the thiol molecules occur simultaneously within a single step. This procedure gets modified by using alkanethiol of different chain lengths like aromatic thiol, dialkyl disulfides, reactant ratios for the synthesis of monolayer-protected clusters [18,19]. As the phase transfer agent and stabilizer corresponding to the Brust–Schiffrin technique are chemically different, the obtained nanoparticles often contain nitrogenous surface impurities due to the phase-transferring agent. Another problem with the method is that the stabilizing ligands, such as thiol or amine, must be compatible with all of the reagents, including the reducing (i.e., $NaBH_4$) and phase transfer agents to avoid their influence on the reaction chemistry [20]. Murray's research group in 1998 used hexanethiol instead of dodecanethiol as the stabilizing agent with a 3:1 hexanethiol to Au ratio and carried out the reaction at low temperature to yield a solution of Au nanoclusters with a mean diameter of 1.6 nm and an average composition of $Au_{145}(S(CH_2)_5CH_3)_{50}$ [21]. Leff et al. reported the preparation of amine-stabilized Au nanoparticles by substituting the dodecanethiol with dodecyl amine or oleylamine and increasing the diameter of the prepared Au nanoparticles up to 7 nm [22]. In the ethanol-mediated method, alkyl amine is used as a stabilizer/phase transfer agent and has a distinctive advantage over the Brust–Schiffrin method [23]. In this two-step approach, aqueous solution of metal ions is mixed with an equal volume of ethanol containing dodecyl amine and the mixture is stirred for few minutes. Then toluene is added with continuous stirring before transferring the mixture to a separating funnel. Two immiscible layers are formed within a few minutes. Then transfer of metal salts from the aqueous phase to toluene is completed where ethanol is found to be a very important candidate without which metal ions would not be transferred to the organic phase by the direct mixing of an aqueous metal precursor solution with an organic solvent containing dodecyl amine. The fact that water and ethanol are miscible ensures maximum contact between metal ions and dodecyl amine. A large number of commonly used methods, for example, wet chemistry reduction, seed-mediated growth, coreduction, and solvothermal approaches, have also been developed to produce noble metal nanoparticles after the transfer into the organic solvent. Compared to others, this protocol allows the synthesis of nanocrystals in an organic medium using aqueous soluble metal salts as the starting materials, which are relatively inexpensive and can be easily obtained [24]. This experiment involves the same steps as for transferring metal ions from the aqueous phase to a nonpolar organic medium. A variety of surfactants, namely, sodium acetate, polyvinylpyrrolidone, bis(p-sulfonatophenyl)-phenyl phosphine, and triton X-100, are used for capping metal nanoparticles to stabilize them [25]. Platinum (Pt) nanoparticles have also attracted researchers due to several applicational aspects, and in this regard different synthesis protocols have also been proposed for Pt nanoparticles. Similar to Ag or Au, Pt nanoparticles are synthesized by reduction using H_2PtCl_6 as precursor for platinum using $NaBH_4$ as a reducing agent [26]. The reaction can be summarized as follows:

$$H_2PtCl_6 + NaBH_4 \rightarrow Platinum + Other\ reaction\ product$$

In this reaction process, H_2PtCl_6 is initially dissolved in aqueous or organic media and reducing agent is added with continuous stirring. There are several methods to prepare platinum nanoparticles on a substrate. One of these is the impregnation method, which involves a support material such as Vulcan 72 carbon, which is used for deposition of $PtCl_6^{2-}$ before reduction of the metal precursor to metal nanoparticles. Then chemical reduction within the liquid phase is carried out by reducing agents [27] or using the gas phase method that involves flow of H_2 gas stream at temperatures between 250°C and 600°C [28]. The popular reducing agents used here are hydrazine, borohydride, formic acid, etc. Additionally, the reaction time, kinetics, and mass transfer of reducing agent would influence the nucleation and growth of the nanoparticles. The main limitation of this method is the lack of control of the size of metal nanoparticles and agglomeration in the case of colloidal particles [29]. But the latter problem is overcome by carrying out the reduction in organic media in the presence of stabilizing agents. For example, Pt colloid is synthesized by reduction of H_2PtCl_6 by alcohol in the presence of polymer capping agent [30]. With suitable tuning of the growth parameter, synthesis of nanoparticles with controlled shape and size is possible. In this context, agglomeration of nanoparticles is prohibited by electrostatic repulsion due to surface charges, and additional coating with organic chain molecules also provides steric stabilization. Some examples of the protecting ligands are NR_4^+, PVP, and PVA. Recently, PVP was introduced to stabilize platinum nanoparticles in solution to prevent aggregation where a reaction between alcohol and the metal precursor occurs. Bock and MacDougall [31] suggested a large-scale synthesis method of monodispersed Pt nanoparticles with tunable size and shape, where they are stabilized by glycol serving as a solvent and protecting agent. Alcohol also serves as solvent for dissolving metal precursors as well as surfactants and as a reducing agent to produce Pt colloids. The main drawback of this system is the existence of a protecting agent that hampers a few surface-related activities like catalytics, sensing, etc. Proper washing by an appropriate solvent or by decomposition at elevated temperatures in an inert atmosphere can be employed to remove the organic protecting shell. Before adsorption into the protecting micropores, catalyst support is sometimes crucial to prevent agglomeration into larger metal particles [29]. An alternative route to obtain stabilized Pt nanoparticles with narrow and controllable size distribution is a two-phase method by phase transfer (water to toluene) of $[PtCl_6]^{2-}$ followed by reduction and surface protection with dodecanethiol [32]. In this technique, chemical steps are conducted within the microemulsion step, which serves as a micro- or nanoscale reactor. The introduction of a reducing agent into the microemulsion is accomplished by stirring and the reaction time is found to be in the order of minutes. The size and distribution of the nanoparticles are controlled and improved by a modified two-stage microemulsion where the reducing agent is confined in a separate emulsion. The parameters monitoring the size of the particles in the microemulsion method are water to surfactant ratio, number of surfactants, concentration of precursor solution, temperature, etc. After the reduction step, synthesized nanoparticles are protected by surface molecules from aggregation. The requirement of costly surfactant molecules with extra washing steps may not be economical for a large-scale synthesis and this limits this process to synthesize noble metal nanoparticles [29].

6.3 Electrochemical Method to Synthesize Noble Metal Nanoparticles

During the past decade, another method that has been developed to synthesize colloidal solution of noble metal nanoparticles is the electrochemical method. This method was initially proposed by Reetz and Heilbig in 1994, wherein they dissolved the metallic material anodically and the resulting salt was reduced at a cathode [33]. Ag nanoparticles have been synthesized using this method. Typically, two identical Ag wires which act as electrodes are submerged in deionized water (shown in Figure 6.1), and then a voltage (~5 V) is applied between them to reduce metal salts at the cathode.

Oxidative dissolution of Ag atoms from anode (also called sacrificial anode) in the form of Ag^+ ions occurs if they give off electrons: $Ag^0 - e^- \rightarrow Ag^+$ (at anode). These formed Ag^+ ions migrate toward the cathode where they are reduced into zero-valent Ag atoms: $Ag^+ + e^- \rightarrow Ag^0$. Ag^0 atoms act as nucleation centers for the formation of Ag nanoparticles. In this regard, nucleation and growth of the Ag nanoparticles occur via van der Waals forces. The kinetics of this reaction depend on the applied voltage and is labeled by saturated current. The major problem with this technique is the formation of metal oxide as impurity and deposition of metal nanoparticles on the electrode. H_2O_2 is frequently used to reduce oxide nanoparticles in order to prevent the formation of metal oxides [34], whereas the polarity between electrodes changes periodically to overcome the deposition of nanoparticles on the cathode. Several ionic and neutral surfactants, namely, polyethylene glycol, oligo-derivatives, polysaccharides, and triblock copolymer, are used

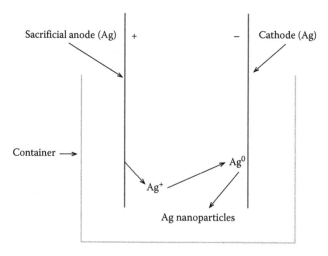

FIGURE 6.1
Schematic diagram of electrochemical synthesis method.

to stabilize nanoparticles in the colloidal solution. It has been shown that the nanoparticle to surfactant ratio plays a significant role. A high ratio between them gives rise to flocculation and growth of the nanoparticles; on the other hand, phenomena such as depletion flocculation occur at a lower ratio. Here, no chemical interaction happens and the phenomenon is reversible as a lower magnitude of formation energy is involved [35]. Not only surfactants but also solvents and electrolytes play a crucial role in determining the shape and size of the nanoparticles. For example, Reetz et al. synthesized Ag nanoparticles using a mixture of acetonitrile and tetrahydrofuran as solvent, but the usage of tetrahydrofuran leads to aggregation of the nanoparticles, and hence larger nanoparticles are formed. Later, Rodríguez-Sánchez et al. synthesized smaller Ag nanoparticles using acetonitrile only [36]. The important point to mention here is that a suitable choice of electrolyte helps to prevent deposition of nanoparticles on the electrode. Tetrabutylammonium bromide or acetate is used in this case. Another important point to mention is that the solvent must be oxygen free in order to avoid formation of Ag_2O. Noble metal nanoparticle synthesis using the electrochemical process has advantages in terms of the easy control of the size and shape of the nanoparticles just by varying applied potential and current. Several electrochemical techniques have also been developed to prepare noble metal nanoparticles using acidic medium. For example, Au nanoparticles are synthesized using $HAuCl_4$ as precursor material, Pt as counter, and working electrodes and calomel as reference electrode [37]. Ag dendritic structure is synthesized by the electrochemical method on ITO-coated glass slides starting from $AgNO_3$, citric acid, and ethanol using two electrode systems [38]. Here, ITO is used as the working electrode while graphite serves as the counter electrode. The synthesis procedure is carried out at 1.5 V for different temperatures to tune the morphology of the sample. One-dimensional growth of the dendritic structure is explained on the basis of favorable growth of the Ag fcc metal along (111) rather than other directions. Wide variation in the morphology of Ag nanostructures has been incorporated by Sivasubramanian et al. when they synthesized Ag nanostructured on ITO-coated glass substrate at constant potential but at different potential without any template or surfactant [39]. This electrodeposition process has been carried out on three electrode systems, where ITO acts as the working electrode and Pt wire and Ag/AgCl serve as the counter and reference electrode, respectively. Morphology significantly depends on equilibrium conditions during the electrodeposition technique. Higher deposition potential creates nonequilibrium within the system; mass transfer toward the electrode increases in comparison to the driving force of crystallization and increased fluctuation results in dendritic structure. On the other hand, at lower potential, slow growth rate produces polygon structure. Nanoporous Au nanostructures have been synthesized electrochemically by Fang et al. for enhanced SERS activity [40]. Rose-like Ag nanoparticles have been synthesized by You et al. using a three-electrodes-based electrochemical technique [41]. Here a screen-printed carbon electrode is used as the working electrode, while Pt and Ag/AgCl are adopted as the auxiliary and reference electrode, respectively. Prior to deposition, the working electrode is preoxidized. The electrochemical reaction is carried out in the voltage range from −0.3 to +0.3 V in aqueous $AgNO_3$ solution (10 mM). The addition of Au nanoparticles here gives a better regular shape of the nanostructure. In addition to single-phase material, noble bimetallic heterostructures have also been synthesized using this technique. For example, Ag nanoparticles are grown on ITO-coated substrate electrochemically from $AgNO_3$, which in the second stage is used as a sacrificial electrode to synthesize AgPt, AgAu, and AgAuPt hybrid structures [42]. AgAuPt hybrid structures are found to be highly sensitive to H_2O_2. In this regard, fractal bimetallic Cu/Ag nanodendritic structures have been synthesized by Li et al. [43] with enhanced Raman scattering. A Pt/Au bimetallic membrane with a leaf-like morphology has been grown by X. Shen et al. for oxygen sensing at room temperature [44]. Its typical synthesis protocol includes synthesis of Pt/Au membrane on ITO glass by the electrochemical method using 25 mM $HAuCl_4$, 25 mM H_2PtCl_2, and 20 g/L PVP solution as precursor materials. After the membrane is formed, the glass slides are immersed in boiling water to separate the membrane from the slide.

6.4 Sonochemical Method to Synthesize Noble Metal Nanoparticles

In addition to the chemical reduction and electrochemical methods, colloidal solution of noble metal nanoparticles is largely synthesized by the sonochemical method using ultrasound that causes a chemical reaction. Its basic principle is that when liquids are irradiated with intensed ultrasound, acoustic cavitation, defined as the combined effect of formation, growth, and implosive collapses of bubbles, occurs and causes the reaction. An ultrasonicator is used to convert electrical signal having frequency of 50 Hz into high-energy ($\sim$100 W/cm^2) and high-frequency (20 kHz to 1 MHz) energy flow. This electrical energy is then transmitted through a piezoelectric transducer that converts electrical energy into mechanical energy. Mechanical vibration is allowed to pass through liquid that alternatively gets compressed and relaxed in the presence of vibration at high frequency. Due to the rapid change of pressure, microscopic bubbles, formed within the liquid, expand during the decompressive phase and implode during the compressive phase. Bubbles grow in the solution and, after achieving a maximum size, they collapse resulting in high pressure and temperature. When a large number of such events occurs within a few microseconds, a high level of energy is released in the liquid as a cumulative effect. Due to high energy, several chemical reactions including reduction of metallic salts become feasible. The chemistry of this technique includes H and OH radicals, generated by ultrasound irradiation, responsible for the reduction of metal ions into their atomic state. In 1987, Gutierrez et al. reduced Ag^+ in aqueous solution using 1 MHz ultrasound [45]. Later, in 1992, Nagata et al. prepared Ag nanoparticles after reducing Ag^+ ions under Ar atmosphere by a 200 kHz ultrasonicator [46]. Pt^{2+} ions have been sonochemically reduced to metallic Pt nanoparticles by Mizukoshi et al. at 200 kHz frequency and 6 W/cm^2 power of an ultrasonicator [47]. They have also investigated the effect of a surfactant like sodium dodecyl sulfate on the stabilization of nanoparticles and radicals involved in the sonochemical reduction process. Prepared nanoparticles are often coated with other surfactants, namely, polyethylene glycol, polyoxymethylene, CTAB, etc., for stabilization. Not only Ag nanoparticles but also Au nanoparticles are synthesized by this technique using a precursor in aqueous as well as alcoholic medium [48]. The initial concentration of metal ions is a crucial parameter for the size of the metal nanoparticles. Lower concentration of metal ions results in smaller size of the metallic nanoparticles. Addition of organic additives is often used to control the reduction rate and size of the nanoparticles. For example, when alcohol is used, due to its hydrophobic nature, it causes accumulation of a high concentration of OH radicals at its interface that enhances the reduction rate of metal ions [49]. In this regard, it has been shown that particle size is inversely dependent on the alcohol concentration and chain length of alkyl [50]. The reason behind this is that alcohol, adsorbed on the metal surface, restricts growth. Additive not only accelerates the reduction reaction but also stabilizes metal nanoparticles in some cases. As an example, Pt nanoparticles, stabilized by polyethylene glycol, sodium dodecyl sulfate, etc., are found to have a diameter of ~3 nm [51]. Ag nanoparticles with a diameter of ~2 nm are synthesized in the presence of polymethylacrylic acid using the sonochemical technique [52]. Another important factor affecting the formation of nanoparticles is purging gas through medium, since a cavitation bubble results from this dissolved gas. In this regard, Okitsu et al. observed the reduction rate of Au(III) in the order N_2 = He < Ne < Ar < Kr [53]. It has been shown that the higher the ratio of C_p/C_v (specific heat ratios) and the lower the thermal conductivity of the purging gas within the liquid, then the higher the temperature of the collapsing bubbles, resulting in a higher reduction rate. In addition, higher solubility of the gas also enhances the reaction rate. The temperature of the solution is another important factor affecting the reduction rate. For example, the reduction rate of Au^{3+} ions is found to be increased between 2°C and 20°C; above this, the reduction rate is reduced. Initially, the vapor pressure of water and alcohol are found to be increased, which results in the formation of reducing species for Au^{3+}. Increasing the temperature further, bubbles collapse due to the cushion effect, caused by the presence of an excess number of water and alcohol molecules within the bubbles. Interestingly, the reduction rate is decreased with increasing frequency of ultrasound. The greatest advantage of this rapid technique is not only synthesization of uniform particles with the desired shape, but also synthesis of bimetallic nanoparticles in a single step. As an example, when aqueous solutions of Au^{3+} and Pd^{2+} are sonicated together in the presence of sodium dodecyl sulfate and Ar gas, Au core/Pd shell bimetallic nanoparticles are formed. This method has already been adopted by researchers to synthesize noble metal nanoparticles within a matrix. For example, Pd nanoparticles (5–6 nm in diameter) are synthesized in silica matrix [54]. In this synthesis method, initially silica is prepared by heating tetraethylorthosilicate at 700°C. Then it is immersed in isopropanolic $PdCl_2$ solution for a long duration. After sufficient immersion, the mixture is ultrasonicated at 40 kHz and 100 W output power for 120 minutes. To avoid oxidation, Ar gas is used to purge through the mixture. Not only colloidal particles of noble metal, but also noble metal nanoparticles have been synthesized on glass slides by the sonochemical method. In this typical technique, glass slides have been immersed vertically in a mixture of water–ethylene glycol (10 volume %) solution and $AgNO_3$ (0.05 M) solution [55]. After purging with Ar gas for 1 hour to remove O_2/air, it is ultrasonicated at 20 kHz frequency and 600 W/cm² power at room temperature. As confirmed by scanning electron microscope, micron-sized Ag particles formed on the glass slides.

6.5 Radiolysis Method to Synthesize Noble Metal Nanoparticles

From the above discussion, it is clear that most of the bottom-up approaches for synthesizing noble metal nanoparticles have been developed on the basis of chemical reduction (in aqueous or nonaqueous medium) using a reducing agent in the presence of a stabilizing agent. In addition to these, another method, based on radiolytic reduction, has also been developed by researchers to synthesize colloidal noble metal nanoparticles. Its advantage lies in the fact that the method relies on very mild experimental conditions such as ambient pressure and room temperature. Its basic principle involves reduction of noble metal ions in the absence of oxygen by hydrated ions due to high negative redox potential. A large number of hydrated electrons and atoms are produced by radiolysis of aqueous solutions by radiations including electron beam, X-ray, gamma ray, and UV radiation. And having high redox potential $(E^0(H_2O/H^{\bullet}) = -2.87$ and $E^0(H^+/H^{\bullet}) = -2.3$ $V_{NHE})$ easily reduces metal ions into zero-valent atoms. The reaction mechanism can be written as

$$H_2O \xrightarrow{\text{Radiation}} e^-_{aq}, H^{\bullet}$$

$$M^+ + e^-_{aq} \rightarrow M^0$$

$$M^+ + H^{\bullet} \rightarrow M^0 + H^+$$

The formed atoms act as nucleation centers and coalescence, further giving nanoparticles. Binding energy between two metal atoms or atoms with ions is stronger than atom–solvent or atom–ligand bond energy, and thus dimerization between metal atoms occurs, which can be written as

$$M^0 + M^0 \rightarrow M_2$$

$$M^0 + M^+ \rightarrow M_2^+$$

M_2^+ is reduced further to form the nucleation center. The competition for reduction between M^+ and M_2^+ depends on the formation of reducing agent. It should be mentioned here that M_2^+ favors cluster growth rather than nucleation. The size of the formed nanoparticles depends on their coalescence and could be controlled by stabilizing agents. Briefly, nanoparticles with a large fraction of high-energy surface atoms are less stable and they interact with each other by van der Waals force, giving large particles. Here, the stability of the nanoparticles is achieved by balancing the attractive force and repulsive force induced by the stabilizing agent. Such a stabilization procedure is called "electrostatic stabilization" or "steric stabilization." Stabilizing agents containing –NH$_2$, –COOH, –OH as a functional group have high electron affinity, and they easily interact with metal nanoparticles to stabilize them after forming a covalent bond. But this technique for stabilization reduces the surface activity of the nanoparticles in many optical, catalytic, and sensing activities. To overcome this problem, a new method has been developed using a long chain polymer like PVP that restricts particle agglomeration, and this is known as the steric hindrance method. Recently, PVA has been adopted by researchers to stabilize noble metal nanoparticles synthesized by radiation. PVA prevents agglomeration by forming a metal hydroxide cluster by hydrolysis on the surface of the nanoparticles due to the presence of –OH groups, which are capable of absorbing M^+ through secondary bonds and steric entrapment [56]. The reaction between PVA and M^+ can be written as

$$R-OH + M^+ \rightarrow R-O-M + H^+$$

where R–OH represents the PVA monomer. In the absence of any radical scavenger, PVA reacts with M^+ that reduces M^+ into M^0. In this regard, Wang et al. showed that the dendritic structure of Ag becomes isotropic when Ag is synthesized in the presence of ethanol and $C_{12}H_{25}NaSO_4$ [57]. Au nanoparticles have been synthesized by gamma radiation in the presence of aqueous medium including acetone and 2-propyl alcohol. Being an electron scavenger, acetone reduces ions to give nanoparticles. In this context, the pH of the medium plays a significant role in this radiolysis reaction to determine the valence state, to restrict reoxidation, and to prevent unwanted precipitation [58]. The initial concentration of the precursor's concentration is found to play a significant role in the final size of the nanoparticles. Ion association gives a larger size of the particle, and hence particle size increases with initial ionic concentration. With increasing concentration, nanoparticles are aggregated from smaller particles due to collision with each other. Aggregation is generally tuned by viscosity, which subsequently affects the speed of the particle, and hence collision. This is carried out by the addition of polymer molecules that act as a capping agent. Using this radiolysis method, Ag and Au nanoparticles have been synthesized from $AgNO_3$ and $HAuCl_4$ in the range between 7 and 15 nm with concentration of ionic salt of 2×10^{-4} and 2×10^{-3} M, respectively, in the presence of 2-propanol and PVP [59]. It has been shown that at a constant number of nuclei, particle size increases with increasing concentration of ions. The important point is that this method doesn't require any additional reducing agent and no side reactions occur. The size of the nanoparticle is simply controlled by the irradiation dose and amount of ionic salt added in this reaction. The formation of stable and uniformly dispersed noble metal nanoparticles primarily depends on homogeneous nucleation and elimination of excessive reducing agents. The size and crystal structure of the synthesized nanostructure by this method depends on the adsorbed dose.

6.6 Physical Methods to Synthesize Noble Metal Nanoparticles

In physical methods, noble metal nanoparticles are generally synthesized by the evaporation–condensation technique within a tube furnace at atmospheric pressure. The basic principle behind this technique is the vaporization of source material taken in a boat and placed at the center of the furnace; finally, the carrier gas, used to flow through the furnace, is used to condense on a substrate. A schematic diagram of the laser ablation technique is represented in Figure 6.2. Using this process, nanoparticles of

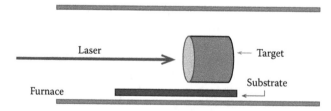

FIGURE 6.2
Schematic diagram for laser ablation technique.

silver and gold have been synthesized. But this method has several limitations. First, as the tube furnace occupies a large space, it consumes a very high amount of energy (~several kilowatts) while raising the temperature of the furnace. Secondly, the tube furnace requires sufficient time to attain stable operating temperature. This evaporation–condensation technique has been modified further by Jung et al. [60]. They prepared silver nanoparticles using the local heating technique by producing a steep temperature gradient in the vicinity of the heater surface. After evaporation–condensation, a large number of noble metal nanoparticles was synthesized by the laser ablation technique using bulk metallic materials in a solution. This method has been developed to synthesize different nanoparticles in the liquid or gas phase starting with solid, liquid, and gaseous precursors. Different experimental setups depending on precursor materials, laser parameter, and ambient conditions have been explored. It should be mentioned that the selection of appropriate parameters is necessary to synthesize the pure phase of the nanomaterial and to increase the yield of the preparation. Primarily, the interaction of the laser with matter plays the role of synthesizing nanomaterials in this technique. Precisely, when light interacts with any gaseous or liquid material, the incident laser light is either absorbed or reflected. The matter then gains energy from the absorbed portion of the laser, resulting in a thermal or chemical reaction within the matter. The mechanisms involved here can be categorized into photothermal, photochemical, and photophysical depending on the physical state of the matter (e.g., solid, liquid, or gas), type of material (conductor, insulator, or semiconductor), laser parameters (wavelength, pulse width, fluence), relaxation times, or thermalization and initial excitation. The photothermal process involves a thermal activation process where a change of phase occurs due to enhancement of temperature and enthalpy of the system. The photochemical process relies on the nonthermal reactions induced by high-energy photons which are strong enough to break the bonds between molecules and generate electrons, atoms, and ions. This laser ablation method has been exploited to synthesize noble metallic nanoparticles, particularly silver nanoparticles.

The main feature of this method is that the characteristic of the synthesized metallic nanoparticles depends on the wavelength of the laser impinging the target material, laser pulse duration (femto-, pico-, nanosecond regime), ablation time duration, and surfactant. It has been observed that the ejection of the nanoparticles strongly depends on the minimum power of the laser, commonly called laser fluence. It has been shown that the mean size of the nanoparticles increased with increasing laser fluence. In addition to laser fluence, laser ablation time also affects the shape and size of the nanoparticles. The interaction of the laser with nanoparticles is responsible for generating nanoparticles with different sizes and shapes [61].

6.7 Template and Hydrothermal-Assisted Synthesis of Noble Metal Nanostructures

One of the great triumphs of nanoscience and nanotechnology is the repetitive synthesis of nanostructured materials to tune different properties. In this regard, one-dimensional nanostructures of noble metal such as nanorods, nanowires, nanobelts, and nanotubes have wide scope of application, especially due to their anisotropic properties such as directional conductivities. Among them, one-dimensional Ag nanostructures exhibit potential application in plasmonic waveguide, surface-enhanced Raman spectroscopy, many biological and chemical sensors, etc. The template-assisted method, being a straightforward method, has attracted researchers due to the precise tuneability of the shape and size of the nanostructure materials. This method can be classified into template-based and template-less methods. The template-based method relies on the preparation of nanostructures within a cylindrical membrane pore corresponding to carbon nanotubes, porous anodic aluminum oxide, block copolymer, etc. On the other hand, the template-less method, which includes a capping agent for directional growth, has been proposed involving seeds and shape-directing capping agents. The template-based method is a straightforward method to synthesize nanostructures. Its basic principle lies in the fact that prefabricated or preexisting templates with definite pore size and shape are used as a structural framework to grow nanostructures within their spatially confined spaces. Based on their nature, templates are divided into two categories: soft templates including biological scaffolds such as peptides, lipids, polymers, micelles, liquid crystals, block polymers, etc., and hard templates like anodic aluminum oxide, polycarbonate membrane, silica spheres, polymer latex colloids, etc. Templates are widely used to grow nanostructures in axial as well as radial directions. Hard templates have an advantage over soft ones in the sense that the latter can fabricate nanostructures with a wide range of narrow distributions. One of the most used hard templates is the anodic aluminum oxide (AAO) template, which is a self-organized material with honeycomb-like structure of highly dense arrays of uniform and parallel pores with a diameter ranging from 5 nm to several hundred nm. The pores are produced by electrochemical oxidation of aluminum in such a way that a balance between growth and dissolution of aluminum oxide exists. Pore parameters like pore organization, pore diameter, interpore distance, wall thickness, and pore length are controlled by anodizing conditions including applied potential, current, pH, electrolyte type, etc. Generally, sulfuric, oxalic, and phosphoric acids are used as electrolyte material during electrochemical etching. Pores are filled with precursor material of required nanostructures and finally are removed by chemical etching or by calcination. As an example, single-crystal Au nanowire has been synthesized by the electrochemical method within polycarbonate and anodic aluminum oxide templates. Recently, graphene oxide has also been used as a template for growth of Au nanowire due to the presence of an abundant oxygen functional group with hexagonal close-packed and face-centered cubic crystal phase [62]. Self-standing corrugated Ag and Au nanorods have been synthesized on an Si-supported AAO template by Habouti et al. for plasmonic application [63]. Prior to synthesis, aluminum in this film was deposited on a Si wafer by DC sputtering and anodization of the aluminum film was carried out by oxalic acid under constant potential. Anodized pores are used to synthesize Ag and Au nanorods taking aqueous solution of $AgNO_3$ (0.22 g is dissolved in 40 mL of water) and $HAuCl_4, 3H_2O$ (0.411 g in 70 mL of water), respectively, as precursor materials. The pH of the solution has been maintained at 2 with the help of an etic acid. Two- and three-dimensional ordered structures, made with hollow spheres of Ag, have been synthesized by Wang et al. using the hard template technique [64]. Beeswax has also been adopted as a hard template to synthesize Ag spheres [65]. Pd nanotubes have been synthesized within polycarbonate foil acting as a template. Initially, foil is irradiated with Au ions to introduce damage in it followed by washing with NaOH. After being activated with $AgNO_3$ solution, foils are plated with 5.45 mM $PdCl_2$, 25 mM L(+)-glutamine, and 3.3 mM L(+)-ascorbic acid, yielding Pd nanotubes [66]. The main disadvantage of hard template techniques is that the chemical etching requires a strong acid (like HF) or strong base (like HCl) or high temperature in the case of calcination. It is often noted that the template removal process, being a complicated one, diminishes the quality of the formed nanostructure [67].

The soft template-based method, on the other hand, is found to have more advantages compared to the hard template method due to easy etching, and has recently been widely used. This method has been used by Zhang et al. to synthesize submicron-sized hollow Ag spheres using PEO-b-PMAA–SDS complex micelles as soft templates [68]. Poly(hydroxyethyl methacrylate) is used as a soft template to synthesize uniformly sized (diameter ~11 nm) spherical Ag nanoparticles [69]. In this context, template-based methods generally help to grow spherical hollow nanostructures, and very few synthesis techniques for nonspherical hollow nanostructures are available. Zhang et al. prepared rhombdodecahedral cages of Ag after combining self-assembly and template-based technique [68]. Chen et al. introduced a facial template-based method to prepare a cubic microbox structure of Ag whose walls consist of net-like Ag nanofibers [70]. Importantly, they observed surface plasmon resonance in the visible and infrared region instead of the normal UV region. In this typical method, initially NaCl crystals are grown within anhydrous ethanol medium and a mixture is prepared after pouring NaCl crystal into anhydrous ethanolic solution containing $AgNO_3$ and PVP. Finally, a mixture of NaCl-saturated solution, ascorbic acid, and NaOH is centrifuged and a grey precipitate of the product is obtained. An Au nanotube has been synthesized using the template-based technique by Ballabh et al. using DNA as a template [71]. The first step of this process included the synthesis of sodium sulfate nanowire that acts as a template for Au nanoparticle formation. Recently, DNA has been used as a soft template to synthesize Ag nanoparticles [72]. In this method, 20 mL of aqueous solution is prepared with 0.25 mM $AgNO_3$ and a varied amount of DNA, and then 0.6 mL of 10 mM $NaBH_4$ is added into the above mixture to reduce Ag^+ ions into Ag^0, and thus Ag nanoparticles form within the DNA template.

In contrast to the template-based methods, recently various nanostructures of noble metal have been grown in the presence of surfactants, and this method has been established as a simple and reliable pathway for the synthesis of well-ordered nanostructures. The basic principle of this method is that the surfactant molecules play the role of soft template and form a concentrated critical self-aggregation of molecules (called micelles). Micelles, present in the solution, are responsible for the generation of different nanostructures. Surfactants are classified into anionic, cationic, and nonionic categories depending upon the hydrophobic or hydrophilic head groups at pH = 7, that is, at neutral pH. Anionic and cationic surfactants interact electrostatically with the precursor materials after dissolving in proper solvent and generate micelles at some critical concentration and undergo phase transition to a crystalline state. Various ionic surfactants like CTAB, SDS, octadecyltrimethylammonium chloride, disodium (2-ethylhexyl) sulfosuccinate, etc., and nonionic surfactants like oleic acid, oleylamine, trioctylphosphine, trioctylphosphine oxide, etc., have been employed to synthesize various nanostructures of noble metals. For example, oleylamine, which acts as a reducing as well as a capping agent, is being widely used for the synthesis of Au nanowire using $HAuCl_4$ as precursor material. The basic mechanism of nanowire formation depends on the formation of $[(Oleylamine)AuCl]_n$ inorganic polymer that serves as the template for the growth of ultrathin uniform nanowire (diameter ~1.6 nm, length ~4 μm). Another explanation is the orientated attachment of Au nanoparticles from which two facets fuse to form a bigger particle [73]. Ascorbic acid is used to remove the amine group, attached to the surface of the Au nanoparticles, and the addition helps to grow nanowire lengthwise in the [111] direction [74]. The main advantage of this method is that it is very simple, involves low cost, and wide variation into the nanostructures is easily achieved by simply varying concentration and composition. But its disadvantage lies in the fact that this method is very complicated to control.

In addition to template-based methods, recently the hydrothermal technique has been widely used to synthesize nanostructure materials due to the advantage of preparation of various nanostructures. This concept was first introduced by Sir Roderick Murchison (1792–1871) while investigating the effect of water at higher temperature and pressure. This method has the potential to produce monodispersed and highly homogeneous nanostructures and nanocomposites. The other advantage of this method is that large single crystals could be synthesized. The hydrothermal method is defined as a heterogeneous reaction in aqueous or nonaqueous medium at higher temperature and pressure, required to dissolve and recrystallize materials that are insoluble under normal conditions. A closely related term, frequently used by chemists, is "solvothermal" to synthesize nanoparticles and it is defined as the chemical reaction occurring in nonaqueous medium near or in supercritical conditions of solvent. The hydrothermal method relies on the supercritical water (water is used as solvent) or supercritical fluid technology (organic materials are used as solvent) to yield synthesis of nanostructures in a controlled manner. Though this method has largely been adopted by scientists, its chemistry is largely unknown. Over the past decade, vast efforts have been given to examine the solution chemistry of solvents in critical, supercritical, and subcritical conditions, pH variation, viscosity, density, and dielectric constant with respect to temperature and pressure, but hydrothermal crystallization is still under debate. Therefore, in the absence of any predictive model, the role of reaction parameters like temperature, pressure, precursor concentration, and time has been speculated empirically to predict the crystallization process. But it has been observed that widely varied nanostructures can be produced just by changing reaction time, and by varying the dielectric constant and density of the solvent with respect to temperature and pressure. In the recent past, nanostructures of various noble metals like Au, Ag, Pt, etc., have been synthesized using the hydrothermal as well as supercritical water technique. For example, Zhu et al. have synthesized the dendritic structure of Ag using anisotropic nickel nanotubes that act as a reducing agent during the hydrothermal reaction [75]. Upon changing the reaction system from nonequilibrium to quasiequilibrium, the dendritic structure is modified into compact crystal. When PVP is added into the reaction, compact particles are formed. Yin et al have synthesized Ru nanoparticles having a triangular shape using $RuCl_3$, H_2O, HCHO (40 wt%), and PVP by the hydrothermal method at 160°C. By changing $RuCl_3$:H_2O:PVP ratio, irregular shaped Ru nanostructures are obtained [76]. Au nanowire (diameter ~50–110 nm) has been synthesized using hexamethylenetetramine and ethylenediamine tetra acetic acid by the hydrothermal method. Octahedral-shaped Au nanoparticles were grown in aqueous medium by Chang et al. [77] using $HAuCl_4 \cdot 3H_2O$, CTAB, and trisodium citrate as precursor. In this typical procedure, the reaction is carried out at 110°C taking 9.7 mL of high-purity water, 0.055 g of CTAB, 250 μL of 0.01 M $HAuCl_4$, and 50 μL of 0.1 M trisodium citrate. Au octahedrals with a size between 31 and 149 nm are generated just by carrying out the reaction at time intervals of 6, 12, 24, 48, and 72 hours. Changing the molar ratio of CTAB: $HAuCl_4$, different polyhedral nanostructures of Au are generated [78]. Ag nanowire (diameter ~53 nm and length ~6 μm) has been synthesized by the hydrothermal technique using 100 μL of $AgNO_3$ solution (0.1 M) and 100 μL of sodium citrate solution (0.1 M) [79]. The reaction was carried out at 130°C for 3 hours. It is observed that the pH of the solution has a significant effect on the dimension of the nanowire. Increasing pH reduces the length of the nanowire, whereas at a much higher pH monodispersed nanosphere (diameter ~58 m) is formed. Recently, B. Bari et al. also developed a hydrothermal method to prepare Ag nanowire (diameter ~45–65 nm and length ~200 µm) starting from $AgNO_3$ (0.02 M, 15 mL), D+ glucose (0.12 g, 5 mL), NaCl (0.04 M, 15 mL), and PVP (1 g, 5 mL) [80]. The reaction was carried out at 160°C for 22 hours. The same research group also prepared Ag nanowire thin film on glass and PET substrates. Two-dimensional arrays of Ag nanoparticles (diameter ~17 nm) have been grown by O. Ayyad et al. by the hydrothermal method [81]. Ag nanoplates were synthesized from a hydrothermal reaction among 0.4 g $AgNO_3$, 10 mL $NaHCO_3$ (0.12 M), and 2.5 g of PVP at 170°C for 5 hours [82]. Ag nanosieve has been synthesized by R.P. Singh et al. using $AgNO_3$, 1,2-benzenedicarboxylate (10 mM), and sodium stearate (10 mM) as precursor materials by the hydrothermal technique, carried out at 250°C [83]. The formation and stabilization of the nanostructure are explained in the way that 1,2-benzenedicarboxylate having carboxylate ligands possesses a preference for metal coordination due to the presence of the carboxylic anion group ($-COO^-$) resulting in the formation of Ag–carboxylate, which exhibits van der Waals interaction between chains due to the presence of the hydrophobic nature of their tails. The combination of sodium stearate and Ag–carboxylate gives control over particle size, stabilizes them, and prevents oxidation. Adams et al. prepared Pd nanostructure on Ti substrate using $PdCl_2$ (5 mM) and ammonium formate (1 M) as precursor materials [84]. The reaction was carried out at 180°C for 2 hours.

Acknowledgment

The author thanks all of the CRC Press team for their suggestions to improve the chapter.

References

1. P. Zijlstra, J.W.M. Chong, M. Gu, *Nature* 459 (2009), 410.
2. E. Abbasi, M. Milani, S. FekriAval, M. Kouhi, A. Akbarzadeh, H. TayefiNasrabadi, P. Nikasa, S.W. Joo, Y. Hanifehpour, K. Nejati-Koshki, *Crit. Rev. Microbiol.* 39 (2014), 1–8.
3. V. Amendola, M. Meneghetti, *Phys. Chem. Chem. Phys.* 11 (2009), 3805–3821.
4. S.M. Landage, A.I. Wasif, P. Dhuppe, *Int. J. Adv. Res. Eng. Appl. Sci.* 3 (2014), 5.

5. H. Wang, X. Qiao, J. Chen, S. Ding, *Colloids Surf. A: Physicochem. Eng. Aspects* 256 (2005), 111–115.
6. J.G. Barrasa, J.M. López-de-Luzuriaga, M. Monge, *Cent. Eur. J. Chem.* 9 (2011), 7–19.
7. C.N. Andhariya, O.P. Pandey, B. Chudasama, *RSC Adv.* 3 (2013), 1127.
8. B. Chudasama, A.K. Vala, N. Andhariya, R.V. Mehta, R.V. Upadhyay, *J. Nanopart. Res.* 12 (2010), 1677–1685.
9. H. Ma, B. Yin, S. Wang, Y. Jiao, W. Pan, S. Huang, S. Chen, F. Meng, *ChemPhysChem* 5 (2004), 68–75.
10. W. Zhang, X. Qiao, J. Chen, *Mater. Sci. Eng. B* 142 (2007), 1–15.
11. K. Mavani, M. Shah, *Int. J. Eng. Res. Technol.* 2 (March 2013), 3.
12. S.D. Solomon, M. Bahadory, A.V. Jeyarajasingam, S.A. Rutkowsky, C. Boritz, L. Mulfinger, *J. Chem. Educ.* 84(2) (2000), 322–325.
13. I. Pastoriza-Santos, L.M. Liz-Marzán, *Langmuir* 15 (1999), 948–951.
14. J. Kuna, P. Wang, L. Yuan, X. Zhou, W. Chen, X. Liu, *J. Mater. Chem. C* 3 (2015), 3522.
15. M.A. Uppal, A. Kafizas, M.B. Ewing, I.P. Parkin, *New J. Chem.* 34 (2010), 2906.
16. S. Kumar, K.S. Gandhi, R. Kumar, *Ind. Eng. Chem. Res.* 46 (2007), 3128–3136.
17. M. Brust, M. Walker, D. Bethell, D.J. Schiffrin, R. Whyman, *Chem. Commun.* (7) (1994), 801.
18. R.L. Donkers, Y. Song, R.W. Murray, *Langmuir* 20 (2004), 4703.
19. V.L. Jimenez, D.G. Georganopoulou, R.J. White, A.S. Harper, A.J. Mills, D. Lee, R.W. Murray, *Langmuir* 20 (2004), 6864.
20. H. Liu, Y. Feng, D. Chen, C. Li, P. Cui, J. Yang, *J. Mater. Chem. A* 3 (2015), 3182.
21. M.J. Hostetler, J.E. Wingate, C.J. Zhong, J.E. Harris, R.W. Vachet, M.R. Clark, J.D. Londono et al., *Langmuir* 14 (1998), 17.
22. D.V. Leff, L. Brandt, J.R. Heath, *Langmuir* 12 (1996), 4723.
23. J. Yang, E.H. Sargent, S.O. Kelley, J.Y. Ying, *Nat. Mater.* 8 (2009), 683.
24. X. Wang, J. Zhuang, Q. Peng, Y. Li, *Nature* 437 (2005), 121; B. Cushing, V.L. Kolesnichenko, C.J. O'Connor, *Chem. Rev.* 104 (2004), 3893.
25. J. Yang, J.Y. Lee, H.P. Too, G.M. Chow, L.M. Gan, *Chem. Lett.* 34 (2005), 354; J. Yang, J.Y. Lee, H.P. Too, S. Valiyaveettil, *J. Phys. Chem. B* 110 (2006), 125.
26. K.W. Park, J.H. Choi, B.K. Kwon, S.A. Lee and Y.E. Sung, *J. Phys. Chem. B* 106 (2002), 1869.
27. S.J. Park, H.J. Jung, C.W. Nah, *Polymer (Korea)* 27 (2003), 46.
28. A.S. Arico, V. Baglio, E. Modica, A.D. Blasi, V. Antonucci, *Electrochem. Commun.* 6 (2004), 164.
29. Md. Aminul Islam, M.A.K. Bhuiya, M. Saidul Islam, *Asia Pac. J. Energy Environ.* 1 (2014), 2.
30. S.M. Humphrey, M.E. Grass, S.E. Habas, K. Niesz, G.A. Somorjai, T.D. Tilley, *Nano Lett.* 7 (2007), 785.
31. C. Bock, B. MacDougall, *Proceedings of the Knowledge Foundation's Fourth International Conference on Nanostructured Materials*, Miami, FL (2003), ed. S. Pan, Knowledge Press, Brookline, MA.
32. M.A. Islam, M.A. Bhuiya, M.S. Islam, *Asia Pac. J. Energy Environ.* 1(2) (2014), 107.
33. M.T. Reetz, W. Helbig, *J. Am. Chem. Soc.* 116 (1994), 1401.
34. L. Blanodn, M.V. Vazquez, D.M. Benjumea, G. Ciro, *Portugaliae Electrochimica Acta* 30(2) (2012), 135.
35. N. Kiratzis, M. Faers, P.F. Luckham, *Colloids Surf. A* 151 (1999), 461.

36. L. Rodríguez-Sánchez, M.C. Blanco, M.A. López-Quintela, *J. Phys. Chem. B* 104 (2000), 9683.
37. H. Ashassi-Sorkhabi, B. Rezaei-moghadam, E. Asghari, R. Bagheri, L. Abdoli, *Phys. Chem. Res.* 3 (2015), 24.
38. M.V. Mandke, S.H. Han, H.M. Pathan, *CrystEngComm* 14 (2012), 86.
39. R. Sivasubramanian, M.V. Sangaranarayana, *CrystEngComm* 15 (2013), 2025.
40. C. Fang, J.G. Shapter, N.H. Voelcker, A.V. Ellis, *RSC Adv.* 4 (2014), 19502.
41. Y.H. You, Y.W. Lin, C.Y. Chen, *RSC Adv.* 5 (2015), 93293.
42. Y. Peng, Z. Yan, Y. Wu, J. Di, *RSC Adv.* 5 (2015), 7854.
43. D. Li, J. Liu, H. Wang, C.J. Barrow, W. Yang, *Chem. Commun.* 52 (2016), 10968.
44. X. Shen, X. Chen, J.H. Liu, X.J. Huang, *J. Mater. Chem.* 19 (2009), 7687.
45. M.S. Gutierrez, A. Henglein, J.K. Dohrmann, *J. Phys. Chem.* 91 (1987), 6687.
46. Y. Nagata, Y. Watanabe, S. Fujita, T. Dohmaru, S. Taniguchi, *J. Chem. Soc. Chem. Commun.* (21) (1992), 1620.
47. Y. Mizukoshi, R. Oshima, Y. Maeda, Y. Nagata, *Langmuir* 15 (1999), 2733.
48. S.A. Yeung, R. Hobson, S. Biggs, F. Grieser, *J. Chem. Soc. Chem. Commun.* (4) (1993), 378.
49. J.H. Bang, K.S. Suslick, *Adv. Mater.* 22 (2010), 1039.
50. R.A. Caruso, M.A. Kumar, F. Grieser, *Langmuir* 18 (2002), 7831.
51. Y. Mizukoshi, E. Takagi, H. Okuno, R. Oshima, Y. Maeda, Y. Nagata, *Ultrason. Sonochem.* 8 (2001), 1.
52. H. Xu, K.S. Suslick, *ACS Nano* 4 (2010), 3209.
53. K. Okitsu, A. Yue, S. Tanabe, H. Matsumoto, Y. Yobiko, Y. Yoo, *Bull. Chem. Soc. Jpn.* 75 (2002), 2289.
54. W. Chen, W. Cai, Y. Lei, L. Zhang, *Mater. Lett.* 50 (2001), 53.
55. N. Perkas, G. Amirian, G. Applerot, E. Efendiev, Y. Kaganovskii, A.V. Ghule, B.J. Chen, Y.C. Ling, A. Gedanken, *Nanotechnology* 19 (2008), 435604.
56. A. Gautam, P. Tripathy, S. Ram, *J. Mater. Sci.* 41 (2006), 3007.
57. S. Wang, H. Xin, *J. Phys. Chem. B* 104 (2000), 5681.
58. Q.M. Liu, T. Yasunami, K. Kuruda, M. Okido, *Trans. Nonferrous Met. Soc. China* 22 (2012), 2198.
59. T. Li, H.G. Park, S.H. Choi, *Mater. Chem. Phys.* 105 (2007), 325.
60. J. Jung, H. Oh, H. Noh, J. Ji, S. Kim, *Aerosol Sci.* 37 (2006), 1662.
61. M. Becker, J. Brock, H. Cai, D. Henneke, J. Keto, J. Lee, W. Nichols, H. Glicksman, *Nanostructured Mater.* 10 (1998), 853.
62. X. Huang, S. Li, S. Wu, Y. Huang, F. Boey, C.L. Gan, H. Zhang, *Adv. Mater.* 24 (2012), 979–983.
63. S. Habouti, M.M. Tempfli, C.H. Solterbeck, M.E. Souni, S.M. Tempfli, M.E. Souni, *J. Mater. Chem.* 21 (2011), 6269.
64. Z. Chen, P. Zhan, Z. Wang, J. Zhang, W. Zhang, N. Ming, C. Chan, P. Sheng, *Adv. Mater.* 16 (2004), 417.
65. Z. Wang, M. Chen, L. Wu, *Chem. Mater.* 20 (2008), 3251.
66. E.M. Felix, M. Antoni, I. Pause, S. Schaefer, U. Kunz, N. Weidler, F. Muench, W. Ensinger, *Green Chem.* 18 (2016), 558.
67. X. Lou, L.A. Archer, Z. Yang, *Adv. Mater.* 20 (2008), 3987.
68. D. Zhang, L. Qi, J. Ma, H. Cheng, *Adv. Mater.* 14(2002), 1499.
69. M. Salsamendi, A.G.P. Cormack, D. Graham, *New J. Chem.* 37 (2013), 3591.
70. B. Chen, X. Jiao, D. Chen, *CrystEngComm* 13 (2011), 204.
71. R. Ballabh, S. Nara, *Indian J. Exp. Biol.* 53 (2015), 828.

72. B. Nithyaja, H. Misha, V.P.N. Nampoori, *Nanosci. Nanotechnol.* 2(4) 2012, 99.
73. X. Lu, M.S. Yavuz, H.Y. Tuan, B.A. Korgel, Y. Xia, *J. Am. Chem. Soc.* 130 (2008), 8900.
74. A. Halder, N. Ravishankar, *Adv. Mater.* 19 (2007), 1854.
75. Y. Zhu, H. Zheng, Y. Li, L. Gao, Z. Yang, Y.T. Qian, *Mater. Res. Bull.* 38 (2003), 1829.
76. A.X. Yin, W.C. Liu, J. Ke, W. Zhu, J. Gu, Y.W. Zhang, C.H. Yan, *J. Am. Chem. Soc.* 134 (2012), 20479.
77. C.C. Chang, H.L. Wu, C.H. Kuo, M.H. Huang, *Chem. Mater.* 20 (2008), 7570.
78. Y. Liu, J. Zhou, X. Yuan, T. Jiang, L. Petti, L. Zhou, P. Mormile, *RSC Adv.* 5 (2015), 68668.
79. Z. Yang, H. Qian, H. Chen, J.N. Anker, *J. Colloid Interface Sci.* 352 (2010), 285.
80. B. Bari, J. Lee, T. Jang, P. Won, S.H. Ko, K. Alamgir, M. Arshad, L.J. Guo, *J. Mater. Chem. A* 4 (2016), 11365–11371, DOI: 10.1039/c6ta03308c.
81. O. Ayyad, D.M. Rojas, P.G. Romero, *Chem. Commun.* 47 (2011), 11285.
82. J. Cao, D. Zhao, X. Lei, Y. Liu, Q. Mao, *Appl. Phys. Lett.* 104 (2014), 201906.
83. R.P. Singh, A.C. Pandey, *Anal. Methods* 3 (2011), 586.
84. B.D. Adams, G. Wu, S. Nigro, A.J. Chen, *J. Am. Chem. Soc.* 131 (2009), 6930.

7

Biological Synthesis of Metallic Nanoparticles: A Green Alternative

Kaushik Roy and Chandan Kumar Ghosh

CONTENTS

7.1 Introduction

Biological synthesis of nanoparticles, as the name suggests, may be defined as the synthesis of different nanoparticles and nanostructures using biological elements like plant materials or microbes. The growing need to develop eco-friendly and economically viable synthesis procedures for metallic nanoparticles prompted scientists to explore the use of microorganisms, plant extracts, and other biomaterials. Involvement of natural elements in this route of synthesis reduces the possibility of environmental hazards. The green synthesis of nanoparticles encourages the use of various microorganisms for preparing nanoparticles of noble metals like gold, silver, palladium, platinum, etc., and also a few semiconducting oxides like zinc oxide, titanium oxide, etc. The desired organisms include algae, fungi, and bacteria. Microbe-mediated procedures are generally followed to produce highly stable metal nanoparticles when relevant aspects like genetic and inheritable properties of organisms, optimum conditions for organism growth, and enzymatic activity, etc., are considered. The use of plant extract is gaining in popularity now due to its ease of preparation and lack of risks associated with handling. Among different biogenic nanoparticles produced so far, medicinal plant extract–mediated nanoparticles have been found to be pharmacologically most active, probably due to the surface attachment of pharmacologically active biomass residues.

7.2 Advantages of Biological Synthesis over Other Routes

A number of reports prevailing in the literature indicate that the synthesis of nanoparticles by chemical approaches is hazardous and expensive. In contrast, biological synthesis methods offer many advantages over other synthetic routes attracting nanoresearchers in the last few years. Biological synthesis procedures have the following advantages:

- *Simplicity*: Biosynthesis procedures are generally single-step and easy.
- *Low cost*: Synthesis of nanoparticles using plants is very cost-effective and may be used as an economic and valuable alternative for large-scale production.
- *Eco-friendliness*: Use of microbes and plant materials makes this synthetic protocol clean and fully eco-friendly.
- *Fast process*: The biological synthesis of nanoparticles is regarded as a rapid and easy-to-scale-up technology.

7.3 Use of Microorganisms for the Preparation of Metal Nanoparticles

Use of microbes for the production of nanoparticles is treated as a useful and eco-friendly biological protocol for preparing metal nanoparticles from the aqueous solution of metal salts. The use of different microorganisms such as bacteria, fungi, and algae for the synthesis of nanoparticles is discussed in detail in the following sections.

7.3.1 Use of Bacteria

In the last decade, many studies have reported the use of prokaryotes to produce metallic nanoparticles. Their abundance in the atmosphere and adaptability to extreme conditions make bacteria a good choice for this purpose.

Bacteria are generally fast-growing, easy to cultivate, and well-manipulated biosystems. He et al. prepared extracellular gold nanoparticles 10–20 nm in size using the photosynthetic bacteria *Rhodopseudomonas capsulata* (He et al., 2007). They proposed the key role of a bacterial enzyme, that is, nicotinamide adenine dinucleotide hydride (NADH)-dependent reductase, for the reduction of gold ions to gold nanoparticles during interaction (see Figure 7.1). They also noted that the variation of pH in the growth medium during reaction may result in the formation of different shapes and morphology of nanoparticles. In another study, gold nanoparticles were prepared by using gram-negative aerobic bacteria *Delftia acidovorans* that excrete peptide delftibactin, which was expected to be the reducing and stabilizing factor for the generation of nanoparticles (Johnston et al., 2013). Stable silver nanoparticles were also synthesized involving bacteria as reported by Kailashwaralal et al. where nanosilver was produced by bioreduction of aqueous silver cations using culture supernatant of nonpathogenic bacteria *Bacillus licheniformis* (Kalishwaralal et al., 2008).

Palladium is another noble metal that is currently used in dehalogenation and hydrogenation reactions as a catalyst. Recently, a relevant study reported that palladium nanoparticles can be synthesized using bacterial species found near alpine sites, which are heavy metal–contaminated zones (Schlüter et al., 2014). Out of all heavy metal–resistant bacteria, *Pseudomonas* cells were found to produce catalytically active palladium nanoparticles used frequently in reductive dehalogenation of chemical congeners.

Saifuddin et al. described an innovative synthetic approach for preparing silver nanoparticles using a combination of *Bacillus subtilis* culture filtrate and simultaneous microwave irradiation in aqueous medium (Saifuddin et al., 2009). In another research report, rapid synthesis of nanosilver from a solution of silver salt was achieved by introducing culture supernatants of *E. coli*, *Enterobacter cloacae*, and *Klebsiella pneumonia* (Shahverdi et al., 2007). *Lactobacillus* strains, when exposed to metal ions, result in the biosynthesis

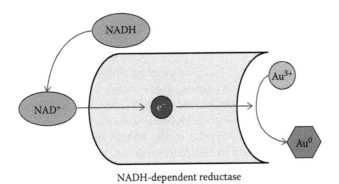

NADH-dependent reductase

FIGURE 7.1
Suggested mechanism for reduction of gold ions into gold nanoparticles. (From He, S. et al., *Mater. Lett.*, 61, 3984, 2007.)

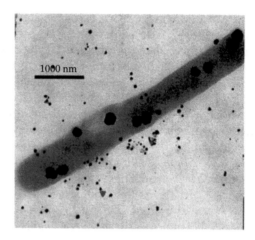

FIGURE 7.2
Synthesis of metallic nanostructures inside and outside bacterial cells. (From Korbekandi, H. et al., *J. Chem. Technol. Biotechnol.*, 87(7), 932, 2012.)

of nanoparticles within the bacterial cells (shown in Figure 7.2). It has been found that exposure of lactic acid bacteria present in the whey of buttermilk to mixtures of silver ions could be used to grow silver nanoparticles (Nair et al., 2002; Korbekandi et al., 2012). Although silver and gold nanoparticles are important due to their antimicrobial abilities, a few studies have also been conducted on other metals such as uranium.

The literature contains many reports on the performance of *Bacillus* species for their excellent capacity for metal bioaccumulation (Kalimuthu et al., 2008; Pugazhenthiran et al., 2009). Pollman and his team explored the capability of *Bacillus sphaericus* to accumulate toxic metals like lead, cadmium, copper, and uranium (Pollmann et al., 2006). The proteins found on the surface layers of bacterial cells were supposed to be responsible for this activity.

7.3.2 Use of Fungi

Fungi are the other type of microorganisms with the potential to produce metal nanostructures when exposed to metal ions during biochemical interactions. A significant amount of research has already been done on the use of fungal species for the preparation of nanometals, as fungi offer some advantages over bacteria. Mycelia of the fungal body have a large surface area for interaction, leading to rapid downstream processing of the bioreduction of metal ions (Mukherjee et al., 2001). Another advantage of this process was reported by Ahmad et al., who prepared extracellular silver nanoparticles using *Fusarium oxysporum* and noted no flocculation of particles even after 30 days of production. The reduction of silver cations and long-term stability of the nanoparticles were attributed to the enzymatic activity of

NADH-dependent reductase (Ahmad et al., 2003). It has also been observed that the fungal cells secrete higher amounts of protein compared to bacteria. Hence, the production of nanoparticles may be amplified when fungal synthetic platforms are involved (Mukherjee et al., 2001).

Trichoderma reesei, a species of soil fungi, was tested to synthesize extracellular silver nanoparticles by Vahabi et al. (2011). The size of the nanoparticles was found to be 5–50 nm (Figure 7.3). The diffraction rings observed from the selected area electron diffraction (SAED) pattern were correlated to the face-centered cubic structure of silver. The use of *T. reesei* seems to have merit over using other fungi as it can be manipulated to produce higher quantities of enzymes up to 100 g/L, which may result in the substantial yield of produced nanoparticles.

Extracellular synthesis of nanoparticles has some advantages like simplicity and low cost, but intracellular synthesis procedures are important for specific environmental applications. An important study performed by Sanghi et al. reported the use of white rot fungus, *Coriolus versicolor*, for the generation and accumulation of intracellular nanoparticles of silver (Sanghi et al., 2009). The unintended metal ions of platinum or copper present in real water samples may be reduced, accumulated, and removed from the environment using this fungus in the same way.

Along with the nanoparticles of noble metals, magnetic nanoparticles have also been reported to be produced by endophytic fungus *Verticillium* sp. and pathogenic fungus *F. oxysporum* (Sun et al., 2002; Bharde et al., 2006). These magnetic nanoparticles are currently being explored in applications like magnetic resonance imaging, magnetic recording, and position sensing.

As with bacteria, fungi also have some noteworthy disadvantages when it comes to safety. Many well-studied fungi such as *F. oxysporum* are pathogenic and therefore might pose a safety hazard. *Trichoderma asperellum* and *Trichoderma reesei* are both fungi that produce Ag nanoparticles when

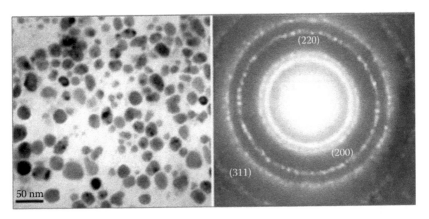

FIGURE 7.3
TEM micrograph and SAED pattern of silver nanoparticles prepared using *Trichoderma reesei*. (From Vahabi, K. et al., *Insciences J.*, 1(1), 65, 2011.)

exposed to silver salts and have been proven to be nonpathogenic, which makes them ideal for use commercially. In fact, *T. reesei* is already being used widely in industries such as food, animal feed, pharmaceuticals, paper, and textiles.

Kaul et al. claimed to synthesize iron nanoparticles from ferric oxide solution using two bacteria, *Alcaligenes faecalis* and *Bacillus coagulans*, and four fungi, *Curvularia lunata*, *Chaetomium globosum*, *Aspergillus fumigatus*, and *Aspergillus wentii* (Kaul et al., 2012). Among all these microorganisms, *C. globosum* was proved to be the most efficient for the synthesis of iron nanoparticles from Fe_2O_3.

Kaul and his team also demonstrated the preparation of magnesium nanoparticles from magnesium salts involving two fungal species, *Pochonia chlamydosporium* and *Aspergillus fumigatus* (Kaul et al., 2012). *A. fumigatus* was able to synthesize Mg nanoparticles extracellularly from magnesium sulfate, whereas *P. chlamydosporium* produced similar particles inside fungal cells while exposed to magnesium chloride. In this way, the fungal strains were selective to different metal salts in response to nanoparticle production.

Raliya and Tarafdar reported biosynthesis of magnesium, zinc, and titanium nanoparticles from standard salts using a group of soil-borne *Aspergillus* species (fungal) including *A. fumigatus*, *A. niger*, *A. tubingensis*, *A. oryzae*, *A. flavus*, and *A. terreus* (Raliya et al., 2014). The isolated strains were collected from arid agricultural lands in Rajasthan, India, and later authenticated. The cell-free culture filtrate was added to different metallic precursor compounds and nanoparticles of different size were obtained as shown in Table 7.1. The extracellular enzymes secreted from the fungal cells are expected to be the key factors that act as reducing and surface-stabilizing agents and further produce metal nanoparticles.

TABLE 7.1

Response to Different Metal Salts for Fungi-Mediated Synthesis

Metal	Precursor Compound	Size of Nanoparticles (nm)
Mg	$Mg(NO_3)_2$	15 ± 0.3
	$MgCl_2$	69 ± 1.6
	$MgSO_4$	96 ± 2.7
	MgO	10 ± 0.8
Zn	$Zn(NO_3)_2$	46 ± 1.4
	$ZnCl_2$	74 ± 0.8
	$ZnSO_4$	88 ± 0.7
	ZnO	30 ± 0.6
Ti	TiO_2 rutile	17 ± 0.7
	TiO_2 anatase	13 ± 0.4

Source: Raliya, R. et al., *Int. Nano. Lett.*, 4(93), 1, 2014.

7.3.3 Use of Algae

The use of algae for preparing metal nanoparticles is a less explored field of research, though a small number of reports are still available on this subject. Singaravelu et al. presented an algal platform involving *Sargassum wightii* for producing extracellular and highly stable gold nanoparticles where almost 95% of production was achieved within only 12 h of incubation (Singaravelu et al., 2007). Another study, by Chakraborty et al., described the use of a panel of algal species including *Cladophora prolifera, Lyngbya majuscule, Spirulina subsalsa, Chlorella vulgaris, Padina pavonica, Rhizoclonium heiroglyphicum, Sargassum fluitans,* and *Spirulina platensis* for the formation and accumulation of gold nanoparticles in aqueous medium (Chakraborty et al., 2009).

A novel algal platform involving freshwater green algae, *Chlorella pyrenoidosa*, has recently been used by Aziz and his team for biosynthesis of silver nanoparticles with a particle size distribution between 5 and 15 nm (Aziz et al., 2015). When the biomass of edible cyanobacteria, *Spirulina platensis*, was exposed to 10^{-3} M aqueous $AgNO_3$, extracellular formation of spherical silver nanoparticles (7–16 nm) resulted in 120 h at 37°C at pH ~ 5.6 (Govindaraju et al., 2008). Possibly, the enzymes generated by the cellular activities of the algal cells seized and reduced the metallic ions and finally transformed them into their elemental form for both silver and gold.

7.4 Use of Plant Extracts for the Synthesis of Metallic Nanoparticles

Microbe-mediated synthesis with bacteria and fungi has been studied extensively in the past few decades for its ability to synthesize metallic nanoparticles. The major disadvantage of microorganism-involved procedures is related to the procurement and maintenance of strain cultures. Procurement of the bacterial and fungal strains is quite expensive in most cases. Maintenance of these strains involves special culture and isolation techniques as well. In addition, proper and cautious handling of microorganisms, specifically the human pathogenic species, is very important. Any improper handling in this case may result in infection and illness. Hence, an important advantage of using plant extracts instead of microorganisms for the synthesis of nanometals is lack of pathogenicity (Pantidos et al., 2014). This advantage is making this protocol very popular nowadays, and researchers across the globe have also begun to use extracts of different parts of a plant body such as leaves, fruits, flowers, roots, and seeds for the production of metal nanoparticles.

7.4.1 Use of Leaf Extracts

Synthesis of nanoparticles using plant-mediated procedures is found to be better and easier than the microbe-mediated methods as it does not involve any maintenance or preservation of microbe cultures. Extracts of various plants have already been explored and found to have an impressive ability to reduce and stabilize metal cations into metallic nanoparticles. Among all types of extracts used so far for biosynthesis of metal nanoparticles, leaf extract is the most common and appealing one. This is mainly because the leaves are considered to be the "kitchen" of a plant body where photosynthesis occurs. Major bioactive molecules like glucose, fructose, chlorophyll, and other components of sap remain present in the leaves. Hence, extracts of leaves mostly contain numerous organic molecules that can play a pivotal role during exposure to metal salts (Mittal et al., 2013). The literature contains numerous reports on preparation of silver and gold nanoparticles using many types of leaf extracts. Several groups reported the green synthesis of silver nanoparticles from silver salts using leaf extract of parsley (*Petroselinum crispum*) (Roy et al., 2015d) (shown in Figure 7.4a), celery (Roy et al., 2015c) (shown in Figure 7.4b), lemongrass (Masurkar et al., 2011), neem (Namratha et al., 2013), lotus (Santhoshkumar et al., 2011), and *Aloe vera* (Medda et al., 2015).

7.4.2 Use of Other Parts Like Root, Flower, Fruit, Etc.

A report on a cost-effective, clean, and fast bioprocess to synthesize Ag nanoparticles after reduction of Ag^+ using apple (*Malus domestica*) fruit extract has been presented by Roy et al. (2014). The fruit extract contains water-soluble hydrocarbons, proteins which appear to be an efficient reducer and stabilizer for metallic nanoparticles. The formation of silver nanoparticles was initially examined by UV-Vis spectroscopy. Different phases and

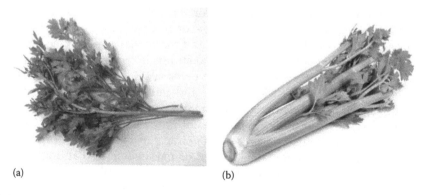

(a)   (b)

FIGURE 7.4
(a) Parsley leaves and (b) celery leaves.

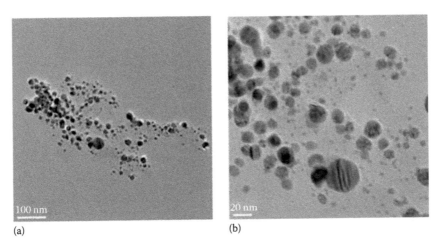

(a)   (b)

FIGURE 7.5
HRTEM images of biogenic silver nanoparticles. (From Roy, K. et al., *Dig. J. Nanomater. Biostruct.*, 9(3), 1137, 2014.)

morphology of silver nanoparticles were analyzed by X-ray diffraction and transmission electron microscope, respectively (shown in Figure 7.5). The fruit extract of apple reduced the silver ions and produced metallic silver nanoparticles with well-defined stability. These nanoparticles are spherical in shape and the average particle size was found to be 20 nm.

Roy et al. reported the biosynthesis of nanosilver using cucumber fruit extract (Roy et al., 2015a). The formation of silver nanoparticles was investigated by scanning the reacting solution under an ultraviolet-visible spectrometer. The extract contains functional organic molecules which are supposed to be efficient reducing and surface-stabilizing agents for synthesis of colloidal particles. The reacting mixture turned from colorless to light brown during incubation, pointing to the production of silver nanoparticles in the medium. The solution color intensified with time and became dark brown after saturation. The recorded absorption spectra of the solution exhibited highest intensity near 450 nm wavelength, which is expected as the characteristic plasmonic peak for silver (see Figure 7.6). These nanoparticles were spherical with average particle size of nearly 10 nm.

In other relevant studies, tuber and root extracts were used for production of stable silver nanoparticles. Velmurugan and his team synthesized gold and silver nanoparticles from metal salts using root extract of common ginger and later studied the antibacterial efficacy of these nanometals toward a few pathogens (Velmurugan et al., 2014). Roy et al. reported green synthesis of silver nanoparticles from silver nitrate using infusion of potato (*Solanum tuberosum*) tuber, which is easily available in the market throughout the year and cheaper than many other vegetables (Roy et al., 2015b). The potato tuber is a potential source of numerous bioactive molecules such as protein, thiamine, amino

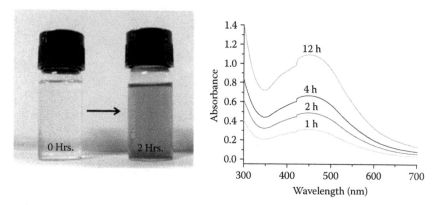

FIGURE 7.6
Color change of reacting solution and UV-Vis spectra at specific intervals. (From Roy, K. et al., *Dig. J. Nanomater. Biostruct.*, 10(1), 107, 2015a.)

FIGURE 7.7
Potato (*S. tuberosum*) tuber and XRD curve of the biogenic silver nanoparticles.

acids, and ascorbic acid. Hence, the infusion of potato reduced the silver ions and stabilized them in the solution. X-ray diffraction pattern was recorded to identify crystal phases of the produced nanosilver (see Figure 7.7). The result of the stability study showed that the size of these biogenic nanoparticles remained unaltered after even a few months of preparation. Such surface stabilization may be the result of biocapping of particle surfaces by amides and amine group molecules (Mendoza-Reséndez et al., 2013).

Another study reports a single-step procedure for preparing silver nanoparticles from silver nitrate solution using fruit extract of *Benincasa hispida* (Roy et al., 2016a), along with the study of its capability for fast detection of hydrogen peroxide. These silver nanoparticles were found to have a nearly spherical shape with an average diameter of 10 nm. The H_2O_2-detecting ability of these biogenic silver nanoparticles was explored and the Ag nanoparticles were found to detect H_2O_2 instantly. A recent study by the same group presents a simple, clean, and eco-friendly synthesis procedure of copper nanoparticles from copper sulfate solution using extract of *Eichhornia crassipes* flowers (see Figure 7.8), along with a hydrogen peroxide detectability study (Roy et al., 2016b). These particles were observed to be spherical with average diameter ~12–15 nm. X-ray diffraction study verified their crystal phases and Fourier transform infrared spectroscopic analysis has confirmed the role of bioactive molecules like lawsone and phenol present in the extract during interaction.

Njagi et al. reported a green synthesis of iron nanoparticles using bran extract of *Sorghum* sp. (Njagi et al., 2011). *Sorghum* is a member of the grass family found mainly in Australia and some parts of Asia. Its bran extract powder is used commercially for preparing alcoholic beverages, syrups, and biofuels. It contains organic compounds like flavonoids, polyphenols, and phenolic acid. This report claims to prepare iron nanoparticles from ferric chloride solution using commercially available *Sorghum* bran extract powder at different temperatures. The field emission scanning electron microscopic and transmission electron microscopic images indicate that the iron nanoparticles are mostly spherical and their size ranges from 40 to 50 nm, as shown in Figure 7.9.

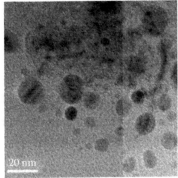

FIGURE 7.8
Eichhornia crassipes flower and its extract-mediated copper nanoparticles. (From Roy, K. et al., Rapid detection of hazardous H_2O_2 by biogenic copper nanoparticles synthesized using *Eichhornia crassipes* extract, *Third International Conference on Microelectronic Circuit and System*, Kolkata, West Bengal, India, 2016b, pp. 203–207.)

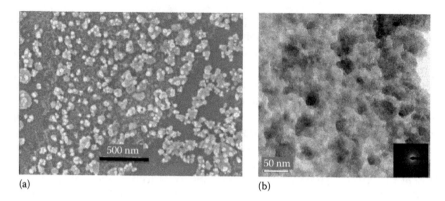

(a)   (b)

FIGURE 7.9
(a) FESEM and (b) TEM images of iron nanoparticles. (From Njagi, E.C. et al., *Langmuir*, 27(1), 264, 2011.)

7.5 Limitations of Biological Synthesis

In spite of the many advantages of biological synthesis discussed already in this chapter, a few limitations of this route still exist today. The green synthesis of metal nanoparticles has significant potential and several advantages compared to conventional physical and chemical routes. However, so far the biological synthesis of nanoparticles has been used in small-scale production only. Another limitation would be the difficulty in controlling the size and morphology of nanoparticles during production, though numerous efforts are now being made to overcome these limitations and scale up the production of metal nanoparticles through eco-friendly biological routes.

7.6 Conclusion: Future Scope of Research

It is well known that metallic nanoparticles have great potentiality for commercial applications. The need for a process to synthesize such nanoparticles in a reliable and green way is becoming stronger. Numerous groups have focused on alternative ways to synthesize nanoparticles as discussed here. Biological systems have been investigated in order to provide a resource-efficient, sustainable, and low-cost platform for nanoparticle synthesis.

The flexibility of synthesis protocols and simple incorporation of nanoparticles into various media encouraged researchers to look further into the mechanistic aspects of antimicrobial, anticancer, and other biological properties of these nanoparticles. Biological synthesis provides an alternative and novel method of conveniently producing nanoparticles involving natural organic agents. These eco-friendly methods can potentially be applied in many areas such as pharmaceuticals, cosmetics, food processing, and medical applications. However, before attributing any commercial relevance to biosynthesis of metal nanoparticles, more examples along with limiting factors need to be considered through further investigations.

References

Ahmad A et al. (2003). Extracellular biosynthesis of silver nanoparticles using the fungus *Fusarium oxysporum*. *Colloids Surf. B: Biointerfaces* 28, 313–318.

Aziz N et al. (2015). Facile algae-derived route to biogenic silver nanoparticles: Synthesis, antibacterial, and photocatalytic properties. *Langmuir* 31, 11605–11612.

Bharde A et al. (2006). Extracellular biosynthesis of magnetite using fungi. *Small* 2, 135–141.

Chakraborty N et al. (2009). Biorecovery of gold using cyanobacteria and an eukaryotic alga with special reference to nanogold formation—A novel phenomenon. *J. Appl. Phycol.* 21, 145–152.

Govindaraju K et al. (2008). Silver, gold and bimetallic nanoparticles production using single-cell protein (*Spirulina platensis*). Geitler. *J. Mater. Sci.* 43, 5115–5122.

He S et al. (2007). Biosynthesis of gold nanoparticles using the bacteria *Rhodopseudomonas capsulata*. *Mater. Lett.* 61, 3984–3987.

Johnston CW et al. (2013). Gold biomineralization by a metallophore from a gold-associated microbe. *Nat. Chem. Biol.* 9, 241–243.

Kalimuthu K et al. (2008). Biosynthesis of silver nanocrystals by *Bacillus licheniformis*. *Colloids Surf. B: Biointerfaces* 65, 150–153.

Kalishwaralal K et al. (2008). Extracellular biosynthesis of silver nanoparticles by the culture supernatant of *Bacillus licheniformis*. *Mater. Lett.* 62, 4411–4413.

Kaul RK et al. (2012). Magnesium and iron nanoparticles production using microorganism and varying salts. *J. Mater. Sci. Poland* 30, 254–258.

Korbekandi H et al. (2012). Optimization of biological synthesis of silver nanoparticles using *Lactobacillus casei* subsp. *casei*. *J. Chem. Technol. Biotechnol.* 87(7), 932–937.

Masurkar SA et al. (2011). Rapid biosynthesis of silver nanoparticles using *Cymbopogan citratus* (Lemongrass) and its antimicrobial activity. *Nano-Micro Lett.* 3, 189–194.

Medda S et al. (2015). Biosynthesis of silver nanoparticles from *Aloe vera* leaf extract and antifungal activity against *Rhizopus* sp. and *Aspergillus* sp. *Appl. Nanosci.* 5, 875–880.

Medina-Ramirez I et al. (2009). Green synthesis and characterization of polymer-stabilized silver nanoparticles. *Colloids Surf. B: Biointerfaces* 73(2), 185–191.

Mendoza-Reséndez R et al. (2013). Synthesis of metallic silver nanoparticles and silver organometallic nanodisks mediated by extracts of *Capsicum annuum* var. *aviculare* (piquin) fruits. *RSC Adv.* 3, 20765–20771.

Mittal AK et al. (2013). Synthesis of metallic nanoparticles using plant extracts. *Biotechnol. Adv.* 31, 346–356.

Mukherjee P et al. (2001). Fungus-mediated synthesis of silver nanoparticles and their immobilization in the mycelial matrix—A novel biological approach to nanoparticle synthesis. *Nano Lett.* 1, 515–519.

Nair B et al. (2002). Coalescence of nanoclusters and formation of submicron crystallites assisted by *Lactobacillus* strains. *Cryst. Growth Design* 2, 293–298.

Namratha N et al. (2013). Synthesis of silver nanoparticles using *Azadirachta indica* (Neem) extract and usage in water purification. *Asian J. Pharm. Tech.* 3, 170–174.

Njagi EC et al. (2011). Biosynthesis of iron and silver nanoparticles at room temperature using aqueous *Sorghum* bran extracts. *Langmuir* 27(1), 264–271.

Pantidos N et al. (2014). Biological synthesis of metallic nanoparticles by bacteria, fungi and plants. *Nanomed. Nanotechnol.* 5(5), 1–10.

Pollmann K et al. (2006). Metal binding by bacteria from uranium mining waste piles and its technological applications. *Biotechnol. Adv.* 24, 58–68.

Pugazhenthiran N et al. (2009). Microbial synthesis of silver nanoparticles by *Bacillus* sp. *J. Nanopart. Res.* 11, 1811–1815.

Raliya R et al. (2014). Biosynthesis and characterization of zinc, magnesium and titanium nanoparticles: An eco-friendly approach. *Int. Nano. Lett.* 4(93), 1–10.

Roy K et al. (2014). Green synthesis of silver nanoparticles using fruit extract of *Malus domestica* and study of its antimicrobial activity. *Dig. J. Nanomater. Biostruct.* 9(3), 1137–1147.

Roy K et al. (2015a). Single-step novel biosynthesis of silver nanoparticles using *Cucumis sativus* fruit extract and study of its photocatalytic and antibacterial activity. *Dig. J. Nanomater. Biostruct.* 10(1), 107–115.

Roy K et al. (2015b). Photocatalytic activity of biogenic silver nanoparticles synthesized using potato (*Solanum tuberosum*) infusion. *Spectrochim. Acta Part A* 146, 286–291.

Roy K et al. (2015c). *Apium graveolens* leaf extract mediated synthesis of silver nanoparticles and its activity on pathogenic fungi. *Dig. J. Nanomater. Biostruct.* 10, 393–400.

Roy K et al. (2015d). Plant-mediated synthesis of silver nanoparticles using parsley (*Petroselinum crispum*) leaf extract: Spectral analysis of the particles and antibacterial study. *Appl. Nanosci.* 5, 945–951.

Roy K et al. (2016a). Fast colourimetric detection of H_2O_2 by biogenic silver nanoparticles synthesised using *Benincasa hispida* fruit extract. *Nanotechnol. Rev.* 5(2), 251–258.

Roy K et al. (2016b). Rapid detection of hazardous H_2O_2 by biogenic copper nanoparticles synthesized using *Eichhornia crassipes* extract. *Third International Conference on Microelectronic Circuit and System*, Kolkata, West Bengal, India, pp. 203–207.

Saifuddin N et al. (2009). Rapid biosynthesis of silver nanoparticles using culture supernatant of bacteria with microwave irradiation. *E- J. Chem.* 6, 61–70.

Sanghi R et al. (2009). Biomimetic synthesis and characterisation of protein capped silver nanoparticles. *Bioresour. Technol.* 100, 501–504.

Santhoshkumar T et al. (2011). Synthesis of silver nanoparticles using *Nelumbo nucifera* leaf extract and its larvicidal activity against malaria and filariasis vectors. *Parasitol. Res.* 108, 693–702.

Schlüter M et al. (2014). Synthesis of novel palladium(0) nanocatalysts by microorganisms from heavy-metal-influenced high-alpine sites for dehalogenation of polychlorinated dioxins. *Chemosphere* 117C, 462–470.

Shahverdi AR et al. (2007). Rapid synthesis of silver nanoparticles using culture supernatants of *Enterobacteria*: A novel biological approach. *Process Biochem.* 42(5), 919–923.

Singaravelu G et al. (2007). A novel extracellular synthesis of monodisperse gold nanoparticles using marine alga, *Sargassum wightii* Greville. *Colloids Surf. B: Biointerfaces* 57, 97–101.

stanford.edu. (2015). What_are_nanomaterials. Retrieved from the web address: https://web.stanford.edu/dept/EHS/prod/researchlab/IH/nano/what_are_nanomaterials.html. Accessed May 23, 2016.

Sun S et al. (2002). Size-controlled synthesis of magnetite nanoparticles. *Am. Chem. Soc. J.* 124, 8204–8205.

Vahabi K et al. (2011). Biosynthesis of silver nanoparticles by fungus *Trichoderma reesei* (A route for large-scale production of AgNanoparticles). *Insciences J.* 1(1), 65–79.

Velmurugan P et al. (2014). Green synthesis of silver and gold nanoparticles using *Zingiber officinale* root extract and antibacterial activity of silver nanoparticles against food pathogens. *Bioprocess Biosyst. Eng.* 37, 1935–1943.

8

Environmental and Biological Applications of Nanoparticles

Kaushik Roy and Chandan Kumar Ghosh

CONTENTS

8.1 Introduction

Polluting elements or compounds that have a negative impact on environment and human health may be termed pollutants or contaminants. Environmental remediation is the process of removing these materials from different environmental media like water, air, or soil. There are several man-made pollutants present in the environment that resist degradation through natural procedures and disrupt hormonal and other physiological systems in living creatures. Removal of these toxic materials with existing detection and treatment methods is generally found to be time-consuming and expensive. In the last few years, researchers around the world have been focusing on the use of nanomaterials for environmental monitoring and remediation. Recently, some reports have instilled hope as nanomaterials were found to clean up toxic elements and harmful organisms from soil, air, and water bodies. The use of nanomaterials for the remediation of environmental hazards may be termed "nanoremediation."

The field of nanotechnology revolutionized almost all scientific and engineering fields, and environmental remediation is no exception. Nanoparticles (<100 nm) with a high surface-to-volume ratio possess a large reactive surface area compared to bulk materials and have a significant role to play in environmental remediation. One of the most promising and well-developed environmental applications of nanotechnology has been in water treatment and purification where different nanoparticles and structures are used to purify natural and waste water through different mechanisms including degradation of toxic dyes, detection and adsorption of heavy metal ions, removing pathogenic microorganisms, and chemically converting toxic materials into less toxic ones (Prabhakar et al., 2013).

Research in nanotechnology has grown by leaps and bounds and has now entered into the field of biotechnology and medical science. Indeed, the widespread use of metal nanoparticles, from household paints to artificial prosthetic devices, has promised to affect our daily lives quite significantly in the foreseeable future (Holtz et al., 2012). Many high-quality research articles have been published on various methods of synthesis and applications of silver nanoparticles, both in basic science and in more clinically oriented domains. Here, we will provide readers with further and up-to-date understanding of the current and future biotechnology applications of noble metal nanoparticles, with a special focus on silver nanoparticles.

In terms of biological applications, silver and its compounds are known to have efficient antibacterial properties against aerobic and anaerobic bacterial species. Recently, silver nanoparticles have been extensively used to make antimicrobial coatings on textiles, furniture, and wound-dressing materials and have been incorporated in food preservation systems, films, washing machines, air conditioners, water purifiers, and biomedical instruments as well. The continuous progress of nanotechnology has enabled us to utilize these diverse properties of nanoparticles, especially of noble metals. The new horizons of nano research are revolutionary and the available research reports are likely to expose only a small fraction of the tremendous benefits and wide area of application. In this chapter, a few of the environmental and biological applications of nanoparticles will be briefly discussed.

8.2 Photocatalytic Degradation of Dyes

Several studies show that the nanostructures of noble metals such as silver and copper are easy and effective options for photo-induced degradation of organic dyes under ultraviolet or solar irradiation. The dyes, used

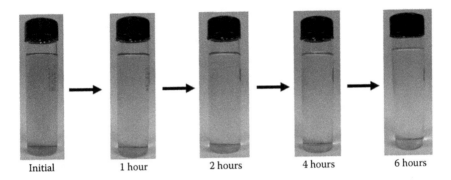

FIGURE 8.1
Change of dye color with exposure time. (From Roy, K. et al., *Appl. Nanosci.*, 5(8), 953, 2015d.)

extensively in the cosmetics, paper, and textile industries, are mostly organic compounds with an azo bond (R–N=N–R') that flow down to canals and rivers through industry outlets and pollute the water considerably. Using the enhanced photocatalytic property of noble metal nanoparticles, these dyes could be degraded before exposure to the environment. Roy et al. reported efficient photocatalytic degradation of organic dyes like methylene blue in the presence of bioengineered silver nanoparticles under solar irradiation (Roy et al., 2015d). When dye solution with colloidal particles was exposed to sunlight with constant stirring, visually, fading of dye coloration was noted after a few hours of exposure, indicating photocatalytic degradation of methylene blue (shown in Figure 8.1).

In another report, the same group of researchers used potato (*Solanum tuberosum*) infusion-mediated silver nanoparticles for photocatalytic degradation of highly toxic methyl orange dye under sunlight (Roy et al., 2015c). The UV-Vis spectra of the dye solution recorded at regular intervals experimentally verify that the corresponding absorbance peak of the dye decreases gradually with time of exposure and eventually reaches a minimum value after the full period of observation (shown in Figure 8.2). In this study, over 75% dye degradation is achieved in the presence of nano-silver after 8 hours of solar exposure. As per the reports, the degradation mechanism is initiated when photons of sunlight striking surface electrons become excited. The oxygen molecules, dissolved in the water, become excited after capturing these electrons and superoxide radicals (O_2^-) are produced. Subsequently, the capture of electrons from the particle surface leaves negatively charged silver nanoparticles that attract anionic dye molecules and finally dye molecules are adsorbed on the surface. Finally, (O_2^-) oxidizes the azo bond of the dye molecules adsorbed on the surface of nanoparticles and produces less harmful by-products like NO_3^-, NH_4^+, etc.

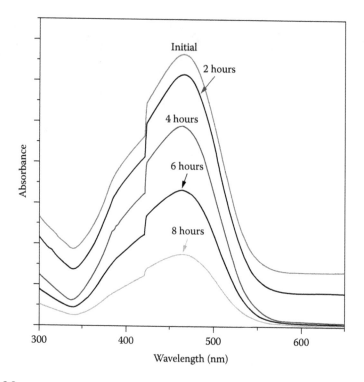

FIGURE 8.2
Gradual decrease of absorbance peak indicating dye degradation.

The chemical reactions of dye degradation generally follow first-order kinetics where reaction rate (R) is given by

$$R = -\frac{d[A]}{dt} = k[A]^1 = k[A]$$

where
  A is the concentration of dye that decreases with exposure time t
  k is reaction rate coefficient (time^{-1})

Simplifying both sides, the following equation may be obtained:

$$\ln\left(\frac{A_0}{A_t}\right) = kt$$

where
  A_0 is the initial concentration
  A_t is the final concentration of dye after t hours of exposure

In addition, the degradation of organic dyes specifically follows pseudo-first-order kinetics where the concentration of photocatalysts (i.e., nanoparticles) remains almost constant throughout the process but the dye concentration decreases gradually. In other cases, where the rate of degradation is proportional to the square of the dye concentration, the reaction would be considered second order and expressed as

$$R = -\frac{d[A]}{dt} = k[A]^2$$

Nanoparticles of metal oxides like titanium dioxide (TiO_2), zinc oxide (ZnO), and tungsten oxide (WO_3) are some other materials to be utilized in water purification technologies due to their prominent abilities to remove chemical and biological pollutants from water (Baruah et al., 2012). A phtotocatalyst is expected to absorb light in visible as well as the near UV part of electromagnetic spectra. Recently, photocatalytic procedures involving semiconducting oxides have drawn attention because of their low cost, eco-friendliness, and sustainability. Titanium dioxide is a leading option for waste water management and other environmental remediations in this case (Chong et al., 2010).

In a few reports, it has been noted that photocatalysts like nanosilver or TiO_2 nanoparticles have been modified to absorb a higher number of photons for overall enhancement of photocatalysis (Theron et al., 2008; Di Paola et al., 2012). These experimental modifications involve doping or attaching with other metals or elements like carbon or nitrogen. At present, the only drawback of this process that hinders its commercial scale-up is the low yield of postrecovery after treatment of sample waters.

Pant et al. prepared silver-impregnated TiO_2/nylon-6 nanocomposite mats that showed the desired features of a filter medium with excellent photocatalytic durability after repeated use (Pant et al., 2011). Nanoparticles of silver were embedded in nanofibers of TiO_2/nylon-6 composite through photo-induced reduction of silver salt under ultraviolet radiation. The photocatalytic property of this composite mat loaded with nanoparticles was investigated on organic methylene blue dye. The results indicate higher efficiency of silver-loaded nanocomposite mats than mats without nanosilver. This material not only performs as an eco-friendly filter media but also permits the nanoparticles to be recycled.

The previous reports on the photocatalytic activity of nanomaterials promoted their applications for water treatment options, including removal of the finest contaminants from water (<300 nm) and introduced pollutant-specific "smart materials" or "reactive surface coatings" that can destroy, transform, or immobilize toxic compounds. Heterogeneous systems using semiconductors like ZnO and TiO_2 are yet to be exposed for efficient waste water treatment through the procedure already discussed. Functionalized photocatalytic membranes incorporating nanoparticles have recently been considered over colloidal suspended particles because of the ease of their removal from real water samples.

8.3 Nanoparticles as Mercury Detector

Mercury is treated as one of the most harmful water pollutants and has received the attention of research communities around the globe. Nanoparticles have the ability to sense trace levels of hazardous materials present in water or soil, resulting in efficient environmental remediation. The equipment used outside the lab for environmental monitoring is complicated and less sensitive to sense trace elements like mercury or lead. Low cost and fast and effective detection systems for sensing pollutants in natural water, waste water, and soil would enhance the efficiency of sensing and clean-up technologies.

One possible method is chemical separation of analytes from real samples and their further concentration to smaller volumes, resulting in easier detection and monitoring of heavy metals like mercury, cadmium, lead, etc. In solid-phase microextraction, solid sorbents in small quantities are directly used to absorb or adsorb the harmful trace elements from air or water. Nanoparticles with a large surface area and high reactivity can be efficient sorbents for microextraction, specifically as self-assembled layers based on mesoporous supports. Mesoporous structure of silica was prepared using a surfactant template through the sol-gel process for developing a self-assembled monolayer with a rigid pore structure and large surface area for interaction. This structure may be effective for sorbing target elements including mercury, chromium, arsenic, etc., and radioactive nuclides like uranium, caesium, and actinides (Addleman et al., 2005).

Suspension of green synthesized silver nanoparticles was used by Roy et al. for selective colorimetric detection of Hg^{2+} in a recent study (Roy et al., 2015b). The dark brown suspension of nanosilver was exposed to various heavy metal ions and only mercury ions were able to decolorize the solution as shown in Figure 8.3. The reason for this activity could be

FIGURE 8.3
Selective colorimetric detection of mercury.

explained in terms of the higher reduction potential of mercury compared to silver. Mercury oxidizes silver during interaction and forms precipitation of silver salt (AgCl), resulting in the decolorization of the dark brown suspension. This visual detection could be verified through UV-Vis spectrometry, and the minimum detectable concentration of Hg^{2+} in this process is around 10 µM.

An important study was carried out by Ojea-Jiménez et al. where colloidal gold nanoparticles were employed as single entities for screening of mercury from multicomponent aqueous solutions with a very low concentration of pollutants (Ojea-Jiménez et al., 2012). Under experimental conditions, sodium citrate was identified as the reducing agent and gold nanoparticles as the catalyst in the reduction of mercury that was trapped in the presence of other cations such as copper and iron. Mercury uptake resulted in amalgam formation, which caused noticeable morphological transformations. The hydrophobicity of obtained amalgam and its consequent expulsion from medium made its recovery easier.

Mercaptosuccinic acid–capped silver nanoparticles (Ag@MSA) were utilized for removing mercury from water samples in a study conducted by Sumesh et al. (2011). In this study, the researchers used silver nanoparticles with a core diameter between 9 ± 2 and 20 ± 5 nm, which were capped by mercaptosuccinic acid (MSA) and supported on activated alumina for removing heavy mercury cations present in waste water. Standard analytical methods like XRD, TEM, and UV-Vis spectroscopy were used to realize the interaction of nanosilver with mercury ions. A desired removal ability of 0.8 g of mercury per gram of Ag@MSA was achieved for the optimum case where contaminant was in the ratio of 1:6 with Ag@MSA. The simplicity of the synthesis procedure, good loading ability on a suitable substrate, and affordable cost could make this process useful for field applications.

Self-propelled autonomous micromachines that take chemical energy from the environment and convert it to kinetic energy are at the forefront of nanotechnology. Great expectations have been placed on nanotechnology for innovative solutions to address challenging environmental problems such as water pollution via mercury contamination, a major issue threatening human health and biodiversity on earth.

8.4 Nanoparticles as Hydrogen Peroxide Sensor

Hydrogen peroxide (H_2O_2), being the simplest peroxide, is famous for its powerful oxidizing ability and is commonly used as a disinfectant, bleaching agent, and for rocket propellants. This is a well-known reactive oxygen species whose production or presence is significant in biological and chemical systems. Hence, development of a sensitive electrochemical H_2O_2 detector has long drawn the attention of researchers. Detection of H_2O_2 is another important factor as its presence causes harmful impacts on human health and other living systems when exposed in even very low concentrations. Biogenic silver nanoparticles were used for sensing hydrogen peroxide in liquid medium by Roy et al. (2016c). The dark brown suspension of silver nanoparticles turned colorless when H_2O_2 was added to it. Generally, the presence of H_2O_2 is detected by UV-Vis spectroscopy. The least detectable concentration was found to be 10^{-3} M in this case. The authors claimed that the reason for this visual change is due to the powerful oxidizing ability of H_2O_2 that oxidizes silver to silver oxide, forming precipitation and finally decolorizing the reacting solution. Another research report by the same group claimed to detect H_2O_2 using suspension copper nanoparticles following a similar procedure of visual detection as shown in Figure 8.4 (Roy et al., 2016a).

Manno et al. studied the synthesis and characterization of starch-stabilized silver nanostructures for sensor applications where silver nanostructures were successfully synthesized through a simple and "green" route using starch as a capping agent (Manno et al., 2008). High-resolution electron microscopy, x-ray diffraction, and UV-Vis absorption suggest the formation of silver nanocrystals with a size below 10 nm, and a self-assembled formation of ribbon-like structures has also been observed. The silver nanostructures are electrodeposited onto suitable substrates with gold inter-digital electrodes realizing amperometric sensors that showed a high sensitivity to hydrogen peroxide.

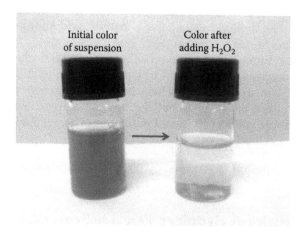

FIGURE 8.4

Visual detection of hydrogen peroxide by suspension of nanocopper. (From Roy, K. et al., Rapid detection of hazardous H_2O_2 by biogenic copper nanoparticles synthesized using *Eichhornia crassipes* extract, *Third International Conference on Microelectronic Circuits and Systems*, Kolkata, West Bengal, India, 2016a, pp. 203–207.)

Miao et al. developed a H_2O_2 sensor based on nanotubes of Ag/SnO_2 composites (Miao et al., 2013). The results of voltammetry and amperometry showed that the Ag/SnO_2 composite-based sensor exhibits a rapid amperometric response to H_2O_2 with a wide range and small detection limit. Hence, it can be used for detecting H_2O_2 in environmental and clinical analytes. Recently, Sophia and team fabricated a sensitive hydrogen peroxide detector based on polyvinylpyrrolidone-stabilized gold nanoparticles (Sophia et al., 2015). The easy and enzyme-free approach of this process may lead to the development of an effective and commercial sensor for hydrogen peroxide.

8.5 Antibacterial Activity of Nanoparticles

The antimicrobial effects of silver ions and their salts are well known historically but the effects of silver nanostructures on microorganisms and their detailed mechanisms had not been revealed clearly until Kim et al. showed the antimicrobial activity of Ag nanoparticles exposed to yeast, *Escherichia coli*, and *Staphylococcus aureus* (Kim et al., 2007). For the antibacterial assay, muller hinton agar plates were used and silver nanoparticles of different concentrations were supplemented in culture media. The results showed significant inhibition of yeast and *E. coli*, but the effect on *S. aureus* was mild. Silver at nanoscale was found to possess less toxicity but better antibacterial efficacy compared to its bulk form. This report strongly suggests that the free radical formation from the metallic surface is primarily responsible for the impressive antibacterial property of nanosilver. Therefore, silver nanoparticles may be used as bacterial growth inhibitors against many bacterial species and can be incorporated in medical equipment and bacteria control systems.

The increased emergence of drug-resistant microorganisms has presented a major challenge to research communities around the world for developing effective drugs. Hence, silver nanoparticles are an easy choice to be utilized to fight and prevent bacterial infections. In a well-documented study, starch-stabilized silver nanoparticles were exposed to a number of human pathogens by Mohanty et al. (2012). The result of this study showed that nanoparticles are more responsive toward gram-positive and gram-negative bacterial species than acid-fast organisms. In addition, these particles were found to interrupt the formation of biofilm and exhibited better bactericidal properties than cationic antimicrobial peptide (LL-37) as well. The large amount of data obtained from this study forecast the future use of silver nanoparticles as a novel and effective template for developing antibacterial agents.

The mechanism for the antibacterial efficacy of silver nanoparticles toward various pathogenic bacteria was initially described by Sondi et al. (2004). They proposed that the reason for the superior antibacterial activity of nanosilver is the release of metallic ions from nanoparticles when the particles come close to the bacterial cells. The cell wall bears small negative charges, thus attracting the nanoparticles to attach on its surface and form "pits" on the wall. In this way, the composition and permeability of the cell wall degrades and cellular fluid begins to leak out. As a consequence, bacterial cells fail to survive for too long and eventually succumb to death. This theory was later supported by another investigative report that showed detailed activity of silver nanoparticles against *E. coli* and *Vibrio cholerae* responsible for gastro-intestinal infections in the human body (Le et al., 2012). The various stages of the bactericidal activity of silver nanoparticles toward *V. cholerae* cells are demonstrated by biological TEM images (shown in Figure 8.5).

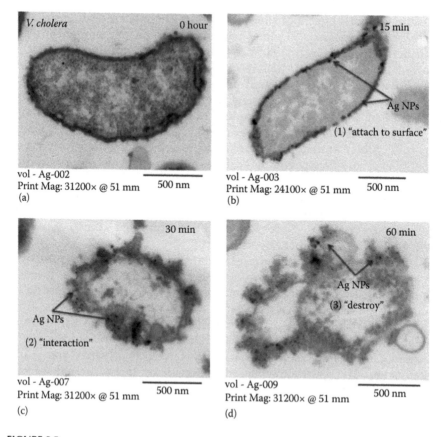

FIGURE 8.5

(a–d) Different stages of interaction of silver nanoparticles with *V. cholerae*. (From Le, A. et al., *Adv. Nat. Sci. Nanosci. Nanotechnol.*, 3(045007), 1, 2012.)

Another study reported the production of silver nanoparticles using gallic acid and demonstrated their efficacy in preventing the growth of clinically important drug-resistant bacteria (Martínez-Gutierrez et al., 2012). The antibacterial impact of silver nanoparticles was investigated on a panel of bacterial strains collected from biomedical devices frequently used in hospital ICUs. The nanoparticles with a diameter of around 24 nm were found to suppress the growth of all tested clinically relevant bacteria, though the efficacy was different toward different species.

Roy et al. reported different activity of plant extract–mediated silver nanoparticles against gram-positive and gram-negative strains (Roy et al., 2015e) as shown in Table 8.1. They considered that the reason for this difference in action may be due to the difference in composition of the cell walls of these two types of bacteria. The peptidoglycan layer in the cell wall of gram-positive bacteria is much thicker than that of gram-negative strains. Therefore, the metallic penetration is easier in the case of gram-negative bacteria and better activity can be observed toward gram-negative bacterial species.

Roy et al. reported in a very recent study that biosynthesized copper nanoparticles also possess appreciable antibacterial efficacy (Roy et al., 2016b). Interestingly, these nanoparticles show different activity against gram-positive and gram-negative bacteria, unlike silver and gold. During the antibacterial assay, the wells are created on solidified agar disks seeded with a specific type of bacteria and different doses of Cu nanosuspension are applied into different wells. The result is shown in Figure 8.6 where the observed inhibition zone (after 24 hours of incubation) is larger for gram-positive *S. aureus* than gram-negative *E. coli*. This may be due to the presence of amine and carboxyl group compounds in the gram-positive bacterial cell wall that undergoes copper-catalyzed amination and decomposition when copper nanoparticles interact with the microorganisms, as proposed by Roy et al. (2016).

TABLE 8.1

Antibacterial Activity of Silver Nanoparticles toward Different Bacteria

Tested Bacteria	Applied Concentrations of Silver Nanoparticles (µg/mL)	Inhibition Zone Diameter (mm)
Staphylococcus aureus	0	0
(gram-positive)	25	6.15
	50	12.50
Klebsiella pneumoniae	0	0
(gram-negative)	25	6.55
	50	13.35
Escherichia coli	0	0
(gram-negative)	25	6.95
	50	14.15

Source: Roy, K. et al., *Appl. Nanosci.*, 5(8), 945, 2015.

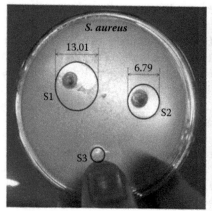
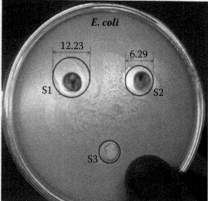

FIGURE 8.6

Inhibition zone on agar disk noted after 24 hours.

8.6 Antifungal Applications

Increased interest and recent explorations with noble metallic nanostructures such as gold and silver nanoparticles have uncovered novel facts about their antifungal abilities. Nanoparticles of noble metals are known for their fungicidal properties along with antibacterial activities, as stated in the previous section. The antifungal properties exhibited by metallic nanoparticles, mostly silver, have made them a favorable option for application in the field of management and treatment of fungal infections in plants as well as the human body. The following scientific reports indicate a few current and important findings regarding the antifungal properties of metallic nanostructures.

Narayanan prepared silver nanoparticles using amino acid and further tested the antifungal activity of nanosilver against *Candida* species (Narayanan et al., 2014). The spherical nanoparticles with a diameter around 8 nm showed positive activity against both *Candida dubliniensis* and *Candida albicans*. This report confirmed that nanosilver-based antifungal agents can be an efficient alternative to common drugs frequently used to resist infections associated with *Candida*.

Roy et al. showed the concentration-dependent activity of biogenic silver nanoparticles against two common fungi: *Aspergillus niger* and *Aspergillus wentii* (Roy et al., 2015a). The conventional disk diffusion method was followed to assess the fungicidal efficacy of the nanoparticles and the result is shown in Figure 8.7. These *Apium graveolens* leaf extract–mediated silver nanoparticles were found to have better activity toward *Aspergillus wentii*.

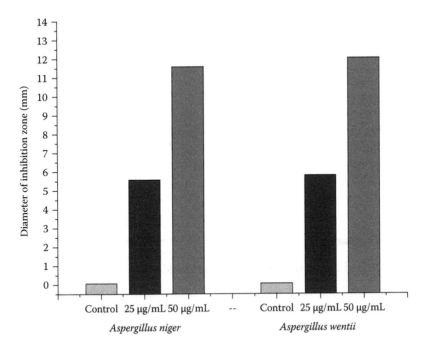

FIGURE 8.7
Antifungal activity of biogenic silver nanoparticles. (From Roy, K. et al., *Dig. J. Nanomater. Bios.*, 10, 393, 2015a.)

The interaction mechanism of silver nanoparticles with fungal cells is not fully realized, though the proposed theory is similar to that of the antibacterial mechanism.

A recent report by Swain at al. confirmed that Au nanoparticles are capable of rendering high antifungal efficacy and have great potential for antifungal therapy (Swain et al., 2016). The team exhibited the synthesis of gold nanoparticles using the root and leaf extracts from *Vetiveria zizanioides* and *Cannabis sativa*, respectively, and compared the two different chemical synthesis methods. The antifungal activity of gold nanoparticles thus observed has been tested for different fungal pathogens using the standard disk diffusion method. Au nanoparticles prepared using the green synthesis method were found to be a more effective antifungal agent and show a reduction in size as compared to other chemical preparation methods, suggesting that the synthesized gold nanoparticles can prove to be an effective antifungal agent. An antifungal study by Lateef et al. showed that *Cola nitida* extract–mediated gold-silver-alloy nanoparticles (Ag–AuNPs) 17–91 nm in size inhibited the growth of *Aspergillus flavus*, *A. fumigatus*, and *A. niger* by 69.51%–100% (Lateef et al., 2016b). The particles also display potent blood anticoagulant and thrombolytic activities, indicative of their potential in the management of blood coagulation disorders.

In another study, zinc oxide (ZnO) nanoparticles have been synthesized using *Lantana aculeata* Linn. Leaf extract and its antifungal activity against plant fungal pathogenesis has been explored (Narendhran et al., 2016). The antifungal activity of ZnO nanoparticles has been determined using the well diffusion method, and this study concluded that the maximum zone of inhibition is observed in *Aspergillus flavus* (21 ± 1.0 mm) and *Fusarium oxysporum* (19 ± 1.0 mm) at 100 μg/mL concentrations. These results clearly indicated the benefits of using ZnO nanoparticles synthesized using biological methods and shown to have antifungal activities. This result promotes the use of ZnO nanoparticles as antifungal agent in environmental aspects of agricultural development.

8.7 Antiviral Activity of Nanoparticles

Currently available data and reported interactions between noble metal nanoparticles (mainly silver) and viral biomolecules strongly indicate that the use of nanomaterials can contribute significantly to the improvement of treatment and prevention of viral infections. However, it is a big challenge for researchers to develop nanostructures or composites that can target specific pathogenic viruses and overcome the drawbacks of commonly used therapies, such as low efficiency and unwanted side effects.

Lara and team studied the antiviral effects of silver nanostructures for the first time and proposed that silver nanoparticles bind with the external membrane of a lipid-enveloped virus to prevent infection, though the interaction of Ag nanoparticles with viruses is a largely unexplored field to date (Lara et al., 2011). Silver nanoparticles were studied specifically in human immunodeficiency virus (HIV) for which the report proposed the mechanism of antiviral activity of the nanoparticles along with the transmission and inhibition of HIV-1 infection in human cervix organ culture.

Another study, by Carja et al., successfully demonstrated the antiviral effect of engineered plasmonic gold and layered double hydroxide self-assemblies (AuNPs/LDHs), taking Hepatitis B virus as a model virus and hepatoma-derived HepG2.2.215 cells for viral replication, assembly, and secretion of virions and subviral particles (Carja et al., 2015). The AuNPs/LDHs substantially reduced the number of viral and subviral particles from cells by almost 80% and showed excellent biocompatibility. The virucidal and cytocompatible properties of AuNPs/LDHs can make it a potential choice for novel and effective therapy for treating Hepatitis B in future.

Osminkina et al. reported that silicon nanoparticles with a size of 5–50 nm may act as effective scavengers for HIV and respiratory syncytial virus (RSV) (Osminkina et al., 2014). This study revealed powerful prevention of viral activity in the presence of silicon nanoparticles with concentrations of 0.01 mg/mL and 0.1 mg/mL for RSV and HIV, respectively. The reason for this effect was explained as the binding of virions with silicon nanoparticles, which is expected to be identical when exposed to other enveloped viruses. Due to a very low cytotoxic concentration of silicon nanoparticles, it may be incorporated in new applications of antiviral therapies.

8.8 Anticancer Activity of Nanoparticles

Cancer nanotechnology is a growing field of research that combines nanotechnology and medical science and finds wide application in molecular diagnosis, bioimaging and targeted chemotherapy, etc. Research on cancer nanotechnology involves nanostructures like nanocrystals and quantum dots that possess enhanced structural, magnetic, and optical properties unlike molecules or bulk materials. Recently, green synthesized silver nanoparticles were exposed to various cancer cell lines to assess their anticancer potentials by Abd-Elnaby et al. (2016). The cytotoxicity of silver nanoparticles was measured on eight types of human tumor cell lines following colorimetric MTT assay, that is, a standard technique for studying anticancer activity. The result showed a concentration-dependent decrease in the percentage of cell viability as the biogenic nanoparticles exhibited a significant amount of anticancer activity after 24 hours of incubation. A few cell lines like HEP-2, CACO, and HELA were most resistant to the cytotoxicity of silver nanoparticles whereas the other cell lines were very sensitive toward it as shown in Table 8.2.

A study by Priyadharshini et al. reported that zinc oxide nanoparticles possess better anticancer properties against prostate carcinoma (PC-3) cell lines than silver nanoparticles (Priyadharshini et al., 2014). Cell viability assays were performed to assess the cytotoxic effects of both silver and ZnO nanoparticles against PC-3 and normal African monkey kidney (VERO) cell line. The inhibitory concentration values were found to be 39.60 µg/mL, 28.55 µg/mL, and 53.99 µg/mL against PC3 cells for silver nanoparticles, ZnO nanoparticles, and aqueous *G. edulis* extract, respectively (after 48 hours of incubation). When verified by the acridine orange/ethidium bromide staining procedure, the percentages of apoptotic bodies were found to be 62% and 70% for silver and zinc oxide nanoparticles, respectively.

Bismuth nanoparticles were explored to centralize radiation used in cancer therapy and the results showed that the nanoparticles can enhance the dose of radiation by 90% to the targeted tumor (Alqathami, 2012). Dhar et al. used gold nanoparticles not only to deliver platinum to cancer tumors but also to minimize the ill effects of platinum therapy at the same time (Dhar et al., 2009). The toxicity of platinum mainly depends on the molecules it is attached to. Therefore, a platinum-containing molecule with low toxicity was chosen to be attached to the gold nanoparticles. When the molecules with

TABLE 8.2

Anticancer Activity of Biosynthesized Silver Nanoparticles against Different Cancer Cell Lines

Conc. of AgNPs (µg/well)	Cell Viability (%)							
	Hepatocellular Carcinoma Cells (HepG-2)	Breast Carcinoma Cells (MCF-7)	Colon Carcinoma Cells (HCT-116)	Prostate Carcinoma Cells (PC-3)	Lung Carcinoma Cells (A-549)	Intestinal Carcinoma Cells (CACO)	Larynx Carcinoma Cells (HEP-2)	Cervical Carcinoma Cells (HELA)
50.00	21.63	36.42	17.97	48.13	38.59	54.31	63.82	56.79
25.00	63.14	70.31	24.85	79.24	74.62	78.29	85.17	74.13
12.50	78.92	84.19	37.28	91.36	87.31	91.45	94.26	85.82
6.25	89.56	89.27	60.35	98.63	94.05	98.73	99.02	92.45
3.13	94.32	94.53	83.51	100.00	98.79	100.00	100.00	98.37
1.56	98.75	97.46	91.84	100.00	100.00	100.00	100.00	100.00
0.00	100.00	100.00	100.00	100.00	100.00	100.00	100.00	100.00

Source: Abd-Elnaby, H.M. et al., *Egyp. J. Aquat. Res.* 42(3), 301–312, 2016, doi: 10.1016/j.ejar.2016.05.004.

nanoparticles reach the sights of tumors, they first interact with the acidic solution, leading to the transformation to the most toxic state, and destroy the cancer cells.

Another important study explored the use of polymer nanoparticles for delivering JSI-124, a molecule commonly used to target cancer tumors (Molavi, 2010). JSI-124 efficiently decreases the ability of cancerous cells possibly by delaying the growth of tumors. This study concluded that the ability of that molecule along with polymer nanoparticles as carrier affects tumor cells significantly and may be utilized for further development of cancer immunotherapy.

Though researchers are trying to develop efficient cancer treatment and therapies, drug resistance is considered to be one of the major difficulties in cancer treatment. Drug-coated polymer nanoparticles and nanospheres were explored to assess their ability to inhibit drug resistance by Wang et al. (2007). In this study, the researchers adopted a novel strategy for inhibiting multi-drug resistance of tumor cells by adding tetraheptyl-ammonium-capped fer-roso ferric oxide (Fe_3O_4) nanoparticles with accumulation of daunorubicin, a well-known anticancer drug. The result indicates the amazing synergistic effect of magnetic Fe_3O_4 nanoparticles on the uptake of daunorubicin in leukemia cells. The interaction between bioactive molecules and magnetic Fe_3O_4 nanoparticles on the cell membrane of leukemia cells probably contributes to the improved effect of cellular uptake, leading to enhanced synergistic effect to inhibit the drug resistance of leukemia K562 cells.

8.9 Nanoparticles for Bioimaging

In the recent past, nanomaterials have become increasingly important in the development of new molecular probes for experimental and clinical in vivo imaging. Nanoparticulate imaging probes have included semiconductor quantum dots, magnetic and magneto-fluorescent nanoparticles, gold nanoparticles and nanoshells, among others. In the past more than 25 years, there have been few fundamental improvements in clinical x-ray contrast agents, since the chemical platform of triiodobenzene has not changed. There are serious limitations imposed by current agents on medical imaging, for example, short imaging times, the need for catheterization in many cases, occasional renal toxicity, and poor contrast in overweight patients. A report by Hainfeld and team first demonstrated that gold nanoparticles may be used to overcome these limitations (Hainfeld et al., 2006). Gold has significantly higher absorption ability than iodine, with less bone and tissue interference. Thus, this method achieves better contrast with a lower dose of X-ray. Nanoparticles clear the blood more slowly than iodine agents, and this permits longer imaging times. In this experiment, 1.9 nm diameter gold nanoparticles were injected intravenously into mice. The standard unit of mammography was used to obtain the images over time. The distribution of gold nanoparticles in biosystems was evaluated using atomic absorption spectroscopy. Retention in the spleen and liver was low with elimination by both kidneys. Tumors and some organs like kidneys were seen with sound clarity and impressive spatial resolution. Blood vessels that are less than 100 μm in diameter were observed to be delineated, so that in vivo vascular casting could be performed. The scientists could clearly identify regions of increased vascularization and angiogenesis. Routine mouse behavior was noted initially with a 10 mg/mL concentration of gold nanoparticles in the blood and no further toxicity was observed after 30 days of injection. Hence, gold nanoparticles can be employed as X-ray contrast agents to overcome the limitations of commonly used iodine-based materials.

In another study, a team of scientists presented another approach utilizing collagen-targeting gold nanoparticles that enable the imaging of myocardial scars via CT scans in a rodent model (Danila et al., 2013). The findings of this study would enable clinicians to judge the recovery potential of myocardium more accurately compared to the current CT scan–based approaches. In myocardial ischemia, the recovery of myocardial function by revascularization procedures depends on the myocardial scar burden and also the extent of coronary disease. Although CT imaging offers superior evaluation of coronary lesions at present, this method does not have the ability to measure the transmural extent of myocardial scars. This study investigates if collagen-targeting gold nanoparticles provide better contrast for CT imaging by effectively targeting myocardial scars. Gold nanoparticles were coated with a collagen-homing peptide, that is, collagen adhesion or CNA35. Myocardial scars were created in the bodies of mice by reperfusion or occlusion of the left anterior coronary artery. Nongated CT imaging was recorded after 30 days. Over 6 hours, CNA35-Au nanoparticles provided uniform and extended opacification of the vascular structures as shown in Figure 8.8a. Focal contrast enhancement was detected in the myocardium in mice with a larger scar burden. This is not apparent within the myocardium of control mice (see Figure 8.8b). Myocardial scar formation and accumulation of nanogold were confirmed by histological staining.

Another team of researchers also studied the use of nanomaterials for CT imaging (Rabin et al., 2006). It is known that the current contrast agents of CT imaging are mainly based on iodinated molecules. They can efficiently absorb X-rays but their rapid pharmacokinetics and non-specific distribution have limited their targeting and microvascular performance. This study suggested the use of polymer-coated bismuth sulfide nanoparticles as injectable agents for recording CT images. This material showed better stability and appreciable X-ray absorption ability along with long circulation times (>2 hours) and safety profile compared to conventionally used iodinated agents. This result proves the efficacy of these polymer-coated nanoparticles for improved in vivo imaging of lymph nodes, liver, and vascular system in mice. These particles and their conjugates are anticipated to be a significant adjunct to bioimaging of cellular targets and other physiological conditions in the near future.

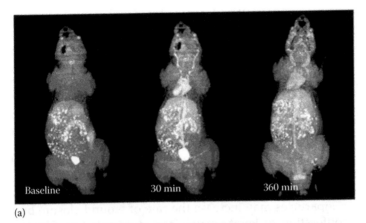

FIGURE 8.8
(a) Prolonged vascular enhancement by Au nanoparticles in vivo. (b) Enhancement of focal contrast of myocardial scar with CNA-35 gold nanoparticles. (From Danila, D. et al., *Nanomed. J.*, 9(7), 1067, 2013.)

8.10 Other Bioapplications of Nanoparticles

A few recent studies have promoted the use of silver nanoparticles in other clinical applications, including wound healing and dressing. A study by Kwan et al. revealed that silver nanoparticles have potential to promote wound healing by accelerated reepithelization and enhanced differentiation of fibroblasts (Kwan et al., 2011). However, the effect of Ag nanoparticles on the functionality of repaired skin is unknown. In this study, the team of scientists explored the tensile properties of healed skin after treatment with nanosilver. Immunohistochemical staining, quantitative assay, and scanning electron microscopy are used to detect and compare collagen deposition, and the morphology and distribution of collagen fibers. The results showed that silver nanoparticles have improved tensile properties and led to better fibril alignments in repaired skin, with a close resemblance to normal skin. Thus, it is suggested that Ag nanoparticles are predominantly responsible for regulating the deposition of collagen, and their use resulted in excellent alignment in the wound-healing process. However, the exact signaling pathway by which these nanoparticles affected collagen regeneration remains open for further investigation.

The effect of silver nanoparticles on different types of skin cells such as keratinocytes and fibroblasts during the process of wound healing was meticulously studied by Liu et al. (2010a). This study concluded that the nanoparticles may increase the rate of wound closure by promoting the proliferation of keratinocytes and differentiating fibroblasts into myofibroblasts.

Silver nanoparticles were found to have anticoagulant and thrombolytic properties too as tested by Lateef et al. recently (Lateef et al., 2016a). In their experiment, the anticoagulant property was evaluated by adding a 0.5 mL suspension of silver nanoparticles (concentration 150 µg/mL) to 5 mL of fresh human blood at room temperature. The thrombolytic property was investigated by spreading the blood clot on clean glass slides and treating it with 0.2 mL of the same nanosuspension. The samples were observed visually as well as through microscopes to capture images of various stages. The microscopic images captured the dispersed red blood cells that prevented blood coagulation after the incubation period. On the other hand, the preformed blood clot on the glass slides was dissolved immediately after addition of nanosuspension, indicating an efficient thrombolytic property of these colloidal metallic nanoparticles as supported by the microscopic views (shown in Figure 8.9). These results further extend the knowledge of the biological properties of silver nanostructures and open up endless possibilities in all scientific research domains.

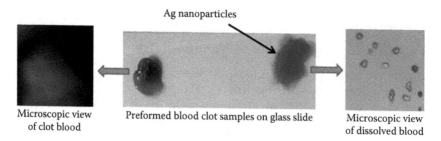

Ag nanoparticles

Microscopic view   Preformed blood clot samples on glass slide   Microscopic view
of clot blood   of dissolved blood

FIGURE 8.9
Thrombolytic activity of silver nanoparticles.

References

Abd-Elnaby HM et al. (2016). Antibacterial and anticancer activity of extracellular synthesized silver nanoparticles from marine *Streptomyces rochei* MHM13. *Egypt. J. Aquat. Res.* 42(3), 301–312. doi: 10.1016/j.ejar.2016.05.004.

Addleman RS et al. (2005). In B. Karn, T. Masciangioli, W. Zhang, V. Colvin, and P. Alivisatos (eds.), Nanostructured sorbents for solid phase microextraction and environmental assay. *Nanotechnology and the Environment: Applications and Implications.* Washington, DC: American Chemical Society, pp. 186–199.

Alqathami M. (2012). Metal nanoparticles may improve cancer treatment. Retrieved from http://www.understandingnano.com/bismuth-nanoparticles-radiation-therapy.html. Accessed November 30, 2012.

Baruah S et al. (2012). Nanostructured zinc oxide for water treatment. *Nanosci. Nanotechnol. Asia* 2(2), 90–102.

Carja G et al. (2015). Self-assemblies of plasmonic gold/layered double hydroxides with highly efficient antiviral effect against the hepatitis B virus. *Nano Res.* 8(11), 3512–3523.

Chong MN et al. (2010). Recent developments in photocatalytic water treatment technology: A review. *Water Res.* 44(10), 2997–3027.

Danila D et al. (2013). CT imaging of myocardial scars with collagen-targeting gold nanoparticles. *Nanomed. J.* 9(7), 1067–1076.

Dhar S et al. (2009). Gold nanoparticles delivery platinum warheads to tumors. Retrieved from http://www.understandingnano.com/cancer-therapy-platinum-gold-nanoparticles.html. Accessed October 30, 2009.

Di Paola A et al. (2012). A survey of photocatalytic materials for environmental remediation. *J. Hazard. Mater.* 211–212, 3–29.

Hainfeld JF et al. (2006). Gold nanoparticles: A new X-ray contrast agent. *Br. J. Radiol.* 79(939), 248–253.

Holtz RD et al. (2012). Nanostructured silver vanadate as a promising antibacterial additive to water-based paints. *Nanomedicine* 8(6), 935–940.

Kim JS et al. (2007). Antimicrobial effects of silver nanoparticles. *Nanomedicine* 3(1), 95–101.

Kwan KH et al. (2011). Modulation of collagen alignment by silver nanoparticles results in better mechanical properties in wound healing. *Nanomed. J.* 7(4), 497–504.

Lara HH et al. (2011). Silver nanoparticles are broad-spectrum bactericidal and virucidal compounds. *J. Nanobiotechnol.* 9(30), 1–8.

Lateef A et al. (2016a). Paper wasp nest-mediated biosynthesis of silver nanoparticles for antimicrobial, catalytic, anticoagulant, and thrombolytic applications. *3 Biotech.* 6(140), 1–10.

Lateef A et al. (2016b). Kolanut (*Cola nitida*) mediated synthesis of silver–gold alloy nanoparticles: Antifungal, catalytic, larvicidal and thrombolytic applications. *J. Cluster Sci.* 27(5), 1561–1577.

Le A et al. (2012). Powerful colloidal silver nanoparticles for the prevention of gastrointestinal bacterial infection. *Adv. Nat. Sci. Nanosci. Nanotechnol.* 3(045007), 1–10.

Liu X et al. (2010). Silver nanoparticles mediate differential responses in keratinocytes and fibroblasts during skin wound healing. *ChemMedChem* 5(3), 468–475.

Manno D et al. (2008). Synthesis and characterization of starch-stabilized Ag nanostructures for sensors applications. *J. Non-Crystal. Solids* 354(52–54), 5515–5520.

Martínez-Gutierrez F et al. (2012). Antibacterial activity, inflammatory response, coagulation and cytotoxicity effects of silver nanoparticles. *Nanomedicine* 8(3), 328–336.

Miao Y et al. (2013). A novel hydrogen peroxide sensor based on Ag/SnO$_2$ composite nanotubes by electrospinning. *Electrochim. Acta* 99, 117–123.

Mohanty S et al. (2012). An investigation on the antibacterial, cytotoxic and antibiofilm efficacy of starch-stabilized silver nanoparticles. *Nanomedicine* 8(6), 916–924.

Molavi O. (2010). Development of a poly(D,L-lactic-*co*-glycolic acid) nanoparticle formulation of STAT3 inhibitor JSI-124: Implication for cancer immunotherapy. *Mol. Pharm.* 7(2), 364–374.

Narayanan KB et al. (2014). Unnatural amino acid-mediated synthesis of silver nanoparticles and their antifungal activity against *Candida* species. *J. Nanopart. Res.* 16(8), 1–11.

Narendhran S et al. (2016). Biogenic ZnO nanoparticles synthesized using *L. aculeata* leaf extract and their antifungal activity against plant fungal pathogens. *Bull. Mater. Sci.* 39(1), 1–5.

Ojea-Jiménez I et al. (2012). Citrate-coated gold nanoparticles as smart scavengers for mercury(II) removal from polluted waters. *ACS Nano* 6(3), 2253–2260.

Osminkina LA et al. (2014). Porous silicon nanoparticles as scavengers of hazardous viruses. *J. Nanopart. Res.* 16(6), 1–10.

Pant HR et al. (2011). Photocatalytic and antibacterial properties of a TiO$_2$/nylon-6 electrospun nanocomposite mat containing silver nanoparticles. *J. Hazard. Mater.* 189(1–2), 465–471.

Prabhakar V et al. (2013). Nanotechnology, future tools for water remediation. *Int. J. Emerg. Technol. Adv. Eng.* 3(7), 54–59.

Priyadharshini RI et al. (2014). Microwave-mediated extracellular synthesis of metallic silver and zinc oxide nanoparticles using macro-algae (*Gracilaria edulis*) extracts and its anticancer activity against human PC3 cell lines. *Appl. Biochem. Biotechnol.* 174(8), 2777–2790.

Rabin O et al. (2006). An X-ray computed tomography imaging agent based on long-circulating bismuth sulphide nanoparticles. *Nat. Mater.* 5(2), 118–122.

Roy K et al. (2015a). *Apium graveolens* leaf extract mediated synthesis of silver nanoparticles and its activity on pathogenic fungi. *Dig. J. Nanomater. Biostruct.* 10, 393–400.

Roy K et al. (2015b). Rapid colorimetric detection of Hg^{2+} ion by green silver nanoparticles synthesized using *Dahlia pinnata* leaf extract. *Green Process Synth.* 4(6), 455–461.

Roy K et al. (2015c). Photocatalytic activity of biogenic silver nanoparticles synthesized using potato (*Solanum tuberosum*) infusion. *Spectrochim. Acta Part A* 146, 286–291.

Roy K et al. (2015d). Photocatalytic activity of biogenic silver nanoparticles synthesized using yeast (*Saccharomyces cerevisiae*) extract. *Appl. Nanosci.* 5(8), 953–959.

Roy K et al. (2015e). Plant-mediated synthesis of silver nanoparticles using parsley (*Petroselinum crispum*) leaf extract: Spectral analysis of the particles and antibacterial study. *Appl. Nanosci.* 5(8), 945–951.

Roy K et al. (2016a). Rapid detection of hazardous H$_2$O$_2$ by biogenic copper nanoparticles synthesized using *Eichhornia crassipes* extract. *Third International Conference on Microelectronic Circuits and Systems*, Kolkata, West Bengal, India, pp. 203–207, ISBN: 978-93-80813-45-5.

Roy K et al. (2016b). Antibacterial mechanism of biogenic copper nanoparticles synthesized using *Heliconia psittacorum* leaf extract. *Nanotechnol. Rev.* 5(6), 529–536. doi: 10.1515/ntrev-2016-0040.

Roy K et al. (2016c). Fast colourimetric detection of H_2O_2 by biogenic silver nanoparticles synthesised using *Benincasa hispida* fruit extract. *Nanotechnol. Rev.* 5(2), 251–258.

Sondi I et al. (2004). Silver nanoparticles as antimicrobial agents: A case study on *E. coli* as a model for gram-negative bacteria. *J. Colloid Interface Sci.* 275(1), 177–182.

Sophia J et al. (2015). Gold nanoparticles for sensitive detection of hydrogen peroxide: A simple non-enzymatic approach. *J. Appl. Electrochem.* 45(9), 963–971.

Sumesh E et al. (2011). A practical silver nanoparticle-based adsorbent for the removal of Hg^{2+} from water. *J. Hazard. Mater.* 189(1–2), 450–457.

Swain S et al. (2016). Green synthesis of gold nanoparticles using root and leaf extracts of *Vetiveria zizanioides* and *Cannabis sativa* and its antifungal activities. *BioNanoScience* 6(3), 205–213.

Theron J et al. (2008). Nanotechnology and water treatment: Applications and emerging opportunities. *Crit. Rev. Microbiol.* 34(1), 43–69.

Wang X et al. (2007). The application of Fe_3O_4 nanoparticles in cancer research: A new strategy to inhibit drug resistance. *J. Biomed. Mater. Res. A* 80(4), 852–860.

9

Nanogenerator: A Self-Powered Nanodevice

Sunipa Roy and Amrita Banerjee

CONTENTS

9.1 Introduction

Living in the present world, we are entering into a period of increasing environmental pollution and of crisis related to the limited fossil fuel and other nonrenewable energy sources. That is why technology is now required to harvest energy from ambient environment. Renewable energy is harvested and converted into useful electrical energy by many devices. Solar, thermal, nuclear, wind, and hydrolytic energy scavenging has paved the path for a new source, namely mechanical energy. Such energy harvesting can be done by a device that has dimensions at the nanoscale and hence is called a nanogenerator (NG).

Mechanical energy has become very useful and highly desired as it is not limited by location, season, weather, or diurnal changes. NGs aim to extract and conserve energy from the environment by making use of various semiconducting and piezoelectric properties of nanodimensional structures. We shall see in later in this chapter that nanotubes and wires of different materials can be used to make such generators. The developmental progress of the NGs is closely related to the various techniques of material growth. Hydrothermal process, physical vapor deposition, thermal evaporation, vapor–liquid–solid, vapor–solid–solid, and chemical approaches are generally followed for the fabrication of the basic core nanostructures. Moreover, each type of material has its own properties and thus holds certain advantages and disadvantages to the nanoscale power devices.

An array of nanowires (NW) or nanotubes are arranged in such a way that the complete structure will function on the principle of energy conversion wherein mechanical energy gets converted to electrical energy. The operating principle of the piezoelectric NG depends on the generation of piezoelectric potential in the NWs when they are dynamically strained under the effect of some external force and transient current flows through the external circuit from the higher- to the lower-potential side of the NWs. The output current of the NG is the sum of the individual currents contributed by all the NWs, while the output voltage is determined by the individual NW.

The ambient environment can provide various types of mechanical vibration in different media and of different frequencies to serve as excitation to the nanoscale power generators. The electrical output obtained from the NGs can be either direct or alternating current of the order of few nanoamperes and the voltage developed is of the order of 10–100 mV. Such power is sufficient to drive other nanodevices such as sensors. Use of nanoscale devices and their miniaturization process was restricted because they had to be powered by traditional batteries, but the use of NGs will now make nanodevices self-powered and also much smaller in size. This will enable the development of more and more battery-less smart systems for future applications. Thus, research continues for the development of cheaper, more portable, environment friendly, and green energy NGs. Figure 9.1 shows the basic piezoelectric nanogenerator circuit.

In certain types of structures, the NGs can also be attached to a very thin metallic cantilever. The cantilever is sensitive to external vibrations so that the NG is fixed on the cantilever where it produces the maximum strain. The use of flexible NGs gives added advantage of harvesting energy from very weak mechanical sources, such as respiratory wind or pulse pressure. Moreover, flexible NGs can be used in various applications without the constraint of any specific shape or size.

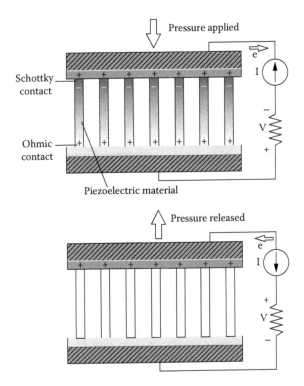

FIGURE 9.1
Schematic of a nanogenerator under strained condition.

9.2 Piezoelectric Property

We have seen in the previous section that electromagnetic energy, thermal energy, mechanical energy, and many other forms of energy found in nature can be harvested and used for powering various devices. Different energy conversion principles are utilized to yield electrical energy from various external excitations. One such conversion of energy wherein mechanical energy gets converted to electrical energy is possible due to a naturally existing property of certain materials called piezoelectric property.

The piezoelectric property is the ability of certain materials to produce electrical output under the influence of applied mechanical force. Piezo is a Greek word that means "pressure." The reverse of this effect also holds equally good. The electrical force given to a piezoelectric material may result in a temporary change in the dimension of the material in response to the signal—that is, the material may get stressed or strained.

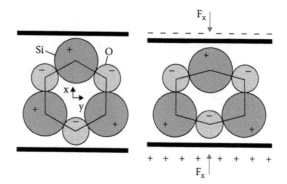

FIGURE 9.2
An unstrained quartz crystal and a strained quartz crystal.

If we take a deeper insight into the structure of the material, the unit cells have a particular arrangement of the anions and cations. When a mechanical stress is applied, the centers of mass of the anions and cations get shifted. This means that the dipoles within the bulk of the material get aligned in a particular direction. The stressed material becomes polarized, and there is resultant appearance of positive charges on one face and negative charges on the opposite face of the material. Inside the material bulk, the opposite charges constituted by the head of one dipole and the tail of the next neighboring dipole get cancelled out. The charge separation produces an electric field and hence a potential gets generated, which is called a piezoelectric potential. It can be seen that the material whose lattice lacks a center of symmetry exhibits piezoelectric property. Examples of piezoelectric materials are quartz, zinc oxide, gallium nitride, and many more.

In Figure 9.2, we can see a unit cell of a quartz crystal under normal and strained conditions. The unit cell is hexagonal. In the first image, the center of mass of the anions (oxygen ions) and that of the cations (silicon) can be seen to coincide with one another. Hence, there is no charge separation and consequently no generation of potential. In the second image, the unit cell is subjected to a force on two opposite surfaces. As a result, there is a slight deformation in the shape of the unit cell, which pushes the center of mass of the anions upward and that of cations downward. A resultant charge separation takes place, exhibiting a potential. If we consider a bulk material sample, then this potential projects itself as the piezoelectric potential.

9.3 Materials

Various types of materials can be used for the production of the NG. Some of them are listed in Table 9.1.

TABLE 9.1

Different Materials Used for NG Device

Material	Type	Structural Details	Electrical Output	Method	Reference
ZnO	Wurtzite	l = 400 nm	−9 mV	Thermal evaporation deposition	[1]
CdS	Wurtzite	d = 100 nm l = 1 μm	−3 mV	Physical vapor deposition	[2]
BaTiO$_3$	Pervoskite	1 cm^2 area, 550 nm	0.4 V, 12 nA	Chemical vapor deposition, wet etching	[3]
In$_2$O$_3$/ZnO core/shell NW	Wurtzite	d = 200 nm l = 2.2 μm	0.902 V	Hydrothermal	[4]
ZnSnO$_3$ + PDMS	Pervoskite + polymer composite	1 cm × 1 cm composite, 500 nm nanocube	20 V, 1 μA cm^{-1}	Not stated	[5]
ZnO/ZnS core/shell NP	Wurtzite Heterostructure	d = 6 nm	4.5 nA	Chemical precipitation	[6]
PVDF + ZnO	Polymer+ Wurtzite composite	d = 80 nm l = 8 μm of NW 30 μm composite thickness	1.3 V, 0.17 μA		[7]
GaN	Wurtzite	d = 500 nm l = 20 μm	1.2 V, 40 nA	Chemical vapor deposition	[8]

9.4 Types of Nanogenerators

There are three categories of NGs, based on their mechanism of conversion of mechanical energy to electrical energy:

1. Piezoelectric
2. Triboelectric
3. Pyroelectric

This section provides a detailed study on each of them.

9.4.1 Working Principle of Piezoelectric Nanogenerator

An NG converts random mechanical energy to electrical energy by using arrays of NWs of piezoelectric materials. When an external strain is applied to the NG, there is a transient flow of electrons to the external load, which is driven by the piezoelectric potential generated between the two contact surfaces on two ends of the vertically grown NWs.

If the conductive tip of an atomic force microscope (AFM) is brought in contact with the NWs and subsequently scanned over all the NWs while maintaining the contact, then many sharp output voltage peaks will be observed across the external load resistance, which must be larger than the internal resistance of the NWs. The output voltage should be continuously measured, and a constant force should be maintained between the AFM tip and the NW surface as the AFM tip is moved over the NWs.

The Schottky contact made between the metallic tip of the AFM and the semiconductor NW results in the flow of the electrons in the external circuit when the NWs are bent under the effect of the constant axial force applied on the NWs. The force is applied along the c-axis. The output voltage peaks may range up to 10 mV. Approximately 40% of the contact made between AFM and NWs results in the production of the output voltage. In any growth process of the NWs, the density of the NWs that are in contact with the AFM tip is $20/\mu m^2$ and the average density of NWs that are capable of generating output voltage when the metallic tip touches them is $8/\mu m^2$. The bottom of the NWs has an electrical contact that must be grounded, and the bottom of the NWs must be fixed.

It has been observed that when the AFM tip touches the stretched end of a particular NW and compresses it, no corresponding output voltage is generated, but an output voltage peak is observed at the time when the AFM tip contact is about to move from the central line of the NW toward the compressed side of it, in the second half of the contact. The magnitude of the electrical output depends significantly on the dimensions of NWs, and the output voltage peak is negative with respect to the ground voltage. The voltage is negative as the current flows from the tip of AFM to the nanowires.

A deeper insight into the structure is required to understand the physical working principle of the piezoelectric energy harvesting process [9]. When a constant force is applied to the top surface of vertically aligned NWs with the help of an AFM, a strain is created. Under the influence of this strain, the outer surface of the NW gets stretched and the inner surface gets compressed. There occurs a difference in strain between the inner and outer surfaces as the outer surface undergoes a positive strain while the inner surface experiences a negative strain. The external strain causes a relative displacement of the anions and cations of the NW material, leading to the generation of an electric field, E_z, directed along the z-direction inside the NW, which is also the c-axis of the NW crystal. This electric field, which is called piezoelectric field, is parallel to the z-axis of the NW at the external surface and antiparallel to the z-axis at the inner wall.

The piezoelectric field eventually gives rise to an electric potential, which varies from V_s^- to V_s^+ between compression and stretching transitions of the top end of the NWs.

During the process of external straining, no output voltage is observed because the relatively displaced ions cannot move freely or recombine as long as the force on the NW is present. The output current is produced only when the pressure is released as the AFM tip is about to transit from the strained NW to the neighboring NW that is about to undergo the same process.

A key measure that must be undertaken to enable energy harvesting is that contacts at the top and bottom of the NWs must be asymmetric: The top contact must be Schottky and the bottom contact must be Ohmic in nature.

As the platinum tip of the AFM scans through the NWs, we can observe two types of electron-transport processes.

The AFM tip first comes in contact with the stretched surface of the NW where the piezo potential V_s is greater than zero. Hence, the platinum tip obtains a potential V_m, which is nearly equal to zero. This leads to a negative bias at the interface of the metal tip–semiconductor NW. The voltage difference $\Delta V = V_m - V_s^+$ is less than zero, which makes the interface behave like a reverse-biased Schottky diode. Under the effect of this reverse bias, only a very small current can flow across the metal–semiconductor interface. This explains the observation that no output current flows during the initial phase of the contact of the AFM tip with the top surface of the vertically grown NW arrays.

The second transport process begins when the AFM tip scans through the stretched surface to the compressed surface during the second half of the contact. At the compressed side, V_s is less than zero, and so $\Delta V = V_m - V_s^-$ is greater than zero. This voltage difference yields a forward-biased Schottky diode at the metal–semiconductor interface, but this time at the compressed side of the NW. Here, the output current will increase from the previous value. According to the characteristics of a forward-biased Schottky diode, electrons will flow from the n-type semiconductor side to the metallic side of the interface.

The NWs have an Ohmic contact with the grounded side. When the AFM tip touches the NW and pushes it, the positive piezoelectric potential is created at the stretched surface. As the AFM tip proceeds with the scanning process, the electrons start flowing from the grounded electrode toward the metal contact tip, but these electrons cannot cross the metal–semiconductor interface owing to the presence of the reverse bias at this junction. When the tip reaches the compressed end, maintaining the constant force on the NW, the local piezoelectric potential continues to decrease, attaining a zero value at the mid-point of the NW top surface, and gradually takes on a negative value at the compressed end. Just as the local potential becomes negative, the electrons that accumulated at the tip flow back to the grounded end through the load. Thus, the electrons follow a closed path of motion through the external circuit, giving rise to a current in the second half of the contact of AFM tip with the NW.

This process of current generation is reversed when the whole experiment is performed under white light illumination. In this case, positive output voltage is observed when the tip comes in contact with the stretched side of the NW (as shown in Figure 9.3).

The basic reason for this phenomenon is that white light reduces the Schottky barrier height that exists at the NW and tip interface.

9.4.2 Working Principle of Triboelectric Nanogenerator

Triboelectric effect occurs due to the contact of two materials composed of different molecules. While the two materials are in contact, electrons are exchanged from one material to the other because a weak chemical bond is created between the two surfaces. As numbers of the surface electrons are not the same for the differently composed material, there is a tendency to make it an equipotential surface when the contact is made by transferring electrons in between them. Naturally, a huge charge imbalance is created between the two.

Now, when the two materials are in separation mode, few bounded atoms of one material keep extra electrons and the other material gives away electrons, depending on the electronegativity effect of the material. At this mode too they retain the imbalance.

When electrons move from one material to another, they leave behind positive charges by generating equal negative charges on the other material. (Scientists have prepared a chart that notes which ones would be positive and which ones would be negative, termed as the triboelectric series.) This imbalance of charge is balanced by producing an electric current in the outer circuitry.

By applying this triboelectric effect, an NG has been developed, which was first demonstrated by Prof. Zhong Lin Wang's research group at Georgia Institute of Technology in the year 2012. They used organic material for the first time.

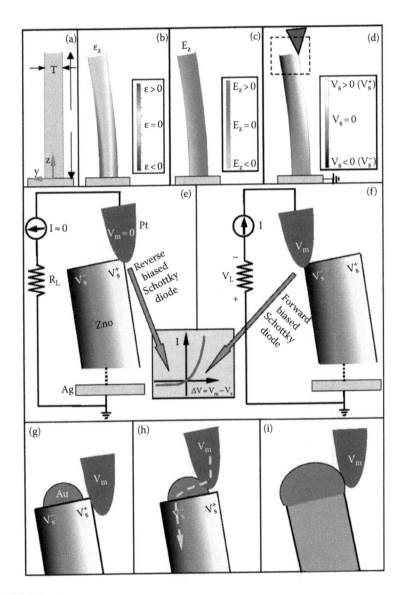

FIGURE 9.3
Simulated result showing the NG exhibiting piezoelectric property to produce current [10]. (a) Schematic of NW and coordinate system, (b) longitudinal strain distribution εz in NW after being deflected by AFM tip from side, (c) corresponding longitudinal induced electric field Ez distribution in NW, (d) potential distribution in the NW as a result of piezoelectric effect, (e and f) contacts between the AFM tip and semiconductor ZnO NW at two opposite ends of the NW showing the reverse and forward biased Schottky rectifying nature, respectively, (g and h) contact of the metal tip with the semiconductor NW having a small gold particle on top, and (i) contact of the metal tip with the semiconductor NW having a large gold particle on top. The charges are gradually leaked off through the compressed side of the NW as soon as the deformation occurs; thus, no potential accumulation occurs. (From Wang, Z.L. and Song, J., *Science*, 312, 242–246, April 2006. With permission.)

The two materials as discussed may be a thin film or bulk material of oppo-site polarity, and they may be of organic or inorganic type. Mostly, the mate-rial of a triboelectric nanogenerator (TENG) is organic and so these NGs are sometimes called organic nanogenerators.

There are three modes of operation present in a TENG.

9.4.2.1 Vertical Contact Separation Mode

As the name suggests, the whole mechanism is based on the contact-making and -breaking of the opposite polarity triboelectric charges on the inside plane of the two materials. When a mechanical force is applied to press the upper portion, the inner planes are in close contact, and, due to the tribo-electric effect as described, charge transfer begins. Due to this phenomenon, one side of the surface becomes much more positive than the other, which gets negatively charged.

Thus, a potential difference is generated across the output drawn from the top and bottom electrodes as shown in Figure 9.4a. As a result, current

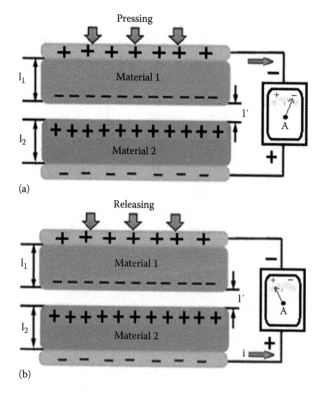

FIGURE 9.4
Triboelectric nanogenerator (a) pressed and (b) released.

will flow through the external load. This phenomenon is repeated, and the periodic change of the induced potential difference generates an alternating current.

When the contact is released, the current starts to flow in the opposite direction. It is better to have one plane as an insulator to prevent the swipe out of the electric charges and to restore it. The whole mechanism is shown in Figure 9.4a and b.

9.4.2.2 Lateral Sliding Mode

The process of triboelectric effect is initiated by the frictional contact between the two planes. There are two types of frictional contact: One is the normal contact, and the other is the lateral sliding mode in which one plane slides over the other. In this process, huge triboelectric charges are accumulated.

When one surface slides over the other, the contact area between the two surfaces changes. As a result, a lateral separation between the two is created, which is also periodic in nature. As discussed in the previous section, this lateral separation induces a charge imbalance, which creates a potential imbalance in the system. This potential imbalance drives the electrons to the external load, and a voltage drop is created. The schematic explains the triboelectrification effect clearly.

It was observed that initially, when the surfaces are in contact with each other fully, tribocharges of opposite polarity are induced as an effect of natural friction (Figure 9.5a). (The polarity difference—positive/negative—is clearly mentioned in the tribo series.) As there is no separation between the two surfaces, charges are balanced and net voltage drop is zero.

When the top surface slides out, the contact area decreases and a charge imbalance occurs in the system. This charge imbalance generates a potential gradient, and current starts flowing from the top to the bottom as depicted in Figure 9.5. The polarity of the voltage drop across the load is such that it would nullify the tribocharge-induced voltage drop across the plate. The flow of current will continue until the top plate is fully separated from the bottom plate as depicted in Figure 9.5b.

Now, when the top plate relapses to its original position, the tribocharges are in contact again. The charges that were induced at the top and bottom electrodes are not annihilated as the in-between layer is di-electric in nature. These excess charges flow through the external load and increase gradually as the contact area increases. Current flows from the top electrode to the bottom one, generating a positive peak, whereas the flow of current from the bottom to the top electrode generates a negative peak as shown in Figure 9.5c. If we plot a graph, it will be periodic in nature, showing zero output when the plates are fully overlapped.

It is very interesting to note that the top plate outside the contact area holds an equal number of negative charges at the top electrode in order to maintain electrostatic equilibrium.

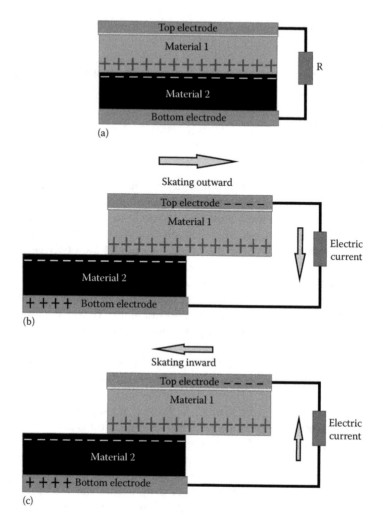

FIGURE 9.5
Triboelectric nanogenerator (a) as it is, (b) slided, and (c) laterally moved.

9.4.2.3 *Single Electrode Mode*

As the name suggests, the single electrode mode TENG has only one electrode [11].

The basic concept is the same, where the entire mechanism is based on coupling of contact electrification and also electrostatic induction. The main difference with lateral mode is that while the lateral mode has two electrodes, top and bottom, this has only one electrode, a bottom electrode.

Among the two layers, the bottom layer is di-electric, whereas the upper layer may be considered as human skin for the ease of understanding as this mechanism is generally fingertip-driven.

When the human skin and the insulating bottom surface are in contact, the charge get transferred between the two surfaces as per normal triboelectric effect. The insulating surface gets negatively charged and the human skin gets positively charged. As already mentioned, the number of tribocharges are equal and opposite when they are in the full-contact mode, to keep the surface electrostatically balanced, and the output voltage/current is zero.

When the top and bottom surfaces are separated, the negative charges on the insulating surface induce an equal amount of positive charges on the bottom electrode (generally indium tin oxide [ITO]). The process is illustrated in Figure 9.6a and b.

As the bottom electrode is grounded, there is a net potential imbalance and current starts flowing from the bottom to the top electrode. If the distance between the two increases, the positive charge induced on the bottom electrode increases to keep it in electrostatic equilibrium.

Please note that separation plays a critical role in obtaining the output voltage or current. We get a response only if the separation is comparable to the thickness of the bottom layer (di-electric). When the separation is higher than the allowed range, then the induced tribocharges are camouflaged, showing no output current. The whole process is elaborately shown in Figure 9.6.

When the upper layer is approaching back, however, the positive charge induced on the bottom electrode (ITO) starts decreasing as a result of the increased electrostatic effect of the two plates. As the bottom electrode is less negative than it previously was, current starts flowing from the ground to the electrode, producing a negative peak at the output. This process continues, until the two surfaces are in full contact, where the output current is zero.

Thus, both a positive and a negative peak are observed in the full cycle duration.

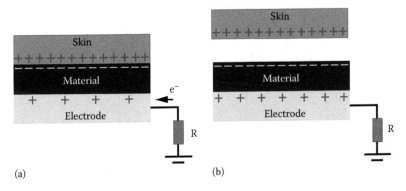

FIGURE 9.6
Triboelectric nanogenerator (a) as it is and (b) separated.

9.4.3 Working Principle of Pyroelectric Nanogenerator

The pyroelectric NG works on the principle of harvesting thermal energy and converting it into electrical energy. If an anisotropic material is subjected to a time-varying temperature fluctuation, the spontaneous polarization of the material changes and flow of electrons takes place to form a current. There can be two types of pyroelectric NGs: primary and secondary.

Primary pyroelectric NGs are generally made of ferrite materials. When such NGs are placed at room temperature, the electric dipoles oscillate about their mean positions by a considerable angular displacement. Under a fixed temperature, polarization induced due to the wobbling of the electrical dipoles is constant, and so the NG does not produce any electrical output (Figure 9.7a). If we allow the temperature to fluctuate, that is, if the surrounding temperature is increased from room temperature to some higher temperature, the dipoles will oscillate by a much higher degree about

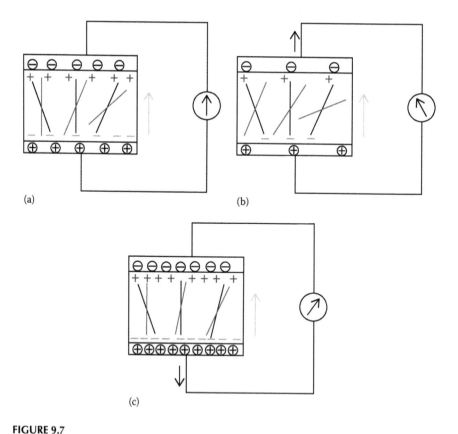

(a)   (b)

(c)

FIGURE 9.7

Pictorial representation of pyroelectric behavior at (a, condition 1: $\frac{dT}{dt} = 0$) fixed temperature, (b, condition 2: $\frac{dT}{dt} > 0$) after heating, and (c, condition 3: $\frac{dT}{dt} < 0$) after cooling.

their aligning axis under the effect of this enhanced heat energy. The spontaneous polarization is reduced due to the higher angular deviation of the dipoles. The induced charges in the electrodes get diminished, resulting in a current flow in the external circuit. In case of a contrasting situation, if the NG is cooled, the spontaneous polarization is increased due to less amount of angular spread, owing to the reduction in the heat energy content. The amount of induced charges increase and the electrons flow in the reverse direction (as shown in Figure 9.7).

The secondary NG will operate under the effect of pyroelectricity as well as piezoelectricity. In this case, the piezoelectricity is induced under the pyroelectric conditions. The secondary process is seen in wurtzite materials. When the NG is subjected to a rise in environmental temperature, it undergoes a strain by thermal expansion. This structural deformation induces a piezoelectric potential across the two surfaces of the material, thereby generating a current flow through the external connection, which is described as the working principle of the piezoelectric NG in Section 9.4.1.

9.5 Types of Excitation

Various types of excitation can be applied to the NG as a compressing force. The NG will convert that specific type of energy to electrical outputs, which can be used to power other nanodevices.

1. *Acoustic source*: A typical NG structure, having a top and a bottom electrode and vertically grown nanowires on the bottom substrate, can be excited by applying a force in the form of sound waves [12]. Sound waves can be used to vibrate the top electrode. When the top electrode is subjected to an acoustic pressure, it compresses the NWs. If the force is applied parallel to the NWs, then a separation of the centers of mass of the positive and negative ions results in the induction of a dipole moment and a generation of electric potential.

   The alternating output corresponds to the positive pulse under compressive strain and to negative pulse under the condition when the NWs are released, that is, when the acoustic pressure is withdrawn.

2. *Thermal source*: Under a particular study, a NG was subjected to the change in the surrounding temperature and an electrical output was recorded in the electrometer [13]. When the temperature around the NG was lowered from 295 to 289 K, a negative open-circuit voltage pulse has been observed under the forward connection. When the connection was reversed, a strong positive pulse was generated when the temperature was made to rapidly increase from

295 to 304 K. This type of NG is called pyroelectric nanogenerator where the Seebeck effect is utilized to convert thermal energy to electrical energy. The NG made of anisotropic material is imposed upon by time-dependent temperature change, which polarizes the anisotropic material and the potential thus created in turn drives the flow of electrons through the Schottky contact established between the NW and the top silver contact.

3. *Ultrasonic vibration*: A conventional NG made of vertically grown ZnO NWs is tested by exciting it with an ultrasonic wave [14]. In the design, the top electrode sheet is corrugated so that the zigzag surface provides scope for better bending of the NWs when they are subjected to some external force. Here, the ultrasonic waves act on the bottom electrode which deflects and bends the NWs, producing a strain. The vertical strain so produced converts the mechanical energy to electrical energy with the help of coupling between piezoelectric and semiconducting properties. The evidence of the electrical output is obtained by external meters connected between the top and bottom electrodes, which record the closed-circuit current and the open-circuit voltage.

There can be numerous other types of excitation. The NGs that are used to power biosensors can generate electricity by the movement of fingers, pressure created by foot while walking, expansion and contraction of diaphragm during respiration, and many more such movements, which can only be limited by human imagination.

9.6 Electrical Contacts

Metal–semiconductor contacts are used in various types of devices. These electrical contacts can be of two types: Schottky contacts and Ohmic contacts.

In case of Ohmic contacts, irrespective of the type of biasing condition applied, electrons can flow across the junctions due to the presence of a very small potential barrier at the interface. On the other hand, in Schottky contacts, electrons are capable of flowing across the junction only under a forward bias and their flow is restricted under a reverse bias. Thus, a Schottky contact acts as a diode, and the current flow is determined by the applied voltage.

In order to elaborate on the above-mentioned contacts, we must understand some basic terms. In metals, all energy levels are filled by electrons up to the Fermi level E_f. The amount of energy $q\Phi_m$ required to remove an electron from the Fermi level to the vacuum is called the work function of the metal. The energy required to extract an electron from the conduction band of a semiconductor to the vacuum level is called the electron affinity. The band diagrams of the two contact types must be studied in order to get a better understanding of the complete physical phenomena taking place within the materials at the surface of contact.

9.6.1 Schottky Barriers

A metal with work function $q\Phi_m$ and a semiconductor of work function $q\Phi_s$ are taken such that $\Phi_m > \Phi_s$. Therefore $E_{fs} > E_{fm}$. When the metal and the semiconductor are brought in contact, at equilibrium condition the Fermi levels on both sides of the contact are perfectly aligned. Charge transfer, that is, the electron flow takes place from the semiconductor to the metal until the Fermi levels are aligned to attain equilibrium. Thus, the electron energy at the side of the semiconductor must be lowered so that the Fermi levels of the two materials are at the same level. A depletion region will be formed consisting of uncompensated donor ions, which will be completely balanced by the electrons on the metallic side of the contact. A downward bending of the conduction and the valence bands takes place, and the contact potential V_o prevents further electron flow from the semiconductor to the metal.

To this type of barrier, if a forward-biasing voltage V is applied, then the contact potential gets lowered by $V_o - V$, which results in the diffusion of electrons from the conduction band of the semiconductor to metal. This gives rise to a forward current. Conversely, when a reverse biasing voltage V_r is applied, the barrier height gets enhanced by $V_o + V$ and the electron flow is hindered. A negligible amount of current flows from the semiconductor to the metal, as shown in Figure 9.8. This type of contact is also called rectifying contact [15].

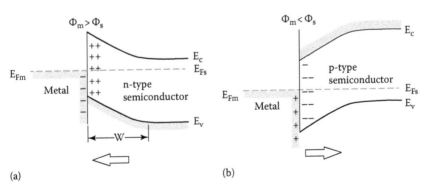

FIGURE 9.8
Schottky contacts when (a) $\Phi_m > \Phi_s$ and (b) $\Phi_m < \Phi_s$.

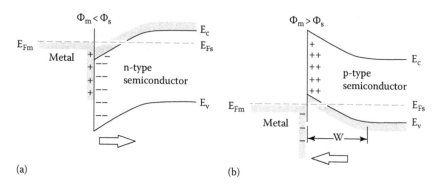

FIGURE 9.9
Ohmic contacts when (a) $\Phi_m > \Phi_s$ and (b) $\Phi_m < \Phi_s$.

9.6.2 Ohmic Barriers

In certain device applications, we may require a metal–semiconductor contact that has a linear I-V characteristic for both types of biasing. Such contacts are called Ohmic contacts. We take a metal and an n-type semiconductor such that $\Phi_m < \Phi_s$. When these two types of materials are brought in contact with each other, the equilibrium condition is achieved by the transfer of electrons from the metal to the semiconductor. This transfer of charge continues till Fermi levels on both sides of the contact get aligned to an equal level. Upward bending of the valence and conduction bands takes place. Positive space charges are present toward the metallic side of the contact and free electrons are accumulated toward the semiconductor side.

Under the application of the forward or the reverse bias, the barrier height faced by the electrons is small enough to be overcome by them and a considerable amount of current can flow in both directions across the junction, which follows a linear relationship with the applied voltage (Figure 9.9). The nature of biasing governs only the direction of the current flow.

In the process of harvesting energy from the NGs, Schottky and Ohmic contacts are required singly or in combination so that the maximum electrical power output can be extracted from the NGs of different designs and materials.

9.7 Alternating-Current Nanogenerator

In the previous sections, we have found that the piezoelectric NGs mostly generate direct current and voltage at the output load.

There are, however, case-specific designs where we can also see alternating current being generated after some mechanical energy gets converted to electrical energy.

ZnO/ZnS core/shell nanoparticles have a special property that the charges in them can be redistributed between the core and the shell, thereby preventing the injected electrons from being captured by electron quenchers [16]. These ZnO/ZnS core/shell nanoparticle electrets are assembled in an electrostatic NG. The NG is then excited by a vibrating metallic tuning fork that produces a sound of frequency 440 Hz. The output that is observed shows that the current through the load is alternating in nature with amplitude of 4.5 nA.

A flexible ZnO nanogenerator, which is fabricated by laterally aligning the nanowires, results in the formation of piezoelectric potential along the diameter of the nanowire when a force is applied perpendicular to the c-axis of ZnO nanowire [17]. An electrometer connected across the flexible NG measures the open-circuit output voltage. The readings of the electrometer clearly show that a positive piezoelectric potential is generated when the NG is bent upward, while a negative potential is generated when the NG is bent in the opposite direction, that is, downward. Thus, it can be said that the direction of bending determines the polarity of the output voltage and hence the voltage is alternating in nature.

Another example of flexible NG is fabricated with several layers grown one after another, as shown in Figure 9.10. Polydimethylesiloxane (PDMS) surface was taken and on top of this layer pyramid microstructures were fabricated. Multi-walled carbon nanotube and piezo-elecric barium titanate were mixed in the PDMS layer which is called the P-BM layer. An arch-shaped ITO electrode was placed on the composite film and an aluminum sheet was placed below the composite layer as the bottom electrode.

A cyclic force was applied to the whole area of the device so that the layers were periodically bent, and as a result, the plates approached closer to one another. The arch-shaped structure thus attained a purely rectangular shape, under the effect of the periodic compression. Now, let us consider the ITO layer as A, the polarized layer as B, and the lower aluminum electrode as C. Initially when the piezoelectric film was in the arched state, due to the bending nature, a negative piezoelectric charge (−qp) was induced in layer A. At this state, there was no friction between the plates and hence no charge transfer took place. As the compression began, the −qp charge was still preserved in layer A, while there was an injection of electrons from aluminum to the PDMS

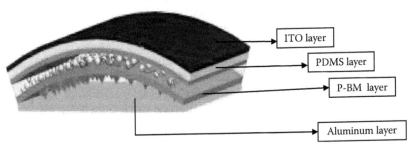

FIGURE 9.10
The schematic of a flexible NG. (From Xue, C. et al., *Nanomaterials*, 5, 36, 2015.)

layer due to friction. This resulted in a positive triboelectric charge (+qt) and piezoelectric charge (qp) remaining in layer C, and a negative triboelectric charge would be maintained in layer B. When the compression was released, the horizontal structure went back to its arch shape so that air gap would be formed between layers B and C, resulting in the creation of a much higher potential of layer C than B. This potential difference led to the flow of positive charges from layer C to A through the external circuit, and that continued till the potential difference was balanced by charge transfer (Δq). Layer A would attain a surface charge of (+Δq−qp), and layer C attained a surface charge of (+qt−Δq). If the compressing force was applied right after this charge buildup had taken place, layers B and C would come in contact with each other and the opposite situation would occur as a positive potential was induced in layer A and the Δq stream of charge now got transferred from layer A to C. Thus, the full cycle of compression and relaxation created alternating electricity under the combined mechanism of piezoelectricity and triboelectricity. The external force was applied at a frequency of 15 to 20 Hz when the performance of the device was measured to be the best, but its performance degraded when the external force frequency was further increased to 25 Hz.

The alternating output can also be generated by using pyroelectric NG, where heat energy gets converted to electrical energy. Under the forward connection, a change of temperature over a particular range results in a positive open-circuit voltage pulse, and for a different temperature change the open-circuit voltage shows a negative pulse.

Sound wave–activated ZnO NGs are capable of producing alternating current. When sound waves are applied on the top surface of the NG and parallel to the NWs, electrons flow from the top to the bottom electrode through the external circuit, whereas when the compression by the sound waves is released from the NG, the electrons flow back to the top electrode via the external circuit. If the same process is made to continue, then repeated cycles of positive and negative voltage will yield an alternating output from the NG.

It has also been reported by other authors that ZnO NGs fabricated by wet chemical method can produce flexible structures [19]. The device thus formed consists of a glass substrate having a deposition of a conductive poly(3,4-ethyl-enedioxythiophene)polystyrene sulphonate (PEDOT) layer. Vertical ZnO NWs are grown on top of the substrate. The final layer of gold-coated PU is deposited over the vertical NWs. Necessary contacts are given to connect the NG with the external measuring circuitry. The flexible NG is rolled down to generate piezo-electric potential. When the electrometer is connected in the forward direction, a positive open-circuit output pulse is recorded, and under the reverse connection of the meter, a negative voltage pulse reading is observed. The positive voltage pulse corresponds to the application of the rolling force on the NG, and the negative pulse corresponds to the releasing of the straining force such that the induced piezo potential gets reversed to neutralize the positive potential. The collective waveforms under both the conditions of strain and release show an alternating power being delivered to the output of the NG.

9.8 Fabrication of Piezoelectric Nanogenerator

There are three types of piezoelectric nanogenerator. This classification fully depends on the orientation of nanostructures. The one-dimensional or c-axis–oriented piezoelectric nanostructures are mechanically much more rugged and also prone to minor variations, which convert small mechanical energy into electrical energy. It is obvious that the force required to mechanically deform the c-axis–oriented nanostructures is comparatively small, which makes it useful in the design of new devices when compared to other forms of nanostructures. Many efforts are being made to enhance the piezoelectricity of the nanostructure by the selection of different piezoelectric materials based on wurtzite structure.

Some of the piezoelectric materials are zinc oxide, cadmium sulfide, barium titanate, gallium nitride, AlGaN, poly vinylidene fluoride, and polytetrafluoroethylene which are very common and have wurtzite structure. For example, ZnO has a hexagonal wurtzite structure. This hexagonal six-faceted structure makes it useful for efficient charge transfer among devices [20,21].

Depending on the configuration of piezoelectric nanostructure, most of the nanogenerators can be categorized into two types: vertical nanowire-integrated nanogenerator (VING) and lateral nanowire-integrated nanogenerator (LING).

9.8.1 Vertical Nanowire-Integrated Nanogenerator

As the name suggests, in VING the nanostructures are vertically oriented, as presented in Figure 9.11. Though the nanostructures are one dimensional, the VING device structure in itself is three dimensional. Generally, it consists of three layers: the bottom electrode, the vertically grown nanostructure, and the top electrode. The device exploits the bottom-up approach. The nanostructure is usually grown on the bottom electrode by various synthesizing techniques, and then the top electrode is deposited using vacuum evaporation techniques.

There are many fabrication techniques of producing these materials. Among these, chemical deposition technique is the cheapest route of producing

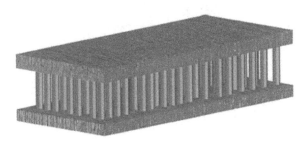

FIGURE 9.11
Schematic of VING.

nanostructures. ZnO can be deposited easily by this route. Hydrothermal route is another cost-effective way of production, which can be done at temperature as low as 100°C and particularly for GaN. It require either microwave or an autoclave for having a vertical c-axis-oriented structures. There are no constraints about the substrate material; nanostructures can be grown easily on any of the substrates, for example, Si, alumina, ITO, fluorine-doped tin oxide (FTO), and borosilicate glass, depending on the applications and annealing temperature.

When it experiences stress/vibration from the environment in any direction, the structure bends. Due to the bending in the nanostructure, strain appears. Because of these stresses on it, a piezo potential is generated on both sides of the nanostructure. Charge generation, accumulation, and transfer are the next consecutive phenomena, and current is finally transported through the top electrode [22].

9.8.2 Lateral Nanowire-Integrated Nanogenerator

LING has a two-dimensional configuration. Like VING, it also consists of three layers: the bottom electrode; the nanostructure, preferably semiconductor; and the intermediate metal electrode. The specialty of this device is that the structures are laterally grown, as illustrated in Figure 9.12. The metal electrode makes a Schottky contact with the semiconductor. As understood from the figure that the diameter (~50 nm) of the nanostructure is much more smaller than the length of the nanostructure (~0.5–1 µm). Eventually, the individual nanostructure is under a pure tensile strain.

All the nanostructures are laterally aligned and integrated on the flexible substrate.

There is another type of nanogenerator which is **Nanocomposite Electrical Generators (NEG)**. NEG also has a 3-D structure: the bottom electrodes, the vertical nanostructure, and the upper layer, which is a polymer matrix used to fill up the in-between void inside the nanostructure. NEGs were found to have a higher efficiency compared to the traditional nanogenerator configuration.

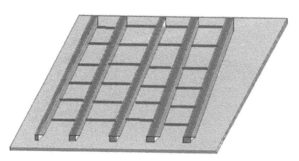

FIGURE 9.12
Schematic of LING.

9.9 Applications

NGs have a wide scope of application, which is mainly dependent on the various principles of energy conversion of the NGs. One major application is to provide continuous power to the large variety of micro- and nanodevices, which operate with the help of relatively low amount of input power. This makes the devices self-sufficient. Since the converted energy is low, the NGs are mainly used as supplementary source of power to the battery.

The NGs are capable of converting the kinetic energy of the wind, waves, and tides to electrical outputs. They can also convert mechanical energy generated from pressure into electrical energy [23]. Thus, implantable biosensors can employ such energy-harvesting devices for the generation of electrical power. Motion of the diaphragm, movements of hands and legs, or rotation of head can provide mechanical strain to the NGs, which in turn will get converted to electric current and voltage. NGs can also provide power to chemical and gas sensors [4].

There is another category of NGs that can harvest vibration energy. The vibrations can be of various types, such as created by voice, walking, clapping, engines, and automobiles. Harvesting energy from such vibrations can be very useful for powering mobile devices as such vibrations exist almost everywhere and at all times.

9.10 Conclusion

Research on NGs continues with the aim to increase the amount of output power so that they can be used for a wider range of applications. In view of this, a special type of NGs is now coming into light, which is of a hybrid structure. Such hybrid structures contain two different piezoelectric materials deposited one on top of the other. The main aim of designing such structures is to get greater amplitude of voltage and current output compared to the traditional designs of NG structures. The hybrid NGs are also capable of harvesting energies from very weak sources. Sources of energy like ocean waves, ripples, and wind, which have low frequency of occurrence and small amplitudes, are also sufficient to excite a hybrid NG to extract measurable values of electrical outputs.

The increasing need for self-powered devices and their varied applications has paved way for opening of diversified research fields. Along with the structural variations of the NGs, research is also carried out on the materials that can be used to fabricate such devices. Implantable biosensors need powering devices that can nontoxically stay inside the body for a long period. Thus, materials need to be judiciously selected. Furthermore, with the material variations, better output may be obtained from the powering devices, number of fabrication steps required may get reduced, and fundamentally new and simpler fabrications may be sufficient to make the NGs with various composite materials. The energy-harvesting systems should be such that they are compatible with the newly emerging cost-effective micro fabrication technologies so that mass production of such devices can be carried out in order to cater to the requirements of different self-powered micro/nanodevices.

Whatever are the new advancements or developments done, we should always keep in mind to use biodegradable and environment friendly materials to save the earth from getting polluted.

References

1. Z. L. Wang, X. Wang, J. Song, J. Liu, and Y. Gao, Piezoelectric nanogenerators for self-powered nanodevices, *IEEE Pervasive Computing*, 7(1), 49–55, January–March 2008.

2. Y.-F. Lin, J. Song, Y. Ding, S.-Y. Lu, and Z. L. Wang, Piezoelectric nanogenerator using CdS nanowires, *Applied Physics Letters*, 92, 022105, 2008.

3. K.-I. Park, S. Xu, Y. Liu, G.-T. Hwang, S.-J. L. Kang, Z. L. Wang, and K. J. Lee, Piezoelectric BaTiO$_3$ thin film nanogenerator on plastic substrates, *Nano Letters*, 10, 4939–4943, 2010.

4. W. Zang, Y. Nie, D. Zhu, P. Deng, L. Xing, and X. Xue, Core–shell In$_2$O$_3$/ZnO nanoarray nanogenerator as a self-powered active gas sensor with high H$_2$S sensitivity and selectivity at room temperature, *Journal of Physical Chemistry C*, 118, 9209–9216, 2014.

5. K. Y. Lee, D. Kim, J.-H. Lee, T. Y. Kim, M. K. Gupta, and S.-W. Kim, Unidirectional high-power generation via stress-induced dipole alignment from ZnSnO$_3$ nanocubes/polymer hybrid piezoelectric nanogenerator, *Advanced Functional Materials*, 24, 37–43, 2014.

6. C. Wang, L. Cai, Y. Feng, L. Chen, W. Yan, Q. Liu, T. Yao et al., An electrostatic nanogenerator based on ZnO/ZnS core/shell electrets with stabilized quasi-permanent charge, *Applied Physics Letters*, 104, 243112, 2014.

7. B. Saravanakumar, S. Soyoon, and S.-J. Kim, Self-powered pH sensor based on a flexible organic–inorganic hybrid composite nanogenerator, *ACS Applied Materials and Interfaces*, 6, 13716–13723, 2014.

8. L. Lin, C.-H. Lai, Y. Hu, Y. Zhang, X. Wang, C. Xu, R. L. Snyder, L.-J. Chen, and Z. L. Wang, High output nanogenerator based on assembly of GaN nanowires, *Nanotechnology*, 22, 475401 (5pp.), 2011.

9. I.-J. No, D.-Y. Jeong, S. Lee, S.-H. Kim, J.-W. Cho, and P.-K. Shin, Enhanced charge generation of the ZnO nanowires/PZT hetero-junction based nanogenerator, *Microelectronic Engineering*, 110, 282–287, 2013.

10. Z. L. Wang and J. Song, Piezoelectric nanogenerators based on zinc oxide nanowire arrays, *Science*, 312, 242–246, 2006.

11. H. Zhang, Y. Yang, X. Zhong, Y. Su, Y. Zhou, C. Hu, and Z. L. Wan, Single-electrode-based rotating triboelectric nanogenerator for harvesting energy from tires, *ACS Nano*, 8, 680–689, 2014.

12. S. N. Cha, J.-S. Seo, S. M. Kim, H. J. Kim, Y. J. Park, S.-W. Kim, and J. M. Kim, Sound-driven piezoelectric nanowire-based nanogenerators, *Advanced Materials*, 22, 4726–4730, 2010.

13. Y. Yang, W. Guo, K. C. Pradel, G. Zhu, Y. Zhou, Y. Zhang, Y. Hu, L. Lin, and Z. L. Wang, Pyroelectric nanogenerators for harvesting thermoelectric energy, *Nano Letters*, 12, 2833–2838, 2012.

14. J. Liu, P. Fei, J. Zhou, R. Tummala, and Z. L. Wang, Toward high output-power nanogenerator, *Applied Physics Letters*, 92, 173105 (3pp.), 2008.

15. C. Periasamy and P. Chakrabarti, Time-dependent degradation of Pt/ZnO nanoneedle rectifying contact based piezoelectric nanogenerator, *Journal of Applied Physics*, 109, 054306 (7pp.), 2011.

16. C. Wang, L. Cai, Y. Feng, L. Chen, W. Yan, Q. Liu, T. Yao et al., An electrostatic nanogenerator based on ZnO/ZnS core/shell electrets with stabilized quasi-permanent charge, *Applied Physics Letters*, 104, 243112, 2014, doi: 10.1063/1.4884366.

17. Y. Chu, L. Wan, G. Ding, P. Wu, D. Qiu, J. Pan, and H. He, Flexible ZnO nanogenerator for mechanical energy harvesting, *14th International Conference on Electronic Packaging Technology*, Dalian, China, 978-1-4799-0498-3/13, 2013.

18. C. Xue, J. Li, Q. Zhang, Z. Zhang, Z. Hai, L. Gao, R. Feng et al., A novel arch-shape nanogenerator based on piezoelectric and triboelectric mechanism for mechanical energy harvesting, *Nanomaterials*, 5, 36–46, 2015.

19. Y.-Y. Cheng, S.-C. Chou, and J.-A. Chang, Development of flexible piezoelectric nanogenerator: Toward all wet chemical method, *Microelectronic Engineering*, 88, 3015–3019, 2011.

20. S.-H. Shin, M. H. Lee, J.-Y. Jung, J. H. Seol, and J. Nah, Piezoelectric performance enhancement of ZnO flexible nanogenerator by a CuO–ZnO p–n junction formation, *Journal of Materials Chemistry C*, 1, 8103–8107, 2013.

21. S. S. Lin, J. H. Song, Y. F. Lu, and Z. L. Wang, Identifying individual *n*- and *p*-type ZnO nanowires by the output voltage sign of piezoelectric nanogenerator, *Nanotechnology*, 20, 365703 (5pp.), 2009.

22. E. Lee, J. Park, M. Yim, Y. Kim, and G. Yoon, Characteristics of piezoelectric ZnO/AlN–stacked flexible nanogenerators for energy harvesting applications, *Applied Physics Letters*, 106, 023901 (5pp.), 2015, doi: 10.1063/1.4904270.

23. X. Chen, S. Xu, N. Yao, and Y. Shi, 1.6 V nanogenerator for mechanical energy harvesting using PZT nanofibers, *Nano Letters*, 10, 2133–2137, 2010.

10

Solar Photovoltaic: From Materials to System

Sunipa Roy and Swapan Das

CONTENTS

10.1 Introduction

Interactions among electrons, holes, phonons, photons, and other particles are required to satisfy the conservation of energy and crystal momentum. A photon with energy near a semiconductor band gap has almost zero momentum. An important process is called radiative recombination, where an electron in the conduction band annihilates a hole in the valance band, releasing the excess energy as a photon. If the electron is at the bottom of the conduction band and the hole is at the top of the valence band, then in the case of a direct band gap semiconductor, this radiative recombination is a preferred phenomenon. In the case of an indirect band gap semiconductor, however, this recombination is not possible as it violates the conservation of crystal momentum, but this can be possible for an indirect band gap material if the process involves absorption or emission of a phonon. In this case, the phonon's momentum must be equal to the difference between the momentums of the electron and the hole. The involvement of a phonon makes the process slower for an indirect band gap semiconductor.

The process of light absorption is just the reverse of radiative recombination.

When light energy of a particular wavelength equivalent to the band gap absorption edge is incident on an indirect band gap semiconductor, it can penetrate much further before being absorbed than on a direct band gap semiconductor. This is the key factor used for photovoltaic devices, and that is why silicon is still used as a solar cell substrate material though it is an indirect band gap semiconductor.

Photoelectric devices convert light energy directly into electrical energy. These devices are self-generating, which means that they require no external power source to deliver the output. Photovoltaic and photoluminescence effects are just the reverse of each other.

Devices that convert electricity to light by exploiting the photoluminescence effect are called emitters (as they emit light), whereas devices that convert light into electricity are called photovoltaic (PV). PV devices include photoemissives (nonsolid-state devices) and photodetectors (solid-state devices).

Electrons are in a higher energy state on the n side, whereas holes are in the lower energy band on the p side. When these electrons–holes recombine, some of the energy generated is given up in the form of heat and light. Generally, compound semiconductors (GaAs, GaP, GaAsP, etc.) respond by releasing greater percentage of energy in the form of light. If the semiconductor material is translucent, light is emitted and the junction becomes a light source. On the other hand, when light energy of a particular wavelength equivalent to the band gap absorption edge and of particular intensity is incident on a semiconductor, the light is absorbed and breaks the covalent bond to generate electron–hole pairs. These carriers cross the junction and generate an electric current, known as photocurrent, flowing through the external circuit.

10.2 Physics of Photovoltaic

Solar cell is the most popular PV device that combines optics with electronics. It converts solar energy into optical energy and is also known as solar energy converter. The energy reaching the earth's surface from the sun is basically electromagnetic radiation, which covers a spectral range of 0.2–0.3 μm.

In 1954, the solar cell was first developed by Chaplin, Fuller, and Pearson, and since then it has shown remarkable progress in the market. Figure 10.1 shows the basic solar cell device structure.

There are two parameters that are ultimately used to characterize a solar cell:

1. *Short-circuit current*: The maximum current at zero voltage. Ideally, if $V = 0$, $I_{sc} = I_L$. Note that I_{sc} is directly proportional to the available sunlight.

2. *Open-circuit voltage*: The maximum voltage at zero current. The value of V_{oc} increases logarithmically with increased sunlight. This characteristic makes solar cells ideally suited to battery charging.

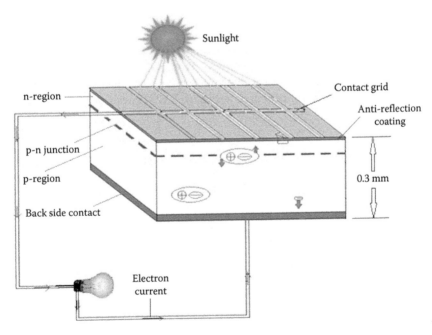

FIGURE 10.1
Solar cell device structure.

The basic physics behind this is as follows:

1. Photons are absorbed and energy is given to electrons in the crystal lattice.
2. Generated free electron–hole pairs flow through the material to produce electricity.
3. Different PV materials have different band gap energies.
4. Photons with energy equal to the band gap energy are absorbed to create free electrons.
5. Photons with less energy than the band gap energy pass through the material.

10.3 Electrical Characteristics of a Solar Cell

A solar cell is basically a p-n junction diode. It requires no bias across the junction to produce electric current. The surface layer of p-type material is made extremely thin to facilitate the incident light to penetrate the junction easily. The device is packaged with glass shutter on the top. When the incident light particles (photons) collide with the valence electrons, they transfer sufficient energy to these electrons to detach them from the parent atoms. Electrons are transferred from the valence band to the conduction band. Free electrons and holes are thus generated on both sides of the junction, and their flow constitutes the minority current IL in the reverse-biased direction. In the absence of light, thermally generated minority carriers constitute the reverse saturation current. Figure 10.2 shows the current–voltage

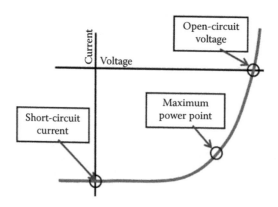

FIGURE 10.2
Current–voltage characteristics curve.

characteristics curve. The corresponding I/V characteristic is described by the Shockley solar cell equation.

The net p-n junction current is described by the Shockley solar cell equation:

$$I = I_L - I_F = I_L - I_S \left[\exp(eV/KT) - 1 \right] \tag{10.1}$$

where
  K is the Boltzmann constant
  T is the absolute temperature
  e is the electron charge
  V is the voltage at the terminals of the cell
  I_S is the diode saturation current

As the diode becomes forward biased due to the voltage drop across the load, the magnitude of the electric field in the space charge region decreases, but it does not go to zero or change direction. The photocurrent is always in the reverse bias direction. This current is directly proportional to the illumination and also depends on the illuminated surface area.

From Figure 10.3, we observe that V_{oc} is the maximum voltage obtained at the load under open-circuit conditions of the diode, and I_{sc} is the maximum current through the load under short-circuit conditions.

The open-circuit voltage V_{oc} is given by the following formula:

$$V_{oc} = \frac{KT}{q} \ln \left(1 + \frac{I_{sc}}{I_s} \right) \tag{10.2}$$

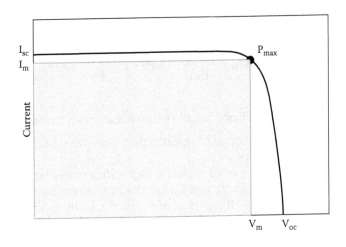

FIGURE 10.3
Maximum power rectangle.

The power delivered by the device can be maximized by maximizing the area under the curve. This can be done by appropriate selection of the load resistor connected across the device.

Both p and n regions of the device are heavily doped. The power rectangle curve shows that to have the maximum power, short-circuit current and open-circuit voltage should be maximum.

There is a term Fill Factor (FF) in the theory of solar cell. The FF is defined as follows:

$$FF = \frac{P_m}{V_{oc}I_{sc}} \tag{10.3}$$

where
P_m is the maximum output power
V_{oc} is the maximum voltage obtained at the load under open-circuit conditions of the diode
I_{sc} is the maximum current through the load under short-circuit conditions

FF depends mainly on V_{oc}/KT. The exact value of FF can be determined by the following equation:

$$FF = \frac{V_{oc} - \ln(V_{oc} + 0.72)}{V_{oc} + 1} \tag{10.4}$$

Equation 10.4 shows that the closer the value to unity the higher is the quality of the solar cell. The value of FF is generally 0.7–0.8.

The conversion efficiency of a solar cell is derived as the ratio of the output electrical power to the incident optical power.

$$\eta = \frac{P_m}{Pin} \times 100\% = \frac{I_m V_m}{Pin} \times 100\% = FF \frac{IL V_{oc}}{Pin} \times 100\% \tag{10.5}$$

The ratio $\dfrac{I_m V_m}{I_{sc} V_{oc}}$ is called FF and is a measure of maximum power from a solar cell. V_m and I_m are the voltage and current at the point of maximum power, and Pin is the incident optical power.

It is important to note that to realize a high-efficiency solar cell, it is not necessary to have only high V_{oc} and I_{sc}, but the FF value should also be very high. A range of 10%–12% efficiency is common for a silicon solar cell.

The actual I-V characteristics of a solar cell may vary from the ideal one and Equation 10.1 can be modified as Equation 10.6. An equivalent two-diode model (as shown in Figure 10.4) is used to present Equation 10.6. The model adheres the resistances R_s and R_{sh} together with photogenerated current IL.

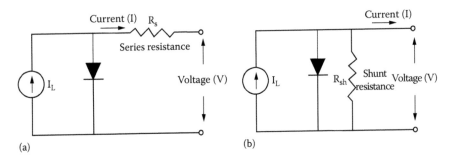

FIGURE 10.4
Non-ideal solar cell with (a) series resistance and (b) shunt resistance.

Considering Figure 10.4, the I-V characteristic of solar cell, when illuminated, can be written as

$$I = I_L - \left[I_{s1} \left\{ \exp\left(\frac{V + IR_s}{kT} \right) - 1 \right\} + I_{s2} \left\{ \exp\left(\frac{V + IR_s}{\eta kT} \right) - 1 \right\} \right] - \frac{V + IR_s}{R_{sh}} \quad (10.6)$$

where

I_{s1} is the reverse saturation current considering the diffusion and recombination of electrons (p region) and holes (n region) and is also the reverse saturation current considering the generation and recombination of electrons and holes in the depletion region

V is the applied voltage

q is the electronic charge

k is the Boltzmann constant

T is the absolute temperature

It is evident from the figure that I_{sh} (current flowing through R_{sh}) reduces the current output (Figure 10.4a) and R_s reduces the output voltage due to the voltage drop across it (Figure 10.4b). The effect of R_s and R_{sh} is obvious if we analyze Equation 10.6 carefully. The ideality factor η is greater than 1, which is reflected in the second part of the equation (exponential term).

How these resistances affect the I-V curve of the solar cell is shown in Figure 10.5.

The ultimate target of a solar cell manufacturer is to maximize efficiency by maximizing absorption and to minimize recombination. To maximize efficiency, parasitic losses should be minimum. FF and efficiency are both decreased by the shunt and series resistance losses.

Shunt resistance appears in a solar cell due to processing defects while its fabrication. On the other hand, series resistance primarily appears due to the movement of current through the solar cell, but the contact resistance between metal contact and silicon is the another source of series resistance. The key impact of

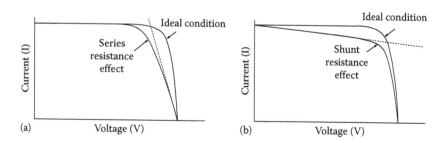

FIGURE 10.5
(a) Series and (b) parallel resistance effect on I-V characteristic.

series resistance is to reduce the fill factor. Higher shunt resistance is always pre-ferred, as low shunt resistance creates an alternating path for the photo current and results in increased power loss. The exact value of shunt resistance can be determined from the slope of the I-V curve near the short-circuit current point. The typical value of the shunt resistance is $1000\ \Omega\,cm^2$ for a commercial solar cell.

10.4 Solar Panel

A solar panel is an indispensable component of a solar substation and is respon-sible for collecting solar radiation and transforming it into electrical energy. It is an array of several solar cells (photovoltaic cells). Figure 10.6 shows the basic solar panel schematic. The arrays can be formed by connecting them in a series or parallel connection, depending upon the energy required.

The most commonly used semiconductor material is silicon. Both crystal-line and amorphous silicon are used for this purpose. Crystalline silicon is of

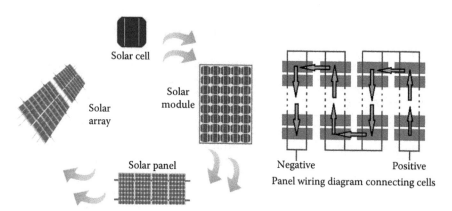

FIGURE 10.6
Solar panel schematic.

two types: Mono-Si and Poly-Si. Mono-Si means that the crystal lattice of the entire sample is continuous, whereas poly-Si means that it is composed of many crystallites of varying size and orientation. The thickness of monocrystalline silicon is 0.17–0.24 mm with a uniform appearance. The maximum efficiency found till date with a monocrystalline solar cell is 22%, whereas a polycrystalline solar cell offers less efficiency (18% maximum). Polycrystalline silicon is composed of different numbers of silicon crystal fused together to form a single crystal. The maximum thickness of polycrystalline silicon is 0.3 mm.

Another promising solar cell material is gallium arsenide. The solar cell fabricated with it is found to be lightweight and more efficient.

Very recently, thin-film solar cells with low cost technology have emerged in the market. The thickness depends on the substrate material as well as the coating used. However, this is at the basic research level and the maximum efficiency observed till date is 10%–12%. The area needs to be explored further.

10.5 Organic Photovoltaic Effect

Organic photovoltaic (OPV) is a very upcoming area of research and there is no existing market for it.

Now, what is OPV?

The PV that deals with conductive organic polymer is termed as organic photovoltaic. It may not always be a polymer but may contain organic molecules. These organic molecules absorb sunlight and convert it to electricity by employing the charge-transport phenomenon.

By changing the functional group content of polymers, the band gap can be manipulated. Organic molecules have higher absorption coefficient, which means that a small quantity of material can absorb a large amount of light.

Compared to organic, the inorganic PV offers greater strength and higher stability. Moreover, the efficiency of the OPV system is less; however, the main advantage associated with it is the low cost involved.

There is no existing market for OPV, which is still at an early stage of research, but due to its inherent advantages such as easy deposition method with large area coverage, it being inexpensive, and device flexibility, it is expected to replace the inorganic PV market very soon.

The present market is very much inclined to OPV because carbon is abundant in nature. Though pure carbon is very difficult to obtain, defect-free carbon deposition in the form of fullerene, carbon nanotube, and graphene is obtained by different researchers. Among them, graphene is successful as a transparent conducting electrode in a PV field.

Among others, optoelectronics and sensor applications are the novel nanodevice applications, which have yielded the greatest scientific breakthrough using single and multilayer graphene technology. Properties such as two-dimensional structures, high light transmittance, and excellent carrier mobility have invited research interest in the application of graphene in photovoltaics. Graphene can also participate in the photo-carrier generation process [12] and could act as an active layer for a p-n junction. Schottky junction solar cells have been assembled by transferring graphene onto various semiconducting substrates, such as Si and GaAs. In spite of the remarkable results recently achieved due to the growing involvement owing to design simplicity and effective photovoltaic property, several basic ambiguities regarding G/Si Schottky junction solar cells are required to be neutralized. Graphene film doping by chemicals, gaseous atmosphere around it, and graphene sheet thickness are some of the studied factors that affect the performance of G/Si solar cells. The motive of such research is to develop a graphene/silicon heterojunction–based solar cell. The result can be a new technology that might open job avenues in industries worldwide.

There are many important optoelectronic properties of graphene.

10.6 Optical Properties of Graphene

Black graphite becomes highly transparent when thinned down to a graphene monolayer. The electrons absorb light in the ultraviolet region but not in the visible region, and hence pure monolayer graphene appears transparent to the human eye. This optical characteristic combined with excellent conductivity of graphene-based materials will make it a replacement of silicon in the near future.

Using layer thickness, we get a bulk conductivity of $0.96 \times 106\ \Omega^{-1}\ cm^{-1}$ for graphene, a value which is higher than the conductivity of copper, which is $0.60 \times 106\ \Omega^{-1}\ cm^{-1}$.

Researchers are more curious about the optical electrical properties of graphene than of the other material as the combination of high film conductivity, optical transparency, and chemical and mechanical stability immediately suggest employing graphene as a transparent electrode for solar cells or liquid crystal and also as a processable, transparent, flexible electrode material [1,2].

10.6.1 Optical Conductance of Graphene

The optical conductivity of a monolayer graphene can be written as

$$\sigma = \sigma_{\pi-\pi*} + \sigma_{\sigma-\sigma*} \tag{10.7}$$

where $\sigma_{\pi-\pi*}$ and $\sigma_{\sigma-\sigma}$ represent conductance from $\pi-\pi*$ and $\sigma-\sigma*$ interband transitions, respectively. σ and π bands transition is not possible as per wave function symmetry. Please note that conductance is in siemens (Ω^{-1}) and conductivity unit is siemens per meter (Ω^{-1}/m).

10.6.1.1 *Optical Conductance of Graphene due to π − π* Transitions*

For the π − π* transitions, optical conductance is a complex number:

$$\sigma_{\pi-\pi^*} = \sigma_{c1} + i\sigma_{i1} \tag{10.8}$$

σ_{c1} is the real part and σ_{i1} is the imaginary part. The real and imaginary conductance of a graphene monolayer is found to be

$$\sigma_{c1} = \sigma_0 \left[\frac{1}{2} + \frac{1}{72} \frac{(\hbar\omega)^2}{\xi^2} \right] \times \left(\tanh\frac{\hbar\omega + 2\beta}{4k_B T} + \tanh\frac{\hbar\omega - 2\beta}{4k_B T} \right) \tag{10.9}$$

$$\sigma_{i1} = \sigma_0 \left[\frac{\beta}{\hbar\omega} \frac{4}{\pi} \left(1 - \frac{2}{9}\frac{\beta^2}{\xi^2} \right) - \log\left| \frac{\hbar\omega + 2\beta}{\hbar\omega - 2\beta} \right| \left(\frac{1}{\pi} + \frac{1}{36\pi}\left(\frac{\hbar\omega}{\xi}\right)^2 \right) \right] \tag{10.10}$$

where
$\xi \sim 3$ eV is the hopping parameter for p_z electrons in a tight binding model
β is the chemical potential with respect to the Dirac point
$k_B T$ is the thermal energy
$\hbar\omega$ is the incident photon energy
σ_0 is the universal optical conductance of a graphene monolayer

$$\sigma_0 = \frac{\pi}{2} \cdot \frac{e^2}{h} = \frac{e^2}{4\hbar} \sim 6.08 \cdot 10^{-5}\ \Omega^{-1} \tag{10.11}$$

At room temperature, we approximate $\sigma_{\pi-\pi^*} \sim \sigma_0$, and the optical conductance of graphene due to π − π* transitions is nearly equal to the universal optical conductance.

Optical conductance is normalized using free space impedance Z_0, and then we have

$$Z_0\sigma_0 = \sqrt{\frac{\beta_0}{\varepsilon_0}} \frac{e^2}{4\hbar} = \frac{1}{c\varepsilon_0} \frac{e^2}{4\hbar} = \pi\alpha \tag{10.12}$$

where, α is the fine structure constant. Optical conductance becomes physical constant when normalized by the impedance of free space.

10.6.1.2 *Optical Conductance of Graphene due to σ − σ* Transitions*

The energy gap in graphene is larger than the visible photon energies, generally 2.25 eV, though Saito [3] proved that the band gap energy of σ − σ* band is 6.0 eV. The transitions in the σ − σ* band result in a phase shift of optical waves while they pass through graphene. This phase shift does not affect the characteristics of graphene as the thickness of few layers of graphene is few angstrom, which is much smaller than the wavelength of the incident photon. Until and unless it has the thickness of hundreds of layers, the phase shift does not matter [4].

$\sigma - \sigma^*$ bands can be expressed by the dielectric constant

$$n = \sqrt{\frac{\varepsilon}{\varepsilon_0}} = \sqrt{1+x} \tag{10.13}$$

where x is the susceptibility and can be approximated as suggested [4]

$$x = \frac{Qe^2}{m\omega_0^2 e_0} \cdot \frac{\omega_0^2}{\omega_0^2 - \omega^2 + j\omega\Delta\omega} \tag{10.14}$$

where
Q is the number of charges per unit volume
ω_0 is the resonant angular frequency
ω is the normal angular frequency of the incident photon

This is a good approximation reported so far.

When $\omega \gg \Delta\omega$, we can simplify Equation 10.14, considering $\varepsilon = \omega\hbar$:

$$x = \frac{e^2}{me_0} \cdot \frac{\hbar^2}{\varepsilon_0^2 - \varepsilon^2} Q \tag{10.15}$$

Say t is the thickness of graphene and m is the mass of the electron. We can define σ as a function of x:

$$\sigma_{\sigma-\sigma^*} = i\omega\varepsilon_0 xt$$

$$= i\frac{e^2}{m} \cdot \frac{\varepsilon\hbar}{\varepsilon_0^2 - \varepsilon^2} Qt \tag{10.16}$$

Qt can be calculated by considering the charge density of $\sigma - \sigma^*$ bands. Graphene has sp² hybridization. With small excitation energy, an electron in the 2s subshell can be promoted into the 2p subshell. This time, the carbon atom hybridizes with three of the orbitals. One s orbital and two p orbitals directed at each other now join together to give molecular orbitals, each containing a bonding pair of electrons. These are sigma bonds. The p_z orbitals are so close that the clouds of electrons overlap sideways above and below the plane of the molecule. A bond formed in this way is called a pi bond, which a very weak bond. The sigma bond as well as pi bond lie in the same plane as the p orbital, at right angles to the s orbital. Any twist in the molecule would break the weak pi bond. In case of graphene, the three sp² hybrid orbitals arrange themselves at 120° to each other in a plane.

Among the eight electrons, four form the 6.0 eV bandgap and are distributed in σ bonds. As there is a weight of half for each polarization, mainly two effective electrons are involved in the 6 eV transition. Therefore, the charge density in $\sigma - \sigma^*$ band is obtained by the following equation:

$$Qt = \frac{2}{|b_1 \times b_2|} = \frac{2}{\frac{\sqrt{3}}{2} b^2} = 3.82 \cdot 10^{19} \text{ m}^{-2} \tag{10.17}$$

Therefore, involvement of $\sigma - \sigma^*$ transitions to overall conductance is

$$\sigma_{\sigma-\sigma^*} = 0.85 \sigma_0 i \tag{10.18}$$

The total conductance of graphene can be expressed as

$$= \sigma_0 [1 + 0.85i]. \tag{10.19}$$

10.6.2 Refractive Index

10.6.2.1 Refractive Index from Optical Conductance

The refractive index of graphene can be calculated from the optical conductivity. The current density per unit of the cross-sectional area can be defined as follows:

$$J = \frac{\sigma_0}{t} F \tag{10.20}$$

where
σ_0 is the optical conductance
t is the thickness of a graphene layer
F is the electric field vector

Ampere's current law states that

$$\nabla \times H = J + \frac{\partial S_0}{\partial_\tau} = \frac{\partial S}{\partial_\tau} \tag{10.21}$$

where $S_0 = \varepsilon_0 F$ and $S = \varepsilon_r \varepsilon_0 F$ are the displacement fields of free space and graphene, respectively. We can rewrite Equation 10.21 as

$$\frac{\sigma_0}{t} F + i\omega\varepsilon_0 F = i\omega\varepsilon_0 \varepsilon_r F \tag{10.22}$$

and we can derive the relative dielectric constant from Equation 10.22 as follows:

$$\varepsilon_r = 1 - i\frac{\sigma_0}{t\omega\varepsilon_0}$$ (10.23)

Using $\sigma_0 = \frac{\pi\alpha}{Z_0}$ and $\omega = \frac{2\pi c}{\lambda}$, we find that

$$\eta = \sqrt{\varepsilon_r} = \sqrt{1 - i\frac{\alpha}{2}\frac{\lambda}{t}}$$ (10.24)

where η is the refractive index and can be expressed in terms of fundamental constants and the ratio of optical wavelength to graphene thickness. Optical conductance per atomic layer is a fixed constant; the refractive index simply depends on the atomic layer's thickness. This is well in agreement that with the increase of optical response, atomic density increases.

Table 10.1 summarizes the refractive index reported so far for incident wavelength of 550 nm. Refractive index of graphene is not well investigated till date; however, the only constraint is that incident angle has to be large enough.

Data for ITO is taken as demonstrated by Reference 10, some data for graphene furnished here is taken from Reference 11.

Pristine graphene has no defect, and the electrons inside the graphene layer move as if they have no mass. Graphene is almost transparent; it absorbs only 2.3% of the light intensity, independent of the wavelength in the optical domain. This number is given by $\pi\alpha$, where α is the fine structure constant. Intrinsic graphene has a breaking strength of 42 N m^{-1}. The hypothetical 2-D steel has a breaking strength of 0.084–0.40 N m^{-1}. Thus, mechanical strength of graphene is more than 100 times greater than that of steel, making it the strongest material in the world. Graphene is a good candidate for high-speed electronics. Using the layer thickness, we get a bulk conductivity of 0.96 × 10^6 Ω^{-1} cm^{-1} for graphene, which is higher than the conductivity of copper (0.60 × 10^6 Ω^{-1} cm^{-1}). Thermal conductivity of graphene at room temperature

TABLE 10.1

Refractive Index of Graphene

Material	Technique	Refractive Index	Reference
Graphene	Picometrology	2.4–1.0i	[5]
Graphene	Reflectivity and transmittance	3.0–1.0i	[6]
Graphene	Reflection contrast, SiO$_2$	2.0–1.1i	[7]
Graphite	Optical reflection	2.675–1.35i	[8]
Graphite	Ellipsometry	2.52–1.94i	[9]

TABLE 10.2

Different Parameter Studies of Monolayer Graphene, Bilayer Graphene, and 100 nm ITO Film

Parameter	ITO	Graphene, Monolayer	Graphene, Bilayer
Transparency on glass (%)	70–85	94	92
Sheet resistance (Ω cm^{-1})	>100	100–30	30–10
Mobility (cm^2 V^{-1} s^{-1})	<50	5,000–20,000	10,000–30,000
Carrier density (cm^{-2})	10^{15}	10^{13}	2–10^{13}

is approximately 5000 W m^{-1} K^{-1}, whereas that of copper is 401 W m^{-1} K^{-1}, which is 10 times lower. The melting point of graphene is estimated at about 6560°F, one of the highest known for any material. No boiling point has yet been established.

Indium tin oxide (ITO) is used as transparent conducting electrodes, but there are many disadvantages associated with ITO: high sheet resistance (>100 = 2), low mobility, and only 70%–80% transparency [10,12]. Due to its lower sheet resistance and high optical transparency graphene has replaced ITO. However, ITO is still used in the market because graphene cannot be deposited through PVD, e-beam evaporation, etc. Defect-free graphene is very hard to get or requires a very hi-end lab and tricky procedure (Table 10.2).

10.7 Solar Cell Device Assembly

We are entering into a period of increasing environmental pollution and crisis of limited fossil fuel and other non-renewable energy sources. Solar PV technology has become an increasingly important energy supply option. Nearly 150 GWs are achieved worldwide till date with the global installed capacity of renewable energy production. Advancement in the microelectronics industry have made the solar PV power plant growth less expensive. This has ultimately improved the solar PV's competitiveness as it has much more scope than have other power-generation options in the market.

The basic need of solar PVs arises from the crisis of electric power supply. The variable grid power and the waste-producing diesel generators have heightened the requirement of solar PVs.

While the majority of the operating solar projects are located in the developed economies, a drop in the prices coupled with unreliable grid power and the high cost of diesel generators has driven fast-growing interest in solar PV technology in the emerging economies as well.

Most of the developing countries have plenty of solar resource, and the governments have made many efforts to encourage the growth of the solar manufacturing companies as solar energy has tremendous benefits for a nation and its economics. Energy is restored, promoting energy security, making the solar PV an emerging field in these days when fossil fuel stores are in the alarming zone.

There is a growing interest of organic PVs as their efficiency has increased remarkably, though more research is needed.

Installation of a solar PV system is tricky but not so time consuming. This is another reason leading to rapidly growing, promising markets with high demand. Further, with consistent development going on in the field, PV market prices continue to fall compared to the prices of other sources of electricity.

A PV plant design is developed with a prefeasibility study as the initial part, which is based on a preliminary energy resource and yield estimates as well as other site-specific requirements and constraints. The plant design is further improved during the feasibility study, which considers site measurements, site topography, and environmental and social considerations. Key design features include the type of PV module used, tilting angle, mounting and tracking systems, inverters, and module arrangement. Optimization of plant design involves considerations such as shading, performance degradation, and trade-offs between increased investment (e.g., for tracking), and energy yield. Usually, the feasibility study also develops design specifications on which the equipment to be procured is based.

PV cell technologies are broadly categorized as either crystalline or thin film. Crystalline silicon (c-Si) cells provide high-efficiency modules. They are subdivided into monocrystalline silicon (mono-c-Si) or multicrystalline silicon (multi-c-Si). Mono-c-Si cells are generally the most efficient, but they are also costlier than multi-c-Si. Thin-film cells provide a cheaper alternative but are less efficient. There are three main types of thin-film cells: Cadmium Telluride (C_dTe), Copper Indium (Gallium) Di-Selenide (CIGS/CIS), and amorphous silicon (a-Si).

The performance of a PV module will decrease over time due to a process known as degradation. The degradation rate depends on the environmental conditions and the technology of the module. Modules are either mounted on fixed-angle frames or on sun-tracking frames. Fixed frames are simpler to install and cheaper, and they require less maintenance. However, sun-tracking systems can increase the yield by up to 45%. Tracking, particularly in areas with a high direct/diffuse irradiation ratio, also enables a smoother power output.

Inverters convert DC electricity generated by the PV modules into AC electricity, ideally conforming to the local grid requirements. They are arranged in either string or central configurations. Central configuration inverters are considered to be more suitable for multi-MW plants. String inverters enable individual string Maximum Power Point Tracking and require less specialized maintenance skills. They also offer greater design flexibility. PV modules and inverters are all subject to certification, predominantly by the International Electrotechnical Commission. New standards are currently under development for evaluating PV module components and materials. The performance ratio (PR) of a well-designed PV power plant will typically be in the range of 77%–86% (with an annual average PR of 82%), degrading over the lifetime of the plant. In general, good quality PV modules may be expected to have a useful life of 25–30 years.

10.7.1 Overview of Grid-Tie Ground-Mounted PV Power Plant

Solar PV modules: Through a silent and clean process that requires no moving parts—through the photovoltaic effect—solar PV modules change solar radiation directly into electricity. The PV effect demonstrates a semiconductor effect, wherein electron movement is initiated when solar radiation falls onto semiconductor PV cells, which in turn yield DC electricity as output. Various types of solar panels are available in the market. To achieve 12–18 V, the required number of solar cells, each of which can have a potential difference of about 0.5 V, are connected in a series or parallel array. Multiple panels are connected together both in parallel and in series to achieve higher current and higher voltage, respectively. Figure 10.7 shows the basic grid-tie ground-mounted PV power plant. Typical components include the following:

Inverters: It is obvious that the electricity produced in a solar panel is DC. However, the electricity that we get from the grid supply is AC, and so for running a common equipment using power from the grid as well as the solar-energy system, it is compulsory to install an inverter to convert the DC of solar system to the AC

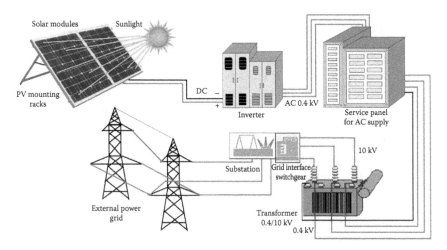

FIGURE 10.7
Overview of a grid-tie ground-mounted PV power plant.

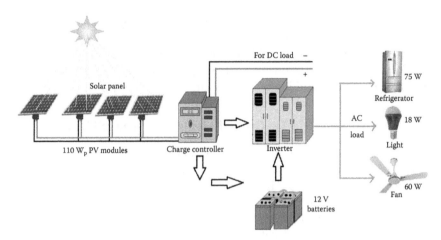

FIGURE 10.8
Overview of an off-grid solar PV system.

of the same level as that of the grid supply. In off-grid system the inverter is directly connected across the battery terminals so that the DC coming from the batteries is first converted to an AC and then served to the equipment. Figure 10.8 shows the basic off-grid solar PV system. In the grid-tie system, the solar panel is directly connected to an inverter that then feeds the grid with the same voltage and frequency power. In the modern grid-tie system, each solar module is connected to the grid through an individual micro-inverter to achieve high-voltage AC from each individual solar panel.

PV mounting systems: Efficiency depends not only on the solar device but also on how scientifically it is mounted. A PV module with a fixed tilt angle is desired. Recently, sun-tracking frames have been developed for better efficiency. This is the only moving part in a solar PV power plant. There are two types of tracking system: single- and dual-axis. Single-axis trackers can change either the orientation or the tilt angle, while dual-axis trackers alter both the orientation and the tilt angle.

AC service panel and step-up transformers: The inverter output requires a further step-up in voltage to reach the AC grid voltage level. Therefore, an AC service panel with circuit breakers and with the option of isolation of the PV power plant is required. The step-up transformer modifies the output obtained from the AC service panel to the required grid voltage (10, 25, 33, 38, or 110 kV, depending on the grid connection point).

Grid connection interface: The grid connection interface plays a critical role, where the electricity is transferred into the grid network. If any problem appears, there should be an option to isolate the PV power plant. The metering points are made external to the PV power plant boundary.

10.7.2 Off-Grid Solar PV System Design

10.7.2.1 Power-Consumption Demand Calculation

The initial stage in designing an off-grid solar PV system is to find out the total power and energy consumption of overall loads that require to be supplied by the solar PV system, as follows:

1. *Determine overall watt-hours per day for each appliance used*: Total watt-hours per day for all appliances is the summation of watt-hours needed for each of them.
2. *Determine overall watt-hours per day wanted from the PV modules*: Multiply the total appliances' watt-hours per day by 1.3 (the energy lost in the system) to get the total watt-hours per day that must be provided by the panels.

10.7.2.2 PV Module Size Calculation

The amount of power produced is different for different sizes of PV modules. The peak watt (W_p) produced depends on the size of the PV module and the climate or geographical location of the place where it is installed.

The 'panel generation factor' (G), varies according to the site location. The size of PV modules can be determined as follows:

1. *Determine overall watt-peak rating needed for PV modules*: Total Watt-peak rating needed for the PV panels to operate the appliances can be obtained by dividing the total watt-hours per day needed from the PV modules by G.
2. *Determine the number of PV panels for the scheme*: On dividing the answer obtained by the above calculation by the rated output watt-peak of the available PV modules, we get a result that on increasing fractionally to next whole number will give us the number of PV panels required for the system.

If more PV modules than the result are installed, the system will perform better and battery life will improve. If fewer PV modules are used, the system may not work at all during cloudy periods and battery life will be shortened.

10.7.2.3 Inverter Size Calculation

It is known that an inverter generates AC power, which is why it is used in a PV system where AC power output is required. Here, the minimum input rating of an inverter should be equal to the total watt of appliances and its minimum output should be equal to the battery voltage. For stand-alone systems, the inverter must be big enough to provide the total amount of watts. In addition, the inverter's input rating should match with the PV array rating for safe operation of the systems connected to the grid.

10.7.2.4 Battery Size Calculation

Selection of a definite variety of battery is bound by certain limitations. Deep-cycle battery, which gets discharged to a low energy and can be rapidly recharged day after day for years, is used. Size of the battery to be used can be determined by the following method:

1. Derive the total watt-hours per day used by appliances.
2. Divide the above result by 0.80 for battery loss.
3. Divide the quotient obtained by the above calculation by 0.6 for depth of discharge.
4. Divide the answer obtained in step 3 by the nominal battery voltage.
5. On multiplying the answer obtained in step 4 with days of autonomy (number of days that the system has to operate when PV panels fail to produce electricity), we get the required ampere-hour capacity of the deep-cycle battery.

$$\text{Battery capacity} \left(\text{Ah} \right) = \frac{\text{Total watt-hours per day used by appliances}}{0.80 \times 0.6 \times \text{Nominal battery voltage}}$$

$$\times \text{Days of autonomy}$$

10.7.2.5 Solar Charge Controller Size Calculation

A solar charge controller is rated with amperage and voltage capacities. The controller's rating must match with the voltage of the PV array, and it should also have enough current handling capacity.

Irrespective of a series or parallel configuration of the controller, the PV input current influences the size of the controller.

As per the standard practice, the size of a solar charge controller can be obtained by multiplying short-circuit current (I_{sc}) of the PV array by 1.30.

Example

A home has the following electrical appliance usage: one 18 W fluorescent lamp with electronic bulk, used 5 h day^{-1}; one 75 W refrigerator that runs 24 h day^{-1}, with a compressor that runs 10 h and is off for 14 h; and one 60 W fan, used for 3 h day^{-1}. The arrangement will be powered by 12V$_{DC}$, 110W$_p$ PV modules.

POWER-CONSUMPTION DEMAND CALCULATION

Appliance use = (18 W × 5 h) + (60 W × 3 h) + (75 W × 10 h) = 1020 W h day^{-1}
Total PV panels energy needed = 1020 × 1.3 = 1326 W h day^{-1}.

PV PANEL SIZE CALCULATION

Assumed panel generation factor G = 3.3
Total W$_p$ of PV panel capacity needed = 1326/3.3 = 401.81W$_p$
Number of PV panels needed = 401.81/110= 3.65 modules
Actual requirement = 4 modules

Hence, this system should be powered by at least four modules of 110W$_p$ PV modules.

BATTERY SIZE CALCULATION

Total appliances use = (18 W × 5 h) + (60 W × 3 h) + (75 W × 10 h)
Nominal battery voltage = 12 V
For one day autonomy
Battery capacity = [{(18 W × 5 h) + (60 W × 3 h) + (75 W × 10 h)} × 1]/ (0.80 × 0.6 × 12)
Total ampere-hours required = 177.08 A h
Thus the battery should be rated 12 V 177.08 A h for one day autonomy.

INVERTER SIZE CALCULATION

Total watt of all appliances = 18 + 60 + 75 = 153 W
For safety, the inverter should be considered 25%–30% bigger size.
The inverter size should be about 190 W or greater.

SOLAR CHARGE CONTROLLER SIZE CALCULATION

PV module specifications

P_m = 110W$_p$
V_m = 16.7V$_{DC}$
I_m = 6.6 A
V_{oc} = 20.7 A
I_{sc} = 7.5 A

Solar charge controller rating = (4 strings × 7.5 A) × 1.3 = 39 A.
Hence, the solar charge controller should be rated 40 A at 12 V or greater.

References

1. I. Meric, M. Y. Han, A. F. Young, B. Ozyilmaz, P. Kim, and K. L. Shepard, Current saturation in zero-bandgap, top gated graphene electrode transistors, *Nat Nanotechnol*, 3, 654–659, Sept 2008.
2. J.-H. Chen, C. Jang, S. Xiao, M. Ishigami, and M. S. Fuhrer, Intrinsic and extrinsic performance limits of graphene devices on SiO_2, *Nat Nanotechnol*, 3, 206–209, Jan 2008.
3. R. Saito, *Physical Properties of Carbon Nanotubes*, Imperial College Press, London, U.K., 1998.
4. B. E. A. Saleh and M. C. Teich, *Fundamentals of Photonics*, John Wiley & Sons, Inc., New York, 1991.
5. X. Wang, Y. P. Chen, and D. D. Nolte, Strong anomalous optical dispersion of graphene: Complex refractive index measured by Picometrology, *Opt Express*, 16, 22105–22112, Dec 2008.
6. M. Bruna and S. Borini, Optical constants of graphene layers in the visible range, *Appl Phys Lett*, 94, 031901, Jan 2009.
7. Z. H. Ni, H. M. Wang, J. Kasim, H. M. Fan, T. Yu, Y. H. Wu, Y. P. Feng, and Z. X. Shen, Graphene thickness determination using reflection and contrast spectroscopy, *Nano Lett*, 7(9), 2758–2763, 2007.
8. E. D. Palik, *Handbook of Optical Constants of Solids*, Academic Press, New York, 1985.
9. G. E. Jellison, J. D. Hunn, and H. N. Lee, Measurement of optical functions of highly oriented pyrolytic graphite in the visible, *Phys Rev B*, 76, 085125, Aug 2007.
10. A. K. Kulkarni, K. H. Schulz, T. S. Lim, and M. Khan, Dependence of the sheet resistance of indium-tin-oxide thin films on grain size and grain orientation determined from X-ray diffraction techniques, *Thin Solid Films*, 345, 273–277, 1999.
11. S. V. Morozov, K. S. Novoselov, M. I. Katsnelson, F. Schedin, D. C. Elias, J. A. Jaszczak, and A. K. Geim, Giant intrinsic carrier mobilities in graphene and its bilayer, *Phys Rev Lett*, 100(1), 016602, 2008.
12. T. Minami, Transparent conducting oxide semiconductors for transparent electrodes, *Semicond Sci Technol*, 20, S35–S44, Mar 2005.

11

Volatile Organic Compound–Sensing with Different Nanostructures

Sunipa Roy and Swapan Das

CONTENTS

11.1 Introduction

Nanotechnology and nanoengineering are key areas of research nowadays. Many researches have been conducted with three-dimensional nanostructures in the area of gas sensing. One-dimensional nanostructured materials including nanotubes, nanorods, nanoflakes, and nanowires have a proven track record in this area with their high sensitivity, and several articles are published every year about the same.

Among different sensor materials, semiconducting metal oxides have drawn much attention during the last few decades due to their unique optical and electrical transport properties, which have potential applications in the field of optics, sensing, and transparent conducting electrodes. To manipulate the materials' intrinsic properties, researchers are still looking for large band-gap intrinsic semiconductor materials. The nanometer regime results in quantum confinement of the photo-generated electron–hole pair, which result in a blueshift in the absorption spectrum.

Among the various structures, some are alike. For example, in rods and wires, the basic structure is the same, just the diameters and lengths vary. In case of wires, the length is 100 times larger than that of a nanorod.

Different types of characterization techniques are used to study the structure as well as properties of a material. Field emission scanning electron microscope (FESEM) is a technique to study the morphology of a thin film. Another, later technique of transmission electron microscopy observation gives 1000 times better resolution than that of FESEM.

Having a unique and very unusual structure, nanomaterial ushered in different opportunities. For example, multifaceted structure facilitates better interaction with the target gas molecules which improved their sensing performance.

This chapter provides an understanding of VOC sensing with chemical sensors. The sensing film may be based on nanotubes, nanorods, nanobelts, and nanowires, or combinations of it.

There are many harmful VOCs in the atmosphere, about which we are not aware. They may cause dizziness, nausea, headache, etc. Hence, detection of VOCs in low-concentration levels is highly desirable. From the societal aspect as well, identifying the VOCs is much needed. The most popular VOC detection device is a breath analyzer.

Due to the presence of a large amount of literature, the following narrative highlights researches published only in the past decades in this demanding field. There is still a need to detect harmful VOCs in a selective manner.

This chapter is divided into three sections. The first section describes the harmful effects of VOCs. The second section presents different nanostructures currently used in this area, and the third section focuses on sensing mechanisms and some important conclusions.

11.2 Global Warming

VOCs are organic substances that can vaporize at room temperature or even below that. They are called organic because they contain a chain of carbon atoms in their molecular structures. VOCs have no color. Some VOCs contain odors, some not. Hydrocarbons (e.g., benzene and toluene), halocarbons, and oxygenates are the different substances found in VOCs, and evaporation of solvents is the source of these. VOCs play a significant role in the creation of the greenhouse effect.

Hydrocarbon mainly specifies methane as it is the most common functional group of any hydrocarbon. Methane is highly combustible and a very important component of VOCs. Its environmental impact is highly dangerous. It contributes to global warming and triggers the production of lower-atmosphere ozone (O_3).

There are two types of ozone: "good" ozone and "bad" ozone. "Good" ozone is natural and stays 10–35 km above from the ground. It protects life from harmful ultraviolet rays. "Bad" ozone is found at the ground level. It is produced by the reaction of VOCs, oxygen, and sunlight. This lower-atmosphere ozone can cause organ infection, breathing problem, and mucous-membrane irritation. It also has a devastating effect on plants and a corrosive effect on metals.

The destructive nature of global warming is that it damages the stratospheric ozone layer, known as the holes in the ozone layer, which creates greenhouse effect.

Tropospheric ozone is another mentionable air pollutant produced by reactions between primary pollutants, VOCs, in the presence of sunlight. This is called a photochemical reaction.

11.3 Source of VOCs

VOCs are carbon-based compounds that evaporate at room (27°C) temperature in the atmosphere. VOCs can be emitted by various sources including soil and building materials. VOCs can aid in developing cancer. The most common VOC found in our daily life is present in our indoor air in the form of formaldehyde, which may cause the sick-building syndrome. Some sources of VOCs are discussed next.

Indoor air: Old buildings are mostly affected by formaldehyde due to the extensive use of plywood rather than solid wood. Pressed wood products like plywood are often used in modern furniture such as computer desks and shelves as they are affordable and durable. This, however, is a major source of VOCs (e.g., formaldehyde) since it is made of glues and other binding material, which is basically carbon polymers.

Formaldehyde concentration above 0.1 ppm initiates an irritation in the respiratory tract; it can even cause nasal cancer and lung cancer.

Paints and varnishes can also boost the VOC content of a building. Paints are chains of carbon polymer. Without carbon base VOCs cannot be made. Oil based paints are more dangerous than water based as it contains higher levels of harmful organic compounds. It should be eliminated from the market.

Chemicals: Toxic chemicals–based products that we use in our daily life make a large source of VOCs. From marker pens to printer ink, they all use carbon polymers.

Diethyl phthalate, which is accountable for neurological problems, is mostly found in daily home necessities. Toluene is a good stain remover but at the same time a notorious nerve-damaging neurotoxin. Hexane and Xylene also damage the nervous system.

Computers are also substantial source of VOCs. Besides the above, there are many other sources of VOCs found in an average building. These include air fresheners, cosmetic products, and detergents.

Residential complexes: When structures are built on or near landfills, methane gas can penetrate the buildings' interiors. This can expose occupants to significant levels of methane. Some buildings (e.g., the Dakin Building) have specially engineered recovery systems below their basements to actively capture such gases and vent them away from the building.

Heavy metals: Heavy metals are metals that have high densities. Mercury, lead, and bismuth are the common examples of heavy metals. Their density ranges from 3.5 to 7 g/cm^3. The two most common heavy metals found in constructions are lead and mercury. Mercury is also used by paint manufacturers as it is an effective antifungal substance.

Natural pollutants: Besides the chemical pollutants mentioned above, numerous natural contaminants carry VOCs, such as volcanic dust, sea salt particles, photochemically and formed ozone. These naturally occurring pollutants may also cause severe illness (e.g., allergic reaction like sneezing, wheezing).

11.4 Operating Principle of a Vapor Sensor

When a vapor sensor or gas sensor is exposed to any VOC, the sensor's resistance primarily decreased due to the release of free electrons or due to the reduction in depletion layer width. The time of this release depends on the temperature as well as the effectiveness of the sensing layer. So when the VOC is ON, the sensor resistance gets saturated and continues to a minimum baseline value. When the VOC is off, the resistance increases, and again it gets saturated to a high value, though minor variations have been observed probably due to the presence of some vapor molecules adsorbed on the sensing layer's surface. The sensitivity depends not only on the temperature but also on the VOC concentration. The other two crucial sensing parameters like response time and recovery time also depend on sensing temperature and VOC concentration.

When VOC concentration increases, more vapor molecules are adsorbed on the sensing layer's surface within a specific period of time. As a result, the number of free electrons increases on the surface. A greater number of electrons means fast electron transport kinetics, and eventually, the sensitivity increases and response time decreases. As the number of electrons is increased, it requires a longer time to wipe out all of them, so the recovery is slower due to slow desorption kinetics. Thus, higher concentration means more time is required to recover.

The effect of using nanostructured material is the lowering of temperature, because of the use of catalytic metal contact, which is of utmost importance. Catalytic noble metal electrode and surface modifiers are solely responsible for the low-temperature dissociation of ethanol molecules just above the sensing layers. Because of the nanostructured surface, the number of sensor nodes are increased and they have excessive free surface energy, which results in greater adsorption of VOC molecules per unit volume. Catalytic metals, owing to their half-filled orbital electrons, induce an exothermic reaction, which is ultimately responsible for lowering the temperature.

11.4.1 Chemical-Sensing Mechanism

11.4.1.1 Ethanol-Sensing Mechanism

As demonstrated by many researchers, the ethanol-sensing mechanism is similar to other gas/vapor sensing mechanisms. Like others, for ethanol sensing, initial oxygen adsorption plays a crucial role in determining the electrical performance of the sensing layer. At some elevated temperature, oxygen molecule dissociates into oxygen ions. When an oxygen ion is adsorbed on the sensing layer's surface, it reduces the concentration of conduction electrons, resulting in an increase in the sensing layer's resistance.

O_2^-, O^{2-}, and O^- are adsorbed on the sensing layer's surface at some elevated temperature. The temperature-dependent chemisorption of oxygen on the surface is shown below:

$$O_2 \, (gas) \leftrightarrow O_2 \, (ads) \tag{11.1}$$

$$O_2 \text{ (ads)} + e^- \leftrightarrow O_2^- \tag{11.2}$$

$$O_2^- \text{ (ads)} + e^- \leftrightarrow 2O^- \text{ (ads)} \tag{11.3}$$

After dissociation, ethanol produces acetaldehyde and hydrogen as mentioned in Equation 11.4:

$$CH_3CH_2OH \rightarrow CH_3CHO + H_2 \tag{11.4}$$

Ethanol may also dissociate in the manner given in Equation 11.5:

$$CH_3CH_2OH \rightarrow C_2H_4 + H_2O \tag{11.5}$$

11.4.1.2 Acetone-Sensing Mechanism

Similarly, at an elevated temperature, oxygen molecules are adsorbed on the sensing layer surface and snatch electrons from the conduction band, producing one of the chemisorbed oxygen species: O_2^-, O^{2-}, or O^-. The production of these species depends fully on the surrounding temperature. O^- is most stable form in the temperature range 300°C–350°C. The reaction kinetics can be dictated as follows:

$$O_2 \text{ (gas)} \leftrightarrow O_2 \text{ (ads)} \tag{11.6}$$

$$O_2 \text{ (ads)} + e^- \leftrightarrow O_2^- \tag{11.7}$$

$$O^{2-} \text{ (ads)} + e^- \leftrightarrow 2O^- \text{ (ads)} \tag{11.8}$$

$$O^- \text{ (ads)} + e^- \leftrightarrow O^{2-} \tag{11.9}$$

11.4.1.3 Methanol-Sensing Mechanism

The methanol-sensing mechanism is similar to the ethanol-sensing mechanism but in a different temperature regime. First, oxygen is chemisorbed on the sensing layer's surface at an elevated temperature. As mentioned above, during the chemisorption, atmospheric oxygen forms ionic species such as O_2^- and O^- by capturing the electrons from the conduction band of the sensing film. The reaction kinetics is as follows:

$$O_2 \text{ (gas)} \leftrightarrow O_2 \text{ (ads)} \tag{11.10}$$

$$O_2 \text{ (ads)} + e^- \leftrightarrow O_2^- \text{ (ads)} \tag{11.11}$$

$$O_2^- \text{ (ads)} + e^- \leftrightarrow O^- \text{ (ads)} \tag{11.12}$$

The reactions between methanol vapor and ionic oxygen can be expressed as

$$CH_3OH \text{ (gas)} + O^- \text{ (ads)} \rightarrow HCOH + H_2O + e^- \tag{11.13}$$

$$HCOH + O \text{ (bulk)} \rightarrow HCOOH + O \text{ (vacancies)} \tag{11.14}$$

$$CH_3OH \text{ (gas)} + O_2^- \text{ (ads)} \rightarrow HCOOH + H_2O + e^- \tag{11.15}$$

Thus, formaldehyde is formed when the sensor is exposed to methanol.

11.5 Challenges in Using Nanotechnology for VOC Detection

The application of nanotechnology can play a significant role in the gas-sensing phenomenon of a thin film. Reduced particle size plays an important role in gas sensing (VOC or others) as the number of nodes to interact with gases increases. This phenomenon has been explained by many researchers and established that a sensor consists of partially sintered crystallites that are connected to their neighbors by necks. These interconnected grains ultimately agglomerate and form a larger boundary that is allied to the neighbors by grain boundaries (GB).

As a basic gas sensing phenomenon, the oxygen is adsorbed on the surface of the sensing layer and pulls the electrons from the surface toward itself due to its strong electron affinity. As the electrons are pulled out from the surface, a depletion layer of thickness L (also called the Debye length) is formed. Now, apart from the grain boundary barrier, an electron has to overcome the depletion layer also as it moves on the surface of the thin film. As a result, there is a decrease in the conductivity of the film. The formula for Debye length is given as:

$$L_D = \sqrt{\left(\frac{\varepsilon kT}{e^2 n}\right)}$$

where
  ε is the dielectric constant of the oxide material
  k is the Boltzmann constant
  T is the temperature
  e is the electronic charge
  n is the electron density

Researchers are continuously trying to unfold the science behind nanoen-gineering and have concluded that several factors are responsible for its superiority:

- Very high surface-to-volume ratio: It is a well-known fact that nano-crystalline materials have much higher surface-to-volume ratio than does any other material, due to the presence of multinodes in the sensing layer. It is already established that when particle size is ~30 nm, 5% of the atoms are on the sensor surface, but when the particle size is reduced to 10 nm, 30% atoms are on the surface. Thus, it is clear that the smaller is the size of the particle, the greater is the percentage. With 3 nm, 50% atoms have been found on the surface. Due to this, we acquire a huge number of dangling bonds on the surface of the material, which in turn increases the surface energy exponentially. The effective sensitive area is increased many-fold, and thus the sensitivity is increased.

- Lowering of operating temperature: The activation energy of gas sorption is lowered due to the availability of a large number of nodes; therefore, the operating temperature of the sensor is also lowered. This is because the same number of gas molecules can be adsorbed on the surface at a lower temperature (e.g., for ordinary thick-film micro-crystalline sensors the operating temperature may be as high as 400°C, but for nanocrystalline materials it can be lowered down to ~150°C). Also notable is the fact that the material shows greater response at the same temperature compared to a microcrystalline thin film.

- Precise control over sensitivity by size reduction: By tailoring the material—that is, by adjusting the particle size, porosity, thickness of the film, etc.—according to our needs, we can increase the sensitiv-ity, selectivity, and response time of a sensor.

11.6 VOC Sensing with Different Nanostructures

VOC detection is one of the key areas of research across the globe. Developing the functional material to be used as the sensing material is a critical task. Alcohol sensors are widely used in industries such as petrochemical, drug, and food. Several nanoforms (nanorods, nanotubes, nanowires, nanoflakes) of semiconducting (group II–VI) metal oxides like ZnO and TiO_2 are depos-ited for the use of solid-state gas sensors. VOCs that are responsible for spoil-age of fruits and vegetables during storage also need to be detected in an efficient manner.

Type of Nanostructure	Material	Method	Precursors Used	Crystallite Size	Sensing Parameter (VOC)	Sensor Operating Temperature (°C)	Conc. (ppm)	Response Magnitude	Res. Time/ Rec. Time (s)	Reference
Nanowires	ZnO	Photolithography to partially etch and evaporation process	ZnO:Ga film, HCl, etching mask, alumina boat, zinc vapor, and argon and oxygen gases	Length and diameter of these laterally grown ZnO NWs were 6 µm and 30 nm	Ethanol	300°C	1500	61[b]	-/-	[1]
	In_2O_3	Carbothermal reduction reaction	Indium oxide and active carbon at 1000°C in flowing nitrogen atmosphere	Diameter from 60 to 160 nm and length ranging from 0.5 to a few micrometers	Ethanol	370°C	100	2.00[b]	10/20	[2]
	V_2O_5	Melt quenching method and hydrothermal reaction	V_2O_5 powder, deionized water, yarns	20–100 nm in diameter and hundreds of micrometer long	Ethanol	330°C	1000	9.09[b]	-/-	[3]
	CuO	Oxidation reaction of copper plate	Copper plate, alcohol	Diameter of 100–400 nm and length of around several micrometers	Ethanol	240°C	1000	1.5[b]	-/-	[4]
	In_2O_3	Transport and condensation method	(In_2O_3 powder, purity 99.999%), alumina substrate, Ar flux	Lateral dimension of the thinnest NWs is about 20 nm. Lateral dimension averages 110 ± 20 nm	Acetone	400°C	25	7[b]	-/-	[5]

(Continued)

Type of Nanostructure	Material	Method	Precursors Used	Crystallite Size	Sensing Parameter (VOC)	Sensor Operating Temperature (°C)	Conc. (ppm)	Response Magnitude	Res. Time/ Rec. Time (s)	Reference
	SnO_2	Vapor–liquid–solid method	Au, Sn, Oxygen	100 nm scale NWs of some microns' length and less than 50 nm width	Formaldehyde	270°C	10	61[b]	90/150	[6]
	V_2O_5	Melt quenching method and hydrothermal reaction	V_2O_5 powder, deionized water, yarns	20–100 nm in diameter and hundreds of micrometer long	Toluene	330°C	1000	2.00[b]	–/–	[3]
Nanorods	SnO_2	Hydrothermal route	7 mL $SnCl_4$ (0.5 mL/L)/NaOH (5 mL/L) + alcohol-water mixture	04–15 nm diameter 100–200 nm length	Ethanol	200°C	300	31.4[b]	1/1	[7]
	ZnO	Hydrothermal process or aqueous solution methods	50 mL zinc chloride dihydrate + ammonium hydroxide (NH_4OH)	80–200 nm diameter	Acetone	300°C	100	30[b]	~5	[8]
	ZnO	Seed-relative hydrothermal method and sol–gel method	$(Zn(Ac)_2·2H_2O)$ and methenamine $(C_6H_{12}N_4)$ (0.02 M)	200–300, 80–100, and 20–40 nm diameters	Formaldehyde	UV- assisted	50	11.7[d]	–/–	[9]
	WO_3	Thermal evaporation	WO_3 and graphite powders	Diameter of ~100 nm	Acetone	300°C	200	108.25[a]	237	[10]
	$Au\text{-}WO_3$		WO_3 and graphite powders, $PdCl_2$ solution	Diameter of ~100 nm	Acetone	300°C	200	131.26[a]	98	[10]
	$Pd\text{-}WO_3$		WO_3 and graphite powders, $HAuCl_4$ solution	Diameter of ~100 nm	Acetone	300°C	200	138.62[a]	120	[10]

(Continued)

Type of Nanostructure Material	Method	Precursors Used	Crystallite Size	Sensing Parameter (VOC)	Sensor Operating Temperature (°C)	Conc. (ppm)	Response Magnitude	Res. Time/ Rec. Time (s)	Reference
Au α-Fe$_2$O$_3$ nanorods	Microwave irradiation method	Alkylamines and tetraethylammonium hydroxide (TEAOH)	6–10 nm diameter and 190–200 nm length	Acetone	270°C	100	45[b]	–/–	[11]
ZnO	Hydrolysis	Zn(Ac)$_2$ and LiOH· Zn(Ac)$_2$, ethanol and water, LiOH, Al$_2$O$_3$ tube, C$_6$H$_{12}$N$_4$, Zn(NO$_3$)$_2$	Diameters of 10–30 nm and lengths about 1.4 μm	Ethanol	350°C	100	18.29[b]	~10/~20	[12]
ZnO	Double-bend FCVA deposition system and chemical synthesis	Pure Zinc, Zn(NO$_3$)$_2$·6H$_2$O, C$_6$H$_{12}$N$_4$	Diameter of 30–50 nm	Ethanol	280°C	250	2.3[b]	16/120	[13]
ZnO	Hydrothermal process	Cetyl trimethyl ammonium bromide (CTAB), deionized water, zinc powder	Diameters of 40–80 nm and lengths about 1 μm	Ethanol	330°C	100	13[b]	–/–	[14]
ZnO	Hydrothermal process or aqueous solution methods	50 mL zinc chloride dihydrate + ammonium hydroxide (NH$_4$OH)	80–200 nm diameter	Benzene	300°C	100	2.5[a]	–/–	[8]

(Continued)

Type of Nanostructure	Material	Method	Precursors Used	Crystallite Size	Sensing Parameter (VOC)	Sensor Operating Temperature (°C)	Conc. (ppm)	Response Magnitude	Res. Time/ Rec. Time (s)	Reference
	In_2O_3	Solvothermal method	10 mL oleic acid, 5 mL n-amyl alcohol, 20 mL n-Hexane, 2 mL 0.5 mol/L $In(NO_3)_3$, and 5 mol/L NaOH solutions	Diameter ~20–50 nm and length more than 100 nm	Ethanol	330°C	50	115[b]	6/11	[15]
	CuO	Hydrothermal method	$Cu(OAc)_2 \cdot 2H_2O$, Cetyl trimethyl ammonium bromide (CTAB), NaOH	80 nm in diameter and 700 nm in length	Ethanol	400°C	100	1.5[b]	–/–	[16]
	ZnO	Chemical bath deposition technique	$C_4H_6O_4Z_n$ and [HMT $((CH_2)_6N_4$, deionized water	~70 nm diameter and ~500 nm length	Ethanol	200°C	1530	89–94[c]	14/70	[17]
Nanotubes	TiO_2	Anodization process	HF (0.5 wt.%), 0.5 wt.% NH4F + ethylene glycol	Diameters in the range of 30–80 nm	Ethanol	200°C	5000	1.6[b]	–/–	[18]
	ZnO	Sonochemical method	6 mL of 1 M $Zn(NO_3)_2$, 6 mL of 10 M NaOH, 15 mL alcohol, 10 mL ethylenediamine	Length and diameter of NTs were 1.5–2 μm and 250 nm, walls were about 30 nm thick	Ethanol	300°C	100	27.5[b]	–/–	[19]

(Continued)

Type of Nanostructure	Material	Method	Precursors Used	Crystallite Size	Sensing Parameter (VOC)	Sensor Operating Temperature (°C)	Conc. (ppm)	Response Magnitude	Res. Time/ Rec. Time (s)	Reference
	4 wt.% Ce-doped α-Fe$_2$O$_3$	Electrospinning method	Fe(NO$_3$)$_3$·6H$_2$O, Ce(NO$_3$)$_3$·6H$_2$O, N,N-dimethyl formamide (DMF) and ethanol, PVP, Mw¾ 1,300,000	Thickness of the sensing film was measured to be about 300 μm	Acetone	240°C	50	21[b]	3/8	[20]
	7 wt.% La-doped α-Fe$_2$O$_3$	Electrospinning method	Fe(NO$_3$)$_3$·6H$_2$O, Ce(NO$_3$)$_3$·6H$_2$O, N,N-dimethyl formamide (DMF) and ethanol, PVP, Mw¾ 1,300,000	Thickness of the sensing film was measured to be about 300 μm	Acetone	240°C	50	26[b]	3	[20]
	TiO$_2$	Electrochemical anodization route	NH$_4$F, ethylene glycol, water	—	BTX	200°C	400	11.06 6.41, and 4.57[c]	36.02, 38.43, 26.70/ 139.95, 125.30, 106.61	[21]

a $(R_a/R_g \, \%)$.
b $(R_a/R_g \text{ or } R_g/R_a)$.
c $([R_a - R_g]/R_a, \%)$.
d $([I_a - I_g]/I_a)$.

11.7 Measurement Techniques

A simple gas measurement setup for VOC sensing is presented in Figure 11.1. Acetone vapors were introduced in the chamber by bubbling IOLAR-grade N_2 (carrier gas) through the acetone kept in a conical bottle. Assuming that the carrier got saturated with the acetone, desired concentrations of vapor were obtained in the sensing chamber by controlling the flow rate of N_2 through acetone and adding IOLAR-grade N_2 (diluents) through a separate gas flow line. The homogeneous mixture carrying the desired percentage of the target vapor was fed into the chamber with a flexible PVC pipe. To measure sensor resistance electrodes were connected to an Agilent U34411A multimeter with Agilent GUI data-logging software (v2.0).

The final concentration of acetone vapor was calculated using this formula:

$$\%\text{Anesthetic concentration} = \frac{\text{SVP} \times \text{Fraction diverted}}{(\text{PBar} - \text{SVP}) \times (1 - \text{Fraction diverted})} \times 100\%$$

Here, SVP is the saturated vapor pressure, which is the partial pressure of a substance in vapor phase at equilibrium with its liquid phase. It increases rapidly as boiling point approaches. PBar is the standard atmospheric pressure (760 mmHg).

It is needed to adjust the fraction of the total fresh gas that is diverted into the vaporizing chamber. Typically, only a small percentage of the total fresh gas flow enters the chamber.

Gas entering the vaporizing chamber is always fully saturated by the time it leaves. Vapor molecules added to the gas molecules entering the chamber increase the volume of the gas leaving the chamber. The extent of this increase in the volume and the amount of vapor picked up per mL of gas entering depends on the volatility of the agent in proportion to atmospheric pressure.

The exact percentage to be diverted depends on the volatility of the agent (SVP), the proportion of the fresh gas diverted into the vaporizing chamber, and barometric pressure.

For finding out the optimum operating temperature of sensing, sensor resistance was measured in presence of air and acetone vapor and the corresponding response magnitude was calculated as a function of temperature. Response magnitude (RM) of the sensor was calculated by using the following formula:

$$RM = \frac{R_a - R_g}{R_a} \times 100\%$$

where

R_g is the sensor resistance in the test gas (i.e., acetone vapor mixed with pure N_2)

R_a is the sensor resistance in air

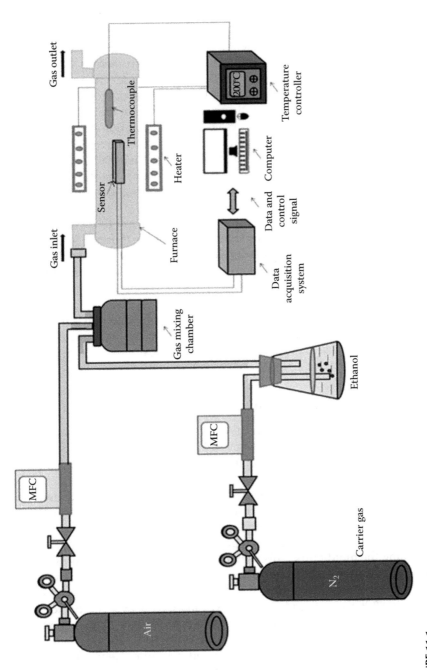

FIGURE 11.1
Gas measurement setup for VOC sensing.

11.8 Protecting the Sensor from the Environment

Miniaturization of electronic devices needs complete integration to protect them from the harsh ambient elements. Therefore, packaging is a challenging issue in the microelectronic industry. A good instrument is that which not only has precision but also protects the inner module with intelligence. For example, a gas sensor needs an open window to make contact with the gases, but at the same time, it should be protected from the external environment. This is a big challenge for sensor encapsulation because when such a small device contains an open window, there is a tendency for it to break up when any external force is applied to it. Generally, a TO5-type package is used for a gas sensor. It is critical to clamp the device exactly below the open window, which is the main functional area of the whole package.

Wire bonding is used to take the output from the sensor. To make this wire bond, the semiconductor (generally) device needs to be clamped judiciously. Generally, steel clamping has been used, though there is a huge clamping force associated with it. Recent developments, however, suggest that the clamping force can be decreased significantly with the soft film compensation technique.

Numerous techniques have been adopted till date by different scientists to meet the present technological demand. Few of them are highlighted here to present an overall idea about the packaging technique.

11.8.1 Principle of Film-Assisted Molding Technology

Boschman's film-assisted molding (FAM) technology supports the market demand of sensor packaging. Moreover, FAM technology is cost-effective and offers excellent performance reported till date. This technology complies with the necessity of encapsulation.

Transfer molding is a fundamental packaging process, which uses an epoxy molding compound (EMC). Initially, the process bore many problems including deformation and wear and tear.

Boschman's FAM technology is a type of transfer molding technology, with a film lining the mold parts. This technology has proven track records in different encapsulation techniques. In this technology, components are released from the mold with a clear surface. FAM technology is generally used in quad flat no leads (QFN)-, ball grid array (BGA)-, and microelectromechanical systems (MEMS)-based sensors with windows open. Figure 11.2a–f show the main process steps of the FAM cycle.

Figure 11.2 depicts the in-taking of a brand-new film. Generally, a Teflon film is preferred as it has high temperature-withstanding capability. By using

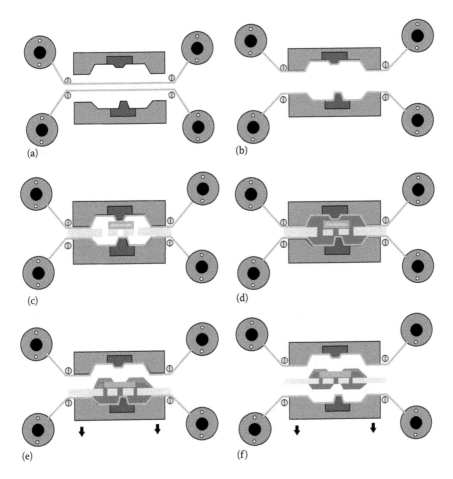

FIGURE 11.2
(a) Fresh film intake. (b) Mold to the film position and suction of film into the mold's inner surfaces by vacuum. (c) Lead frames/substrates loading and closing of the mold. (d) Cavity filling by transferring the molding compound. (e) Opening the molded components. (f) Unloading the molded products.

this film, the encapsulated component never touches the mold, it is protected in every cycle, and the stiction problem is also overcome. The filler constraints are removed. Then, the new film is stuck to the mold's inner surface, after which the substrate is loaded inside the gap and then mold is closed and the mold cavity is filled with EMC. Finally, the mold is opened and the molded components are released. This entire process is one cycle. The vacuum is released before the next cycle starts. The film is forwarded by exactly one length to start the process again with the new one. In this method, no mold cleaning is required.

11.8.2 Applications of The FAM Technology

The FAM technology can be divided as adhesive film technology (AFT) and seal film technology (SFT), both of which are one-film technologies. Two-film technologies also exist, but discussing those is beyond the scope of this chapter.

According to the applications, the adhesive and the seals film have some differences.

Parameter	Seal Film Technology	Adhesive Film Technology
Film material	Teflon	Polyester
Elasticity	High	Low
Film thickness	50–100 μm	50 μm
Melting temperature	362°C	230°C
Film length	110–200 m	100 m
Film width	Application dependent	

AFT is used for one-side molded packages (QFN, single in-line package [SIP], etc.) and to stop compound flash on die pads. The term seal means "vacuum," and SFT is used with vacuum forming into the mold cavities. This is applicable for both sticky and clear compounds. This technique also prevents compound flash, and no ejector is required.

11.8.3 Seal Film Technology

SFT is basically a low-cost encapsulation method whatever may be the application (see Figure 11.3). The film prevents the flow of molding compound through the open window and facilitates the formation of cage-free surfaces. Die encapsulation without compound flash is the key aspect of this technology, which makes it low-cost.

Boschman has developed a film technique so that the film can be used on both sides of a device, with a sealing film on the top and either a sealing or

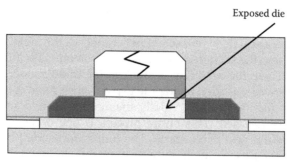

FIGURE 11.3
Encapsulation of exposed die.

an adhesive film on the bottom, depending on the device's application. The technique with sealing film on both sides of the device is particularly useful for lead-frame–mounted chips requiring package structuring on both sides.

11.8.4 Adhesive Film Technology

AFT is mainly used for large-area packages (LLPs, QFN, and array). A temperature-resistant and low-cost polymer is used as an adhesive film (see Figure 11.4). It provides lamination as well as de-lamination in the backend process flow. The adhesive film is temperature-dependent. At low temperatures, it has a little bit of adhesion, but as the temperature rises, the melting characteristics of this layer increase, providing good quality of adhesion to the lead frame and building up resistance against the flow of compound, thus preventing any bleeding of the encapsulated material.

Boschman's AFT is one of the prototypes of this technique. It is used to pre-tape the back side of lead frames.

11.8.5 Release Film Technology

In this technology, the mold area does not come into contact with the mold compound. Release film technology (see Figure 11.2) does not require ejector pins at the film side, the ejection process is smooth even with slurry molding compounds. A release agent can facilitates releasing of the molded device from the space and also decreases the stiction between the molding compounds and the substrate. The film guards the mold surface and the encapsulant. By this technique, contamination due to EMC can be avoided.

11.8.6 Wafer-Level Transfer Molding with FAM

Not only the chip-level but a wafer-level packaging is also possible. M. Brunnbauuer worked on wafer-level packaging for the first time [4]. Encapsulating the chips with the molding compound is the key factor of this packaging technique. Wafer-level packaging holds good in gas sensor array

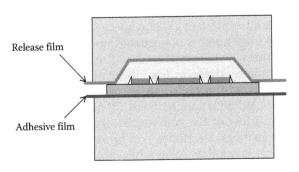

FIGURE 11.4
The adhesive and release film.

FIGURE 11.5
Wafer-level packaging with molded wafer (a) before molding and (b) after de-bonding.

application. Although compression molding is a good candidate for wafer-level molding, transfer molding is preferred for its better performance and reliability. There is a good combination of transfer molding and Boschman Assisted Film technology in MEMS packaging. Figure 11.5 shows the schematic of wafer-level packaging with molded wafer and windows open.

11.9 Conclusion

Nanocrystalline materials are single- or multiphase solids with the grain size of a few nanometers (10^{-9} m), typically less than 100 nm. As the grain size is small, the surface-to-volume ratio is large and there are numbers of GB in the bulk. The high surface-to-volume ratio makes it useful for different applications, for example, optoelectronic devices, biosensors, and nanomachines. Smaller particle size (and higher surface-to-volume ratio) will enhance the sensitivity by increasing the active area of gas sensing in the film *and also* by promoting surface adsorption of gaseous species by increasing the number of dangling bonds. A huge number of fractional atoms reside in the GB, and they form dangling bonds with the target species. The properties of nanocrystalline materials are completely different from those of the polycrystalline one due to the presence of these dissatisfied atoms on the surface. A nanostructured material mainly includes (1) nanoparticles, (2) nanocrystalline materials, and (3) nanothin film. Most of the commercial applications of nanostructured materials include coatings, electrodes, and functional nanostructures.

Lattice mismatch is another issue while designing a sensor. Less lattice mismatch is preferred and there are still so many research needs to be done in future.

References

1. C.L. Hsu, T.J. Hsueh, S.J. Chang, I.C. Chen, Laterally grown ZnO nanowire ethanol gas sensors, *Sens. Actuators B* 126 (2007) 473–477.

2. W.C.C. Xiangfeng, J. Dongli, Z. Chenmou, Ethanol sensor based on indium oxide nanowires prepared by carbothermal reduction reaction, *Chem. Phys. Lett.* 399 (2004) 461–464.

3. S. Wei Jin, L. An, W. Chen, S. Yang, C. Zhao, Y. Dai, Enhancement of ethanol gas sensing response based on ordered V_2O_5 nanowire microyarns, *Sens. Actuators B* 206 (2015) 284–290.

4. A.G.P. Raksa, T. Chairuangsri, P. Mangkorntong, N. Mangkorntong, S. Choopun, Ethanol sensing properties of CuO nanowires prepared by an oxidation reaction, *Ceram. Int.* 35 (2009) 649–652.

5. A. Vomiero, S. Bianchi, E. Comini, G. Faglia, M. Ferroni, N. Poli, G. Sberveglieri, In_2O_3 nanowires for gas sensors: Morphology and sensing characterisation, *Thin Solid Films* 515 (2007) 8356–8359.

6. I. Castro-Hurtado, J. Herrán, G.G. Mandayo, E. Castaño, SnO_2-nanowires grown by catalytic oxidation of tin sputtered thin films for formaldehyde, *Thin Solid Films* 520 (2012) 4792–4796.

7. X.Y. Xue, Y.J. Chen, Y.G. Wang, T.H. Wang, Synthesis and ethanol sensing characteristics of single crystalline SnO_2 nanorods, *Appl. Phys. Lett.* 87 (2005) 233503–233505.

8. Y. Zeng, T. Zhang, M. Yuan, M. Kang, G. Lu, R. Wang, H. Fan, Y. He, H. Yang, Growth and selective acetone detection based on ZnO nanorod arrays, *Sens. Actuators B* 143 (2009) 93–98.

9. L. Peng, J. Zhai, D. Wang, Y. Zhang, P. Wang, Q. Zhao, T. Xie, Size-and photo-electric characteristics-dependent formaldehyde sensitivity of ZnO irradiated with UV light, *Sens. Actuators B* 148 (2010) 66–73.

10. S. Kim, S. Park, C. Lee, Acetone sensing of Au and Pd-decorated WO_3 nanorod sensors, *Sens. Actuators B* 209 (2015) 180–185.

11. L.M.P. Gunawan, J. Teo, J. Ma, J. Highfield, Q. Li, Z. Zhong, Ultra high sensitivity of $Au/1D$ α-Fe_2O_3 to acetone and the sensing mechanism, *Langmuir* 28 (2012) 14090–14099.

12. X.-N.Y. Li-Jian Bie, J. Yin, Y.-Q. Duan, Z.-H. Yuan, Nanopillar ZnO gas sensor for hydrogen and ethanol, *Sens. Actuators B* 126(126) (2007) 604–608.

13. M.Z. Ahmad, A.Z. Sadek, K. Latham, J. Kita, R. Moos, W. Wlodarski, Chemically synthesized one-dimensional zinc oxide nanorods for ethanol sensing, *Sens. Actuators B* 187 (2013) 295–300.

14. Y.C.J. Xu, Y. Li, J. Shen, Gas sensing properties of ZnO nanorods prepared by hydrothermal method, *J. Mater. Sci.* 40 (2005) 2919–2921.

15. Y.C.J. Xu, J. Shen, Ethanol sensor based on hexagonal indium oxide nanorods prepared by solvothermal methods, *Mater. Lett.* 62 (2008) 1363–1365.

16. C. Wang, X.Q. Fu, X.Y. Xue, Y.G. Wang, T.H. Wang, Surface accumulation conduction controlled sensing characteristic of p-type CuO nanorods induced by oxygen adsorption, *Nanotechnology* 18 (2007) 145506–145510.

17. S. Roy, N. Banerjee, C.K. Sarkar, P. Bhattacharyya, Development of an ethanol sensor based on CBD grown ZnO nanorods, *Solid-State Electron.* 87 (2013) 43–50.

18. E.S.N. Kilinc, Z.Z. Ozuturk, Fabrication of TiO_2 nanotubes by anodization of Ti thin films for VOC sensing, *Thin Solid Films* 520 (2011) 953–958.
19. Y.-J. Chen, C.-L. Zhu, G. Xiao, Ethanol sensing characteristics of ambient temperatures on chemically synthesized ZnO nanotubes, *Sens. Actuators B* 129 (2008) 639–642.
20. C. Liu, H. Shan, L. Liu, S. Li, H. Li, High sensing properties of Ce-doped α-Fe_2O_3 nanotubes to acetone, *Ceram. Int.* 40 (2014) 2395–2399.
21. K. Dutta, P.P. Chattopadhyay, C.-W. Lu, M.-S. Ho, P. Bhattacharyya, A highly sensitive BTX sensor based on electrochemically derived wall connected TiO_2 nanotubes, *Appl. Surf. Sci.* 354 (2015) 353–361.

12

Nanoscience with Graphene

Angsuman Sarkar

CONTENTS

12.1 Introduction

In the 1930s, Landau and Peierls claimed that a 2D crystal in a strict sense is thermodynamically unstable and hence cannot exist [1,2]. This was supported by Mermin et al. with experimental observations [3]. Two-dimensional crystals such as a one-atom-thick layer melt at lower temperatures as their thickness decreases. Therefore, a thin film containing a dozens of atomic-layer-thick material becomes unstable. For these reasons, layers having a thickness of one atom, also known as atomic monolayers, which were until now considered as the fundamental element of 3D substrates, normally synthesized epitaxially on 3D substrates. Until 2004, 2D materials without a 3D base were presumed not to exist. However, in 2004, graphene was first discovered and demonstrated experimentally, and other 2D crystals such as isolated and unattached monolayer boron nitride were also discovered [4]. Due to their high crystalline quality, 2D crystals are typically found lying on substrate materials in liquid suspension, which is not crystal and where carrier can move without scattering for thousands of interatomic distances. Subsequent experimental results have verified and validated that the charge carriers are nothing but Dirac fermions having zero mass [5]. As a result, carbon-based materials such as graphene and carbon nanotubes (CNT) have recently been rewarded with high research attention. The objective of this chapter is to describe the advancements and developments of nanoscience using graphene. The electronic and physical structure of graphene is of primary importance for the analysis of graphene-based devices or applications. Moreover, an analysis of the electronic band structure of graphene leads toward the understanding and derivation of CNT. This chapter concludes with the application of graphene nanoribbons (GNR).

Geim and Novoselov demonstrated graphene for the first time in 2004 [4]. Graphene is considered to be a promising candidate for next-generation post-CMOS material. However, there is a long way to go for commercial production using graphene. This can be attributed to reliability issues along with production of high-quality samples with controllable bandgap in a scalable fashion. Despite this, people continue to research on graphene, incorporating different novel techniques, thoughts, and schemes. Graphene has many crucial properties, such as tremendous mechanical strength characterized by the high values of its Young's modulus, fracture strength, and thermal conductivity. It is possible to produce 2D films with graphene. Moreover, at room temperature (300 K) it exhibits the quantum Hall effect [6]. Furthermore, graphene demonstrates unusual electronic properties like massless Dirac fermions with linear dispersions and transport energy gaps, and supports ballistic transport of charge carriers with an electric field that is ambipolar in nature [7]. Due to these exclusive properties, graphene has garnered attention and expectation for scientific as well as application concerns.

Carbon is a group IV element that produces many molecular compounds and crystalline solids. There are four valence electrons in carbon. The interaction among those electrons produces various carbon allotropes (modified structure of an element in the same phase of matter, such as in different solid forms). In elemental carbon atoms, the 2s and 2p orbitals are occupied by valence electrons. Figure 12.1 shows that for the formation of a crystal by gaining excitation energy from its neighbor nucleus, one 2s electron moves to p_z orbital. Interaction or mixing among these 2s and 2p orbitals between neighboring atoms leads to hybridization and new hybridized orbitals. Various types of hybridization result in various forms of carbon allotropes as shown in Table 12.1.

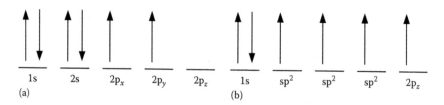

FIGURE 12.1
Electrons, their relative spins, and hybrid orbitals in (a) elemental carbon and (b) graphene.

TABLE 12.1

Graphene Allotropes

Allotropes	C_{60} Bucky Balls	Carbon Nanotubes	Graphene	Graphite	Graphite
Dimension	0D	1D	2D	3D	3D
Electronic properties	Semiconductor $E_g = 1.9$ eV	Metallic or semiconductor $E_g = \sim 0.3–1.1$ eV	Semi-metallic zero-gap semiconductor	Metallic	Insulator
Hybridization	sp²	sp²	sp²	sp²	sp³
Structure	Spherical	Cylindrical	Planar sheet	Stacked planar sheet	Crystalline

There are three sp² hybridized orbitals in graphene, resulting from the interactions between 2s with the $2p_x$ and $2p_y$ atomic orbitals. As a result of these sp² hybridizations, three σ-bonds were produced, which are considered to be the strongest covalent bonds. This is the reason graphene and CNT offer great mechanical strength. The remaining $2p_z$ electrons are weakly bound to the nucleus and form the covalent bonding called π-bonds. It is worth mentioning that in contrast to the σ-bonds, where the electrons are localized along the plane connecting the carbon atoms, in a π-bond the electrons are delocalized where the distribution of electrons cloud is normal to the plane connecting the carbon atoms. The electronic properties of graphene are primarily determined by these delocalized electrons.

Graphene is a planar allotrope of carbon. In graphene, carbon atoms are present in a honeycomb structure with covalent bonding. If graphene is wrapped into a sphere, it will produce Bucky balls. If it is rolled into a sheet, it will produce CNTs. If several such sheets of graphene are stacked, they will produce graphite. Graphene nanoribbon (GNR) is a small piece of graphene sheet cut as a ribbon.

12.2 Lattice Structure of Graphene

Figure 12.2 shows the direct lattice structure of graphene by using a ball-and-stick model. In the honeycomb structure, balls denote the carbon atoms and sticks denote the σ-bonds. The bond length between carbon–carbon atoms is $a_{C–C} \approx 1.42$ Å. The honeycomb lattice can be considered a Bravais lattice with a basis of two atoms, shown as A and B in Figure 12.2, contributing a total of two π– electrons per unit cell, determining the electronic properties of graphene. The Bravais lattice with the corresponding primitive cell

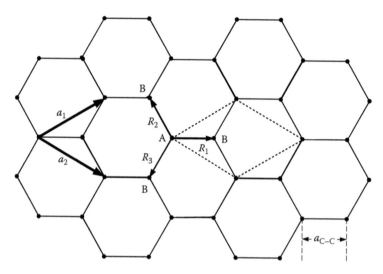

FIGURE 12.2
Honeycomb lattice structure of graphene showing the primitive unit cell as a parallelogram, with dotted line having a basis of two atoms A and B.

is considered as an equilateral parallelogram with side $a = \sqrt{3}a_{C-C} \approx 2.46$ Å. The primitive unit vectors defined in Figure 12.2 is given by

$$a_1 = \left(\frac{\sqrt{3}a}{2}, \frac{a}{2}\right), \quad a_2 = \left(\frac{\sqrt{3}a}{2}, \frac{a}{-2}\right) \quad \text{with } |a_1| = |a_2| = a$$

The distance of separation between a type A atom and the nearest neighbor type B atom is given by

$$R_1 = \left(\frac{a}{\sqrt{3}}, 0\right), \quad R_2 = (-a_2 + R_1) = \left(-\frac{a}{2\sqrt{3}}, -\frac{a}{2}\right),$$

$$R_3 = (-a_1 + R_1) = \left(-\frac{a}{2\sqrt{3}}, \frac{a}{2}\right) \quad \text{with } |R_1| = |R_2| = |R_3| = a_{C-C}$$

Figure 12.3 depicts the reciprocal lattice structure of graphene, which is also a honeycomb lattice, but it is rotated 90° with respect to the direct lattice. The lattice vectors in the reciprocal lattice structures are given by

$$b_1 = \left(\frac{2\pi}{\sqrt{3}a}, \frac{2\pi}{a}\right), \quad b_2 = \left(\frac{2\pi}{\sqrt{3}a}, -\frac{2\pi}{a}\right) \quad \text{with } |b_1| = |b_2| = \frac{4\pi}{\sqrt{3}a}$$

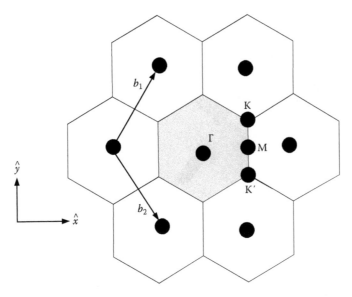

FIGURE 12.3
Reciprocal lattice structure of graphene. Shaded hexagon shows the first Brillouin zone marked with high symmetry points Γ, M, and K located at the center, midpoint of the side, and corners of the shaded hexagon, respectively. The K′-point which is equivalent to the K-point for most of the purposes is also located at the corner.

The shaded hexagon in Figure 12.3 with sides $b_{BZ} = |b_1| / \sqrt{3} = 4\pi/3a$ and area $= 8\pi^2/\sqrt{3}a^2$ depicts the important Brillouin zone. Γ-point, M-point, K-point, and K′-point are the points in the Brillouin zone where a high value of symmetry is observed, and are shown at the center, midpoint of the side, and corners, respectively.

12.3 Energy Dispersion and Zero-Bandgap Semiconductor Property of Graphene

Figure 12.4 shows the energy dispersion throughout the Brillouin zone of graphene. The upper and lower halves correspond to the conduction (π^*) band and the valence (π) band, respectively. As the band edges touch each other at Fermi energy (E_F), no bandgap exists for graphene. Therefore, graphene is considered a zero-bandgap semiconductor. It is worth mentioning that in typical metals, E_F lies in the conduction band and in conventional semiconductors, it lies in the finite bandgap. The E_F will depart from its equilibrium 0 eV value either under the non-equilibrium conditions caused by

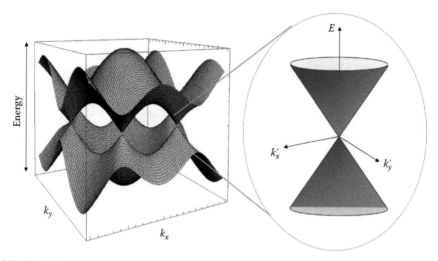

FIGURE 12.4
The nearest neighbor tight-binding band structure of graphene. The magnified figure shows the linear energy dispersion of graphene at the K-point, known as Dirac cone.

the application of electric or magnetic fields or due to extrinsic conditions caused by the presence of impurity atoms. It is known that a linear relationship between the energy and momentum or wavevector defines a massless relativistic particle. As shown in Figure 12.4, a linear dispersion is clearly visible from the band structure of graphene, which results in graphene being a massless (zero effective mass) particle. There are six K-points, as shown in Figure 12.4, known as Dirac points, where the bottom edge of the conduction band and the top edge of the valence band touch each other, producing the famous 3D Dirac cone with linear dispersion, as shown in Figure 12.4.

12.4 Density of States and Carrier Density

Density of states (DOS) is related to the density of mobile electrons or holes present in a material at a given temperature. It is expressed mathematically by

$$g(E)dE = \frac{2g_z dA}{(2\pi)^2/\Omega}$$

where
    g_z is the spin degeneracy, equal to 2 for graphene as six equivalent K-points are shared by three hexagons
    Ω is the area of lattice
    dA, which is the differential area in the k-space, is given by $2\pi k dk$ by considering a circle of constant energy in k-space with radius k

Therefore, DOS can be expressed as

$$g(E) = \frac{2}{\pi}\left|k\frac{dk}{dE}\right| = \frac{2}{\pi}\left|k\left(\frac{dE}{dk}\right)^{-1}\right|$$

as dispersions around K-point are linear and expressed by

$$E(k)_{linear}^{\pm} = \pm\hbar\upsilon_F|k| = \pm\hbar\upsilon_F\sqrt{k_x^2 + k_y^2} = \pm\hbar\upsilon_F k$$

where the reduced plank constant is denoted by $\hbar$ and υ_F is the Fermi velocity. Substituting these in $g(E)$, we obtain

$$g(E) = \frac{2}{\pi(\hbar\upsilon_F)^2}|E| = \beta_g|E|$$

where β_g is a material constant with value equals to 1.5×10^{14} eV^{-2} cm^{-2}. DOS becomes zero when E approaches E_F as there is no bandgap in graphene. This is why graphene is considered as a semi-metal in contrast to metals, which have a large DOS when E approaches E_F.

In order to find the carrier density at a given temperature, the number of states occupied per unit area is determined using the Fermi-Dirac distribution $f(E_F)$.

In order to find the carrier density of electrons, it is possible to write

$$n = \int_0^{E_{max}} g(E)f(E_F)dE = \frac{2}{\pi(\hbar\upsilon_F)^2}\int_0^{E_{max}}\frac{E}{1+e^{(E-E_F)/k_BT}}dE$$

for an intrinsic (undoped) graphene sheet, using Fermi energy 0 eV, the Fermi integral results in $\pi^2/12$. Accordingly, the carrier density is expressed as

$$n_i = \frac{\pi}{6}\left(\frac{k_BT}{\hbar\upsilon_F}\right)^2 \approx 9\times10^5 T^2 \left(\text{electrons cm}^{-2}\right)$$

It is worth mentioning that carrier density here is a square function of temperature, in contrast to conventional semiconductors where carrier density is an exponential function of temperature.

For extrinsic graphene with doped impurities, carrier density is expressed by

$$n_e \cong \frac{\lambda}{\pi}\left(\frac{E_F}{\hbar\upsilon_F}\right)^2$$

where $\lambda = 1.1$ is a fitting parameter.

12.5 Graphene Nanoribbons

GNRs are nanodimensional rectangles made from graphene sheets with widths in the range of a few to tens of nanometers. A long length provides them with a high aspect ratio, and they are considered quasi-1D material. They exhibit metallic or semiconductor-like characteristics and are under intense investigation for exploring their electrical, mechanical, optical, and quantum-mechanical properties [8]. GNRs provide a departure from the conventional graphene sheets by opening the bandgap. This opening of the bandgap unlocks their potential for using them as transistors or for other applications. However, the controllability of the bandgap and the creation of the bandgap in a routine manner is still under thorough investigation. This is the reason that GNRs require intense investigation and are a topic for modern research.

GNRs can be classified into two types—armchair GNR (aGNR) and zig-zag GNR (zGNR)—named after the cross sections at their edges, as shown in Figure 12.5. Moreover, by counting the number of aGNR (N_a) or zGNR (N_z) chains in the width, it is possible to label or identify the GNRs. Figure 12.5 shows the steps to count the number of chains in a 9-aGNR and 6-zGNR. The width of the GNR can also be expressed in terms of the number of chains and is given by

$$\text{aGNR, } w = \frac{N_a - 1}{2}a,$$

$$\text{zGNR, } w = \frac{3N_z - 2}{2\sqrt{3}}a$$

where a is the graphene lattice constant ≈ 2.46 Å.

In a GNR, the nanodimensional width results in quantum confinement of electrons, which restricts the electron's movement along the length of the nanoribbons, in contrast to a graphene sheet where electrons can achieve higher mobility due to the significantly less collisions during its movement in a 2D plane (1D movement confinement). As a result of this quantum confinement and edge effects, there is a significant difference between the electronic band structure of a GNR and a graphene sheet. Son et al. reported the band structure of GNRs by using first principles or tight-binding schemes. Their results reveal that bandgaps are present in zGNRs and are inversely proportional to the nanoribbon width [9]. In a similar manner, aGNRs also provides property of inverse dependence of electronic bandgaps to the width and to the number of armchair chains as well.

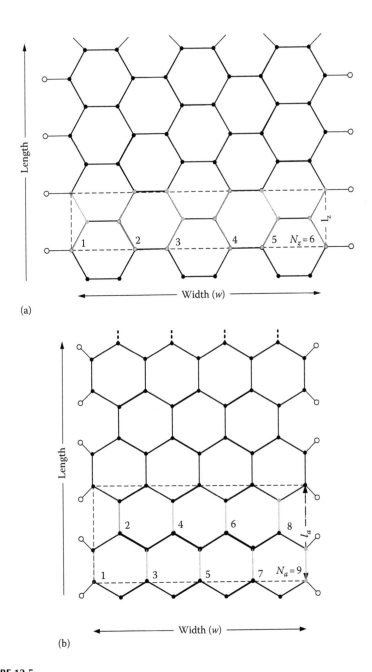

FIGURE 12.5
The finite-width honeycomb lattice structure of (a) 6-zGNR (b) 9-aGNR. The primitive unit cell is shown with dashed box. The open circles at the edges denote passivation atoms, like hydrogen. The gray line denotes the zigzag or armchair chains to determine N_z or N_a, respectively.

A useful semi-empirical expression of width w dependence of bandgap E_g is given by

$$E_g \approx \frac{\alpha}{w + w_o}$$

where α and w_o are fitting parameters. From experimental observations, the value of these fitting parameters is obtained as $\alpha = 0.2$–1 eV and $w_o = 1.5$ nm [10,11].

The GNR behaves like graphene as their bandgap vanishes when the width of the GNR becomes larger than 50 nm.

12.6 Brief Review of the Synthesis of Graphene

The most primitive way of the synthesis of graphene is to synthesize it from graphite and is known as a top-down synthesis approach. Figure 12.6 shows the stripping of graphene layers from a graphite pencil used in everyday life. There are many other top-down synthesis approaches such as stripping of graphite using a scotch tape. In contrast, in a bottom-top approach, the process starts at the atomic scale, with growth achieved by building atom by atom till the desired thickness is obtained. The examples of bottom-top approaches are chemical vapor deposition (CVD), Fischer–Tropsch synthesis, ion implantation, and pyrolysis.

On the other hand, efforts that have been made to produce graphene can be classified in two different approaches: chemical and physical methods.

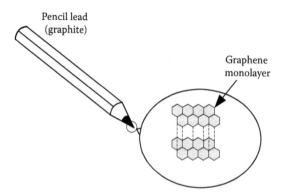

FIGURE 12.6
During writing, the graphite present in a pencil tip can place few monolayers of graphene in the paper. It is worth mentioning that a few monolayers of graphene may not be sufficient for visibility of the writing [12]. (From Ferrari, A.C. et al., *Phys. Rev. Lett.*, 97(18), 187401, 2006.)

It is worth mentioning that it is possible to address the challenges for controllable production of large-scale graphene sheets by employing some variant of the physical or chemical methods.

A most important property of graphene is that it possesses a very high mobility ($1 \times 10^4 - 1.5 \times 10^4$ cm^2 V^{-1} s^{-1}) for carrier transport, which can open up new horizons of performance improvement of electronic devices [13].

In spite of all these processes, exfoliating graphene flakes from graphite and then depositing them on SiO$_2$/Si substrate was considered a basic and standard procedure for the production of graphene sheets. However, the main problems associated with this technology are the deficiency in the controllability of the size and crystalline quality, thus making it a non-scalable technique, suitable for fundamental studies but not for large-scale production. On the other hand, chemical processes are drawing great attention for large-scale production of graphene [14]. Employing dispersion and exfoliation of graphite or its associated intercalated compounds, it is possible to produce graphene by means of chemical route [15]. However, it is worth mentioning that unlike the direct-exfoliation technique, chemical modification destructs considerable graphene electronic structures, thus affecting its distinctive properties. Graphene can be produced from graphite oxide by (1) use of exfoliation or (2) employment of chemical reduction [16,17]. However, it is worth mentioning that when process of reduction is over, graphite flakes tend to agglomerate and then finally precipitate. To avoid agglomeration, poly(4-styrenesulfonic acid) (PSS) is preferred. However, use of polymers such as PSS may deteriorate the electronic property and charge transfer in graphene [18,19]. Moreover, the chemical processes require several steps and 3–5 days of time [20] to insert the intercalants into graphite oxides. Unlike these oxidation and reduction procedures, Coleman et al. demonstrated direct graphite exfoliation using a suitable organic solvent, such as N-methylpyrrolidone (NMP), to produce high-quality graphene monolayer sheets [21] without using intercalants. Recently, in a green approach, hydrothermal dehydration method was proposed [22]. Thermal route to prepare graphene was proposed by Zheng et al. [23]. Nuvoli et al. used HMIH for graphite exfoliation to produce a few layers of graphene [24].

On the other hand, usage of super critical fluids (SCFs), with their unique properties, to produce high-quality graphene is drawing much attention [25–27]. However, it requires high temperature and pressure. Recently, it was demonstrated that when graphene acts together with an electron donor such as aniline and tetrathiafulvalene (TTF) and acceptor molecules such as nitrobenzene and tetracyanoethylene (TCNE), it causes changes in the electronic properties such as shift of the Raman G band [28,29]. Recently, Jang et al. exhibited the SCF exfoliation and subsequent surface modification [30].

Graphene synthesis using CVD started in 2008, in which, depending upon the conditions, it was possible to obtain graphene with thicknesses of a single layer to a few layers [31–44]. In an arc-discharge technique, employed by Volotskova et al. [45], in an extreme high voltage, graphene was produced in a manner similar to welding process.

In the Fischer–Tropsch process, a mixture of carbon monoxide and hydrogen was set to react chemically for obtaining carbon nonmaterials [46,47].

Hibino et al. reported that graphene sheets in SiC, which are epitaxially produced, suffer from a wide thickness variations in their graphene layers [48]. On the other hand, Oshima et al. demonstrated that it is possible to adjust the thickness of graphene sheets, on which the bandgap depends, and to adjust important electronic-property-determining factors such as the width of graphene nanoribbons by choosing the substrate, and regulating the temperature of the process of annealing and time of exposure of the gases to be deposited in a typical CVD or surface segregation techniques [49,50]. By altering the substrate temperature to control the carbon segregation, growth of graphene sheets with a single to a few layers on Ni (111) substrate was demonstrated by Odahara et al. [51].

In an imaginative idea, it has been reported recently that graphene can be obtained by unzipping single-layer CNTs [52]. Employment of ion implantation techniques has also been reported to obtain high-quality graphene [53,54]. Recently, laser synthesis photolysis techniques have become attractive [55].

Recently, carbon nanowalls (CNWs) are drawing a lot of attention for their usage in next-generation electronic devices due to their excellent carrier motilities and low-resistance electrodes. CNWs are vertically standing graphene sheets on substrates. A large surface-to-volume ratio and very high aspect ratio makes CNWs useful in industrial applications [56–58]. The performance of CNWs depends on their electrical properties and morphology, and hence it is essential to make them controllable. The systems using plasma-enhanced chemical vapor deposition (PECVD) technology is the most common one used for controlled synthesis of CNWs [56,58–61]. However, due to the complex growth processes inside the plasma, the growth mechanisms of PECVD process were poorly understood.

12.7 Applications of Graphene

Due to continuous scaling of silicon-based MOSFETs, it has been foreseen that in order to continue device scaling, new materials are required. Coincidentally, stable planar graphene was developed in 2004 and has since become the topic of interest. Two-dimensional monolayer graphene exhibited many interesting properties that can be utilized for device performance improvement. According to recent literatures, a graphene-based transistor exhibited a cut-off frequency equal to 100 GHz [62].

In another study, a high channel mobility equal to 20,000 cm^2 V^{-1} s^{-1} was achieved by employing a top-gated structure [63]. However, the use of graphene is somewhat poor in digital logic design [64] due to its semi-metal nature and corresponding high OFF-state leakage current. However, in radio-frequency applications, where OFF-state current is not the most important issue, graphene-based transistors provide excellent performance. The high standby leakage power is the result of the deficiency of adequate bandgap. Therefore, researchers are putting their effort to introduce sufficient bandgap in graphene in order to convert it into a semiconducting material. It is known that from the application of an external electric field on bilayer graphene an energy bandgap will appear. However, for adequate bandgap, a large electric filed is required. On the other hand, it is easier to integrate semiconducting 1D GNR using existing technologies in order to use their potential in device applications such as MOSFETs [11,65–67], resonant-tunneling diodes (RTDs) [68], and quantum dots [69].

Recently, there has been a rapidly growing interest in the use of graphene as a foundation material to build nanoelectromechanical systems (NEMS), due to its stiffness and light weight. The NEMS resonators provide low internal mass and ultrahigh frequencies as compared to Si-based nanotubes. Therefore, it is expected that further research of graphene-based NEMS can provide answers to challenges like individual atom inertial sensing [70]. On the other hand, results using graphene-ferroelectric memories [71] and lab-on-a-chip development [72] are highly encouraging. Moreover, from terahertz to far infrared optical detector and emitter development, use of graphene has become a major research interest [73,74].

12.8 Conclusions

The discovery of a single atomic sheet of graphene ignited intense research activities in order to explore the interesting properties of 2D graphene. The superior electron–hole mobility results from its conical energy dispersion relation near the Dirac point, and it make graphene a suitable future candidate for electronic device applications. A wide variety of graphene-based devices and problems associated with them have been discussed. It is clear that band opening by 1D confinement is a major achievement and is a significant stride in the direction of the development of ultralow-power, high-frequency, graphene-based devices. Furthermore, a clear understanding of the growth mechanism, controlled growth technology, and modeling of notable properties of graphene is mandatory for its widespread usage in devices and nanoscience applications.

References

1. Peierls, R., 1935. Quelques propriétés typiques des corps solides. *Annales de l'Institute Henri Poincare*, 5(3): 177–222.
2. Landau, L.D., 1937. Zur Theorie der phasenumwandlungen II. *Physikalische Zeitschrift der Sowjetunion*, 11: 26–35.
3. Mermin, N.D., 1968. Crystalline order in two dimensions. *Physical Review*, 176(1): 250.
4. Novoselov, K.S., Jiang, D., Schedin, F., Booth, T.J., Khotkevich, V.V., Morozov, S.V., and Geim, A.K., 2005. Two-dimensional atomic crystals. *Proceedings of the National Academy of Sciences of the United States of America*, 102(30): 10451–10453.
5. Novoselov, K.S.A., Geim, A.K., Morozov, S., Jiang, D., Katsnelson, M., Grigorieva, I., Dubonos, S., and Firsov, A., 2005. Two-dimensional gas of massless Dirac fermions in graphene. *Nature*, 438(7065): 197–200.
6. Dikin, D.A., Stankovich, S., Zimney, E.J., Piner, R.D., Dommett, G.H., Evmenenko, G., Nguyen, S.T., and Ruoff, R.S., 2007. Preparation and characterization of graphene oxide paper. *Nature*, 448(7152): 457–460.
7. Ohta, T., El Gabaly, F., Bostwick, A., McChesney, J.L., Emtsev, K.V., Schmid, A.K., Seyller, T., Horn, K., and Rotenberg, E., 2008. Morphology of graphene thin film growth on SiC (0001). *New Journal of Physics*, 10(2): 023034.
8. Geim, A.K. and Novoselov, K.S., 2007. The rise of graphene. *Nature Materials*, 6(3): 183–191.
9. Son, Y.W., Cohen, M.L., and Louie, S.G., 2006. Energy gaps in graphene nanoribbons. *Physical Review Letters*, 97(21): 216803.
10. Han, M.Y., Özyilmaz, B., Zhang, Y., and Kim, P., 2007. Energy band-gap engineering of graphene nanoribbons. *Physical Review Letters*, 98(20): 206805.
11. Li, X., Wang, X., Zhang, L., Lee, S., and Dai, H., 2008. Chemically derived, ultrasmooth graphene nanoribbon semiconductors. *Science*, 319(5867): 1229–1232.
12. Ferrari, A.C., Meyer, J.C., Scardaci, V., Casiraghi, C., Lazzeri, M., Mauri, F., Piscanec, S. et al., 2006. Raman spectrum of graphene and graphene layers. *Physical Review Letters*, 97(18): 187401.
13. Novoselov, K.S., Geim, A.K., Morozov, S.V., Jiang, D., Zhang, Y., Dubonos, S.V., Grigorieva, I.V., and Firsov, A.A., 2004. Electric field effect in atomically thin carbon films. *Science*, 306(5696): 666–669.
14. Choucair, M., Thordarson, P., and Stride, J.A., 2009. Gram-scale production of graphene based on solvothermal synthesis and sonication. *Nature Nanotechnology*, 4(1): 30–33.
15. Park, S. and Ruoff, R.S., 2009. Chemical methods for the production of graphenes. *Nature Nanotechnology*, 4(4): 217–224.
16. Hong, W., Xu, Y., Lu, G., Li, C., and Shi, G., 2008. Transparent graphene/PEDOT–PSS composite films as counter electrodes of dye-sensitized solar cells. *Electrochemistry Communications*, 10(10): 1555–1558.
17. Li, D., Mueller, M.B., Gilje, S., Kaner, R.B., and Wallace, G.G., 2008. Processable aqueous dispersions of graphene nanosheets. *Nature Nanotechnology*, 3(2): 101–105.
18. Hummers Jr., W.S. and Offeman, R.E., 1958. Preparation of graphitic oxide. *Journal of the American Chemical Society*, 80(6): 1339–1339.

19. Stankovich, S., Dikin, D.A., Piner, R.D., Kohlhaas, K.A., Kleinhammes, A., Jia, Y., Wu, Y., Nguyen, S.T., and Ruoff, R.S., 2007. Synthesis of graphene-based nanosheets via chemical reduction of exfoliated graphite oxide. *Carbon*, 45(7): 1558–1565.

20. Li, X., Zhang, G., Bai, X., Sun, X., Wang, X., Wang, E., and Dai, H., 2008. Highly conducting graphene sheets and Langmuir–Blodgett films. *Nature Nanotechnology*, 3(9): 538–542.

21. Hernandez, Y., Nicolosi, V., Lotya, M., Blighe, F.M., Sun, Z., De, S., McGovern, I.T. et al., 2008. High-yield production of graphene by liquid-phase exfoliation of graphite. *Nature Nanotechnology*, 3(9): 563–568.

22. Zhou, Y., Bao, Q., Tang, L.A.L., Zhong, Y., and Loh, K.P., 2009. Hydrothermal dehydration for the "green" reduction of exfoliated graphene oxide to graphene and demonstration of tunable optical limiting properties. *Chemistry of Materials*, 21(13): 2950–2956.

23. Zheng, J., Di, C.A., Liu, Y., Liu, H., Guo, Y., Du, C., Wu, T., Yu, G., and Zhu, D., 2010. High quality graphene with large flakes exfoliated by oleyl amine. *Chemical Communications*, 46(31): 5728—5730.

24. Alzari, V., Nuvoli, D., Scognamillo, S., Piccinini, M., Gioffredi, E., Malucelli, G., Marceddu, S., Sechi, M., Sanna, V., and Mariani, A., 2011. Graphene-containing thermoresponsive nanocomposite hydrogels of poly(N-isopropylacrylamide) prepared by frontal polymerization. *Journal of Materials Chemistry*, 21(24): 8727–8733.

25. Rangappa, D., Sone, K., Wang, M., Gautam, U.K., Golberg, D., Itoh, H., Ichihara, M., and Honma, I., 2010. Rapid and direct conversion of graphite crystals into high-yielding, good-quality graphene by supercritical fluid exfoliation. *Chemistry: A European Journal*, 16(22): 6488–6494.

26. Serhatkulu, G.K., Dilek, C., and Gulari, E., 2006. Supercritical CO_2 intercalation of layered silicates. *The Journal of Supercritical Fluids*, 39(2): 264–270.

27. Johnston, K.P. and Shah, P.S., 2004. Making nanoscale materials with supercritical fluids. *Science*, 303(5657): 482–483.

28. Rao, C.E.E., Sood, A.E., Subrahmanyam, K.E., and Govindaraj, A., 2009. Graphene: The new two-dimensional nanomaterial. *Angewandte Chemie International Edition*, 48(42): 7752–7777.

29. Subrahmanyam, K.S., Ghosh, A., Gomathi, A., Govindaraj, A., and Rao, C.N.R., 2009. Covalent and noncovalent functionalization and solubilization of graphene. *Nanoscience and Nanotechnology Letters*, 1(1): 28–31.

30. Jang, J.H., Rangappa, D., Kwon, Y.U., and Honma, I., 2011. Direct preparation of 1-PSA modified graphene nanosheets by supercritical fluidic exfoliation and its electrochemical properties. *Journal of Materials Chemistry*, 21(10): 3462–3466.

31. Dervishi, E., Li, Z., Shyaka, J., Watanabe, F., Biswas, A., Umwungeri, J.L., Courte, A., Biris, A.R., Kebdani, O., and Biris, A.S., 2011. The role of hydrocarbon concentration on the synthesis of large area few to multi-layer graphene structures. *Chemical Physics Letters*, 501(4): 390–395.

32. Ago, H., Ito, Y., Mizuta, N., Yoshida, K., Hu, B., Orofeo, C.M., Tsuji, M., Ikeda, K.I., and Mizuno, S., 2010. Epitaxial chemical vapor deposition growth of single-layer graphene over cobalt film crystallized on sapphire. *ACS Nano*, 4(12): 7407–7414.

33. Park, H.J., Skakalova, V., Meyer, J., Lee, D.S., Iwasaki, T., Bumby, C., Kaiser, U., and Roth, S., 2010. Growth and properties of chemically modified graphene. *Physica Status Solidi (b)*, 247(11–12): 2915–2919.

34. Dervishi, E., Li, Z., Shyaka, J., Watanabe, F., Biswas, A., Umwungeri, J.L., Courte, A., Biris, A.S., and Biris, A.S., 2010. Large area graphene sheets synthesized on a bi-metallic catalyst system. *Nanotechnology, 1*: 234–237.
35. Lee, B.J., Yu, H.Y., and Jeong, G.H., 2010. Controlled synthesis of monolayer graphene toward transparent flexible conductive film application. *Nanoscale Research Letters, 5*(11): 1768.
36. Xu, Z., Li, H., Cao, G., Cao, Z., Zhang, Q., Li, K., Hou, X., Li, W., and Cao, W., 2010. Synthesis of hybrid graphene carbon-coated nanocatalysts. *Journal of Materials Chemistry, 20*(38): 8230–8232.
37. Bhaviripudi, S., Jia, X., Dresselhaus, M.S., and Kong, J., 2010. Role of kinetic factors in chemical vapor deposition synthesis of uniform large area graphene using copper catalyst. *Nano Letters, 10*(10): 4128–4133.
38. Kholmanov Kholmanov, I.N., Cavaliere, E., Cepek, C., and Gavioli, L., 2010. Catalytic chemical vapor deposition of methane on graphite to produce graphene structures. *Carbon, 48*(5): 1619–1625.
39. Wang, X., You, H., Liu, F., Li, M., Wan, L., Li, S., Li, Q. et al., 2009. Large-scale synthesis of few-layered graphene using CVD. *Chemical Vapor Deposition, 15*(1–3): 53–56.
40. Malesevic, A., Vitchev, R., Schouteden, K., Volodin, A., Zhang, L., Van Tendeloo, G., Vanhulsel, A., and Van Haesendonck, C., 2008. Synthesis of few-layer graphene via microwave plasma-enhanced chemical vapour deposition. *Nanotechnology, 19*(30): 305604.
41. Cambaz, Z.G., Yushin, G., Osswald, S., Mochalin, V., and Gogotsi, Y., 2008. Noncatalytic synthesis of carbon nanotubes, graphene and graphite on SiC. *Carbon, 46*(6): 841–849.
42. Nakajima, T., Koh, M., and Takashima, M., 1998. Electrochemical behavior of carbon alloy C × N prepared by CVD using a nickel catalyst. *Electrochimica Acta, 43*(8): 883–891.
43. Nakajima, T. and Koh, M., 1997. Synthesis of high crystalline carbon-nitrogen layered compounds by CVD using nickel and cobalt catalysts. *Carbon, 35*(2): 203–208.
44. Rümmeli, M.H., Kramberger, C., Grüneis, A., Ayala, P., Gemming, T., Büchner, B., and Pichler, T., 2007. On the graphitization nature of oxides for the formation of carbon nanostructures. *Chemistry of Materials, 19*(17): 4105–4107.
45. Volotskova, O., Levchenko, I., Shashurin, A., Raitses, Y., Ostrikov, K., and Keidar, M., 2010. Single-step synthesis and magnetic separation of graphene and carbon nanotubes in arc discharge plasmas. *Nanoscale, 2*(10): 2281–2285.
46. Tan, K.F., Xu, J., Chang, J., Borgna, A., and Saeys, M., 2010. Carbon deposition on Co catalysts during Fischer–Tropsch synthesis: A computational and experimental study. *Journal of Catalysis, 274*(2): 121–129.
47. Swart, J.C.W., Van Steen, E., Ciobíċă, I.M., and Van Santen, R.A., 2009. Interaction of graphene with FCC–Co (111). *Physical Chemistry Chemical Physics, 11*(5): 803–807.
48. Hibino, H., Kageshima, H., and Nagase, M., 2010. Epitaxial few-layer graphene: Towards single crystal growth. *Journal of Physics D: Applied Physics, 43*(37): 374005.
49. Oshima, C. and Nagashima, A., 1997. Ultra-thin epitaxial films of graphite and hexagonal boron nitride on solid surfaces. *Journal of Physics: Condensed Matter, 9*(1): 1.

50. Tanaka, T., Tajima, A., Moriizumi, R., Hosoda, M., Ohno, R., Rokuta, E., Oshima, C., and Otani, S., 2002. Carbon nano-ribbons and their edge phonons. *Solid State Communications*, 123(1): 33–36.

51. Odahara, G., Otani, S., Oshima, C., Suzuki, M., Yasue, T., and Koshikawa, T., 2011. In-situ observation of graphene growth on Ni (111). *Surface Science*, 605(11): 1095–1098.

52. Baraton, L., He, Z., Lee, C.S., Maurice, J.L., Cojocaru, C.S., Gourgues-Lorenzon, A.F., Lee, Y.H., and Pribat, D., 2011. Synthesis of few-layered graphene by ion implantation of carbon in nickel thin films. *Nanotechnology*, 22(8): 085601.

53. Garaj, S., Hubbard, W., and Golovchenko, J.A., 2010. Graphene synthesis by ion implantation. *Applied Physics Letters*, 97(18): 183103.

54. Mwakikunga, B.W., Forbes, A., Sideras-Haddad, E., Erasmus, R.M., Katumba, G., and Masina, B., 2008. Synthesis of tungsten oxide nanostructures by laser pyrolysis. *International Journal of Nanoparticles*, 1(3): 185–202.

55. Mwakikunga, B.W., Forbes, A., Sideras-Haddad, E., and Arendse, C., 2008. Optimization, yield studies and morphology of WO_3 nano-wires synthesized by laser pyrolysis in C_2H_2 and O_2 ambients—Validation of a new growth mechanism. *Nanoscale Research Letters*, 3(10): 372.

56. Wu, Y., Qiao, P., Chong, T., and Shen, Z., 2002. Carbon nanowalls grown by microwave plasma enhanced chemical vapor deposition. *Advanced Materials*, 14(1): 64–67.

57. Hiramatsu, M., Shiji, K., Amano, H., and Hori, M., 2004. Fabrication of vertically aligned carbon nanowalls using capacitively coupled plasma-enhanced chemical vapor deposition assisted by hydrogen radical injection. *Applied Physics Letters*, 84(23): 4708–4710.

58. Takeuchi, W., Ura, M., Hiramatsu, M., Tokuda, Y., Kano, H., and Hori, M., 2008. Electrical conduction control of carbon nanowalls. *Applied Physics Letters*, 92(21): 213103.

59. Kasuga, T., Hiramatsu, M., Hoson, A., Sekino, T., and Niihara, K., 1998. Formation of titanium oxide nanotube. *Langmuir*, 14(12): 3160–3163.

60. Wang, J.J., Zhu, M.Y., Outlaw, R.A., Zhao, X., Manos, D.M., Holloway, B.C., and Mammana, V.P., 2004. Free-standing subnanometer graphite sheets. *Applied Physics Letters*, 85(7): 1265–1267.

61. Sato, G., Morio, T., Kato, T., and Hatakeyama, R., 2006. Fast growth of carbon nanowalls from pure methane using helicon plasma-enhanced chemical vapor deposition. *Japanese Journal of Applied Physics*, 45(6R): 5210.

62. Lin, Y.M., Dimitrakopoulos, C., Jenkins, K.A., Farmer, D.B., Chiu, H.Y., Grill, A., and Avouris, P., 2010. 100-GHz transistors from wafer-scale epitaxial graphene. *Science*, 327(5966): 662–662.

63. Liao, L., Bai, J., Qu, Y., Lin, Y.C., Li, Y., Huang, Y., and Duan, X., 2010. High-κ oxide nanoribbons as gate dielectrics for high mobility top-gated graphene transistors. *Proceedings of the National Academy of Sciences*, 107(15): 6711–6715.

64. Schwierz, F., 2010. Graphene transistors. *Nature Nanotechnology*, 5(7): 487–496.

65. Obradovic, B., Kotlyar, R., Heinz, F., Matagne, P., Rakshit, T., Giles, M.D., Stettler, M.A., and Nikonov, D.E., 2006. Analysis of graphene nanoribbons as a channel material for field-effect transistors. *Applied Physics Letters*, 88(14): 142102.

66. Liang, G., Neophytou, N., Lundstrom, M.S., and Nikonov, D.E., 2007. Ballistic graphene nanoribbon metal-oxide-semiconductor field-effect transistors: A full real-space quantum transport simulation. *Journal of Applied Physics*, 102(5): 054307.

67. Liang, G., Neophytou, N., Lundstrom, M.S., and Nikonov, D.E., 2008. Contact effects in graphene nanoribbon transistors. *Nano Letters*, 8(7): 1819–1824.
68. Teong, H., Lam, K.T., Khalid, S.B., and Liang, G., 2009. Shape effects in graphene nanoribbon resonant tunneling diodes: A computational study. *Journal of Applied Physics*, 105(8): 084317.
69. Wang, Z.F., Shi, Q.W., Li, Q., Wang, X., Hou, J.G., Zheng, H., Yao, Y., and Chen, J., 2007. Z-shaped graphene nanoribbon quantum dot device. *Applied Physics Letters*, 91(5): 053109.
70. Robinson, J.T., Zalalutdinov, M., Baldwin, J.W., Snow, E.S., Wei, Z., Sheehan, P., and Houston, B.H., 2008. Wafer-scale reduced graphene oxide films for nanomechanical devices. *Nano Letters*, 8(10): 3441–3445.
71. Zheng, Y., Ni, G.X., Toh, C.T., Zeng, M.G., Chen, S.T., Yao, K., and Özyilmaz, B., 2009. Gate-controlled nonvolatile graphene-ferroelectric memory. *Applied Physics Letters*, 94(16): 163505.
72. Schedin, F., Geim, A.K., Morozov, S.V., Hill, E.W., Blake, P., Katsnelson, M.I., and Novoselov, K.S., 2007. Detection of individual gas molecules adsorbed on graphene. *Nature Materials*, 6(9): 652–655.
73. Dubinov, A.A., Aleshkin, V.Y., Ryzhii, M., Otsuji, T., and Ryzhii, V., 2009. Terahertz laser with optically pumped graphene layers and Fabri–Perot resonator. *Applied Physics Express*, 2(9): 092301.
74. Ryzhii, V., Ryzhii, M., Satou, A., Otsuji, T., Dubinov, A.A., and Aleshkin, V.Y., 2009. Feasibility of terahertz lasing in optically pumped epitaxial multiple graphene layer structures. *Journal of Applied Physics*, 106(8): 084507.

13

Nanoscale Silicon MOS Transistors

Soumya Pandit

CONTENTS

13.1 Introduction

The feature size of a MOS transistor has been subjected to scaling for more than a few decades. This is governed by the Moore's law, according to which the number of devices in a chip is approximately doubled every 18 months. There are three primary benefits of technology scaling: (1) Enhanced integration of components on a given silicon area. (2) Cost reduction—by making the size of the transistors and the interconnects smaller, more circuits can be fabricated on a piece of silicon wafer and therefore each circuit becomes cheaper. (3) Short transit time of current carriers leading to faster operation of the transistors. However, at the cost of these benefits, the scaling down of the feature size of the MOS transistors results in a number of higher-order effects that significantly degrade the performance of the MOS transistors and hence play a critical role in determining the overall circuit performance. These are collectively referred to as the short-channel effects of the MOS transistors. This chapter provides an introductory overview of the major underlying physical phenomena involved in the scaled MOS transistors fabricated over a Si wafer. The various approaches including alternative device architectures for reducing such effects are also introduced.

13.2 Scaling Rules

The reduction of the dimensions of MOS transistors is referred to as scaling. Introduction of a new technology node leads to a 50% reduction in circuit size, as per Moore's law, which in turn dictates a 0.7 times reduction of line width such that the total area reduction is $0.7 \times 0.7 \approx 0.49$. Primarily, there are three types of scaling: constant field scaling, constant voltage scaling, and generalized scaling. The basic idea of scaling is that a large device is scaled down by a factor α (>1) to produce a smaller device with similar behavior.

13.2.1 Constant Field Scaling

In this approach, the applied voltages and the dimensions (both horizontal and vertical) of a transistor device are scaled down by a scaling factor α (>1). The substrate doping concentration is increased by the same factor α. The electric field pattern inside the transistor device thereby remains the same as it was in the original device. This is referred to as the constant field scaling. It may be noted that the substrate doping concentration needs to be increased by the same scale factor α in order to keep Poisson's equation invariant with respect to the scaling. The circuit speed increases in proportion to the factor α and the circuit density increases by α^2; the power dissipation per circuit is reduced by α^2 without increasing the power density. The scaling relations are shown in Table 13.1.

TABLE 13.1

Scaling of MOSFET Device and Circuit Parameters

Physical Parameter	Constant Electric Field Scaling	Generalized Scaling	Constant Voltage Scaling
Device dimensions (L, W, t_{ox})	$1/\alpha$	$1/\alpha$	$1/\alpha$
Electric field	1	E	α
Voltage	$1/\alpha$	ε/α	1
ON current	$1/\alpha$	ε/α	α
Doping	α	$\varepsilon\alpha$	α^2
Area	$1/\alpha^2$	$1/\alpha^2$	$1/\alpha^2$
Capacitance	$1/\alpha$	$1/\alpha$	$1/\alpha$
Gate delay	$1/\alpha$	$1/\alpha$	$1/\alpha^2$
Power dissipation	$1/\alpha^2$	ε^2/α^2	A
Power density	1	ε^2	α^3

13.2.2 Generalized Scaling

The threshold voltage of a MOS transistor is assumed to scale down by the same scaling factor α, in proportion to the power supply voltage. However, in silicon technology, parameters such as energy gap and work function do not change with scaling; hence, in general, the threshold voltage does not scale. This leads to nonscaling of the subthreshold slope and OFF current. In addition, the power supply voltage was seldom scaled in proportion to the feature size. To accommodate this trend, more generalized scaling rules have been created, in which the electric field is allowed to increase by a factor ε. If it is assumed that the electric field is increased by the factor ε while the device dimensions in all directions are scaled down by the factor α, then in the generalized scaling, the potential or voltage will change by a factor ε/α. To keep Poisson's equation invariant, the doping concentration should be scaled up by $\varepsilon\alpha$. The generalized scale rules are also shown in Table 13.1. A serious issue related to the generalized scaling is that the power density is increased by a factor ε^2.

13.2.3 Constant Voltage Scaling

In this approach, the device dimensions (both horizontal and vertical) are scaled down by the factor α. The operating voltages are, however, kept constant. This means that the electric field patterns within the device will increase. The threshold voltage remains constant. The power per transistor increases by α and the power density per unit area increases by α^3. This large increase in power density may eventually cause serious reliability problems for the scaled transistors. In reality, the CMOS technology evolution has followed a combination of constant field scaling and constant voltage scaling.

13.2.4 International Technology Roadmap for Semiconductors

The International Technology Roadmap for Semiconductors (ITRS) is a set of documents produced by a group of semiconductor-industry experts. This roadmap identifies future probable technical challenges. The industry and the research community will subsequently work together to overcome these challenges. The ITRS is sponsored European Semiconductor Industry Association (ESIA) and other semiconductor associations from Japan, Korea, Taiwan and the United States. The 2013 Edition of the ITRS is now released.

The ITRS working group has classified the CMOS process technology into two types depending upon the type of applications: high performance (HP) and low power (LP). The HP process technology refers to the technology required for the chips of high performance at the cost of high power dissipation, such as microprocessor units for desktop PCs and advanced computing servers. On the other hand, the LP technology refers to the process technology required for battery-operated electronic applications, such as mobile system-on-chip, medical electronics, and laptops. Typical values of some important process parameters for the HP logic and LP logic applications based on the ITRS are provided in Tables 13.2 and 13.3, respectively.

13.3 Short-Channel MOS Transistor

When the source-drain distance of a MOS transistor is of the same order of magnitude as that of the depletion layer width of the source-drain junction with the channel, the transistor is considered to be a short-channel transistor. In a short-channel transistor, the electric field in the channel under the gate of a short-channel MOS transistor is two dimensional, that is, the components in the vertical and horizontal directions are comparable. This is in contrast to the gradual channel approximation in a long-channel MOS transistor, where the lateral field gradient is considered to be negligible compared to the vertical field gradient. The electric field in the channel of a short-channel MOS transistor is significantly high, leading to a number of phenomena related to high electric field, such as the hot-electron effect.

The reduction of the channel length of the MOS transistor leads to several effects such as a reduction in the threshold voltage, drain-induced barrier lowering, subthreshold leakage current, mobility degradation, velocity saturation, punch-through effect, increase in parasitic resistance and capacitance, and hot-electron effect. These are collectively referred to as the short-channel effects.

TABLE 13.2

ITRS 2013: High-Performance Logic Technology Requirement

Year of Production	2013	2017	2021	2024	2028
L_g: Physical gate length (nm)	20	13.9	9.7	7.3	5.1
L_{ch}: Effective channel length (nm)	16	11.1	7.8	5.8	4.1
V_{dd}: Power supply voltage (V)					
Bulk/SOI/MG	0.86	0.80	0.74	0.69	0.64
EOT: Equivalent oxide thickness					
Bulk/SOI/MG	0.80	0.67	0.56	0.49	0.41
Channel doping (10^{18}/cm³)					
Bulk	6	9			
SOI/MG	0.1	0.1	0.1	0.1	0.1
I_{off} (nA/µm)					
Bulk/SOI/MG	100	100	100	100	100
$I_{d,sat}$: NMOS drive current (µA/µm)					
Bulk	1348	1267			
SOI					
MG	1670	1660	1450	1170	900
$V_{t,lin}$ (V)					
Bulk	0.306	0.378			
SOI					
MG	0.219	0.264	0.295	0.319	0.364
$V_{t,sat}$ (V)					
Bulk	0.190	0.230			
SOI					
MG	0.174	0.191	0.214	0.233	0.278
R_{sd}: Total parasitic series source/drain resistance (Ω-µm)	188	156			
Bulk					
SOI	128	124	112	123	128
MG					
$C_{g,total}$: Total gate capacitance (fF/µm)					
Bulk/SOI/MG	1.10	1.03	0.93	0.77	0.60
CV^2: NMOS dynamic power indicator (fJ/µm)					
Bulk/SOI/MG	0.82	0.66	0.51	0.36	0.24
τ = CV/I: NMOS intrinsic delay (ps)					
Bulk	0.705	0.650			
SOI					
MG	0.569	0.496	0.477	0.451	0.423
$I_{d,sat}$ (n-channel)/$I_{d,sat}$ (p-chanel)	1.25	1.20	1.15	1.12	1.07

TABLE 13.3

ITRS 2013: Low-Power Logic Technology Requirement

Year of Production	2013	2017	2021	2024	2028
L_g: Physical gate length (nm)	23	16.0	11.1	8.5	5.9
L_{ch}: Effective channel length (nm)	18.4	12.8	8.9	6.8	4.7
V_{dd}: Power supply voltage (V)					
Bulk/SOI/MG	0.86	0.80	0.74	0.69	0.64
EOT: Equivalent oxide thickness					
Bulk/SOI/MG	0.80	0.67	0.56	0.49	0.41
Channel doping ($10^{18}/\text{cm}^3$)					
Bulk	5.0	8.4			
SOI/MG	0.1	0.1	0.1	0.1	0.1
I_{off} (nA/µm)					
Bulk	10	50			
SOI					
MG	10	10	20	20	50
$I_{d,sat}$: NMOS drive current (µA/µm)					
Bulk	490	422			
SOI					
MG	643	574	537	395	295
$V_{t,lin}$ (V)					
Bulk	0.619	0.647			
SOI					
MG	0.483	0.507	0.507	0.520	0.519
$V_{t,sat}$ (V)					
Bulk	0.528	0.530			
SOI					
MG	0.446	0.461	0.446	0.460	0.468
R_{sd}: Total parasitic series source/ drain resistance (Ω-µm)	188	156			
Bulk					
SOI	128	124	112	123	128
MG					
$C_{g,total}$: Total gate capacitance (fF/µm)					
Bulk/SOI/MG	1.21	1.14	1.03	0.89	0.69
CV^2: NMOS dynamic power indicator (fJ/µm)					
Bulk/SOI/MG	0.90	0.73	0.57	0.42	0.28
$\tau = CV/I$: NMOS intrinsic delay (ps)					
Bulk	2.128	2.159			
SOI					
MG	1.622	1.587	1.556	1.557	1.493
$I_{d,sat}$ (n-channel)/$I_{d,sat}$ (p-channel)	1.27	1.22	1.18	1.14	1.10

13.4 Threshold Voltage of a Short-Channel Transistor

The threshold voltage of a long-channel MOS transistor is independent of the channel length and the magnitude of the applied drain voltage. In contrast, the threshold voltage of a short-channel MOS transistor depends upon the channel length and the magnitude of the applied drain voltage. The long-channel theory of the MOS transistor is derived on the basis of the assumption that the depletion charge under the gate is controlled only by the vertical electric field. However, in a short-channel MOS transistor, the built-in potential of the source/drain and the drain voltage considerably affect the threshold voltage. There are two distinct phenomena related to the threshold voltage of a short-channel MOS transistor: a threshold voltage roll-off and a drain-induced barrier-lowering effect. These are qualitatively discussed below.

13.4.1 Threshold Voltage Roll-Off

For a long-channel MOS transistor, the effect of the source and drain depletion regions on the depletion region under the gate is small, so that the depletion region under the gate contributes to the vertical electric field, which in turn determines the voltage across the oxide and thus the threshold voltage V_T. However, as the channel length is reduced, the source/drain depletion regions (due to n^+p junctions between source/drain and substrate for n-channel MOS transistor) contribute greatly to the formation of the channel depletion region.

In a short-channel transistor, the depletion charge under the gate is induced by the gate together with the source and the drain so that the channel charge can be considered to be shared by the gate, the source, and the drain. The charge sharing is illustrated in Figure 13.1. The field lines originating from

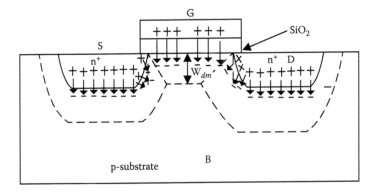

FIGURE 13.1
Charge sharing in a short-channel transistor leading to a reduction in the threshold voltage.

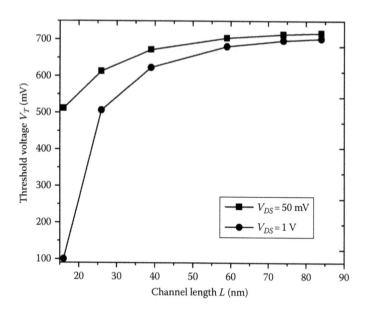

FIGURE 13.2
V_T roll-off characteristics.

the positive charges, which terminate on the negative acceptor ions, have two origins: (1) the positive charge induced on the gate electrode and (2) the positive donor ions within the depletion layer of the n⁺ region of the source and the drain. Therefore, only part of the channel depletion region charge is imaged on the gate charge. It is therefore apparent that a lesser gate charge density (and lower gate voltage) is required to induce inversion in a short-channel MOS transistor. This implies that the threshold voltage of a short-channel MOS transistor is reduced compared to that of a long-channel transistor. The curve representing the reduction of V_T with a decreasing channel length is known as V_T roll-off. This is shown in Figure 13.2. We observe that the roll-off is enhanced at a high drain voltage, which is discussed below.

13.4.2 Drain-Induced Barrier-Lowering Effect

In a MOS transistor, the current carriers, in general, face a potential energy barrier between the source and the channel while moving from the source region to the drain region. For a long-channel NMOS transistor, increasing the gate voltage reduces this potential barrier and eventually allows the flow of electrons under the influence of the channel electric field due to the applied drain voltage.

On the other hand, in a short-channel MOS transistor, as the depletion layer from the drain region encroaches into the channel, the drain-induced electric field reduces this potential barrier between the source and the channel. This may be explained by the fact that a deeper depletion region is associated with a larger surface potential, which in turn corresponds to a lowering of the potential energy barrier for the electrons. The consequence of this barrier lowering is easy transport of the current carriers from the source to the drain region, that is, an increase in the drain current and a decrease in the threshold voltage. The potential energy barrier is further lowered due to the high drain voltage. Consequently, the threshold voltage gets reduced with the increasing drain voltage for short-channel transistors. Thus, the drain current is controlled not only by the gate voltage but also by the drain voltage. This phenomenon is known as drain-induced barrier-lowering (DIBL) and is attributed to the electrostatic coupling between the channel and the drain. This is illustrated in Figure 13.3. In this figure, A represents the amount of barrier lowering because of a reduction of the channel length (from 200 to 22 nm) and B represents the amount of barrier lowering for an increase of the drain bias (from 50 mV to 0.8 V) for a short-channel MOS transistor. The variation of the threshold voltage V_T with the applied drain bias is shown in

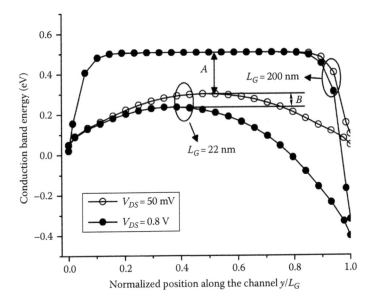

FIGURE 13.3
Illustration of barrier lowering due to a reduction of the channel length as well as due to a high drain bias.

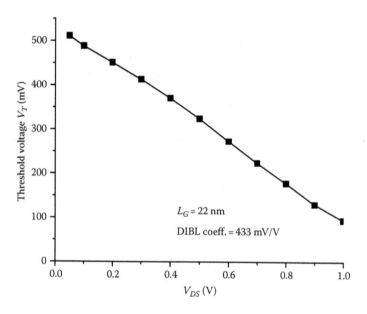

FIGURE 13.4
Reduction of the threshold voltage with the application of drain bias.

Figure 13.4. The strength of DIBL is usually measured by DIBL coefficient, which is defined as follows:

$$\text{DIBL} = \frac{dV_T}{dV_{DS}} = \frac{V_T\left(V_{DS} = 0.05 \text{ V}\right) - V_T\left(V_{DS} = 0.8 \text{ V}\right)}{0.75 \text{ V}} \tag{13.1}$$

A high value of the DIBL coefficient is an indication of poor short-channel behavior.

13.5 Carrier Mobility Reduction

The reduction of mobility of the inversion carriers is an important issue in nanoscale MOS transistors. There are two reasons for this reduction of carrier mobility: (1) vertical electric field and (2) lateral electric field. These are briefly discussed below.

13.5.1 Mobility Dependence on Vertical Electric Field

The electrons in an n-channel MOS transistor suffer from several kinds of scattering mechanisms while traversing from the source to the drain through the channel. These are the (1) Coulomb scattering, (2) phonon scattering, and (3) surface roughness scattering. The Coulomb scattering occurs due to the interaction of inversion carriers with ionized impurities, interface state charges, and fixed oxide charges. The contribution of Coulomb scattering is significant in the weak and moderate inversion mode of operation. The lattice vibrations in silicon at the interface emit and absorb phonons while exchanging energy with the carriers, resulting in lattice scattering. The discontinuities in the lattice structure at the Si-SiO$_2$ interface make the movement of the inversion carriers irregular in the channel. The surface irregularities act as scattering centers, which significantly reduce the carrier mobility.

These three scattering mechanisms degrade the inversion carrier mobility, which are related through Matthiessen's rule. In a nanoscale MOS transistor, due to high normal electric field, surface roughness scattering is the dominant mechanism in reducing the carrier mobility. A simple estimation of the surface mobility of the carrier is given by

$$\mu_s = \frac{\mu_0}{1+\theta\left(V_{GS}-V_T\right)+\theta_B V_{SB}} \tag{13.2}$$

where
μ_0 is the zero-field carrier mobility
θ, θ_B are two empirical parameters, whose values need to be extracted from simulation/experimental data

A typical value for μ_0 is 650 cm^2/V s for the n-channel transistor. For the p-channel transistor, μ_0 is typically smaller by a factor of 3. The parameter θ is of the form β_θ/t_{ox} where t_{ox} is the oxide thickness. $\beta_\theta \approx 5-20$ Å/V. The value of θ_B is usually very small and often neglected in the first-order analysis.

The mobility degradation phenomenon seriously affects the drain current and the transconductance parameter of a MOS transistor. This is shown in Figure 13.5. We observe that beyond a certain gate voltage, the transconductance parameter reduces because of the mobility degradation.

13.5.2 Mobility Dependence on Lateral Electric Field

For a given vertical electric field, the drift velocity of carrier v_d is linearly proportional to the lateral electric field for low values of the electric field. The proportionality constant is given by the surface mobility μ_s. In a nanoscale MOS transistor, the lateral electric field becomes sufficiently high so that the linear relationship is violated. As the lateral electric field is increased, the carrier velocity v_d tends to saturate to v_{dsat}. In a short-channel MOS transistor, the carrier reaches the velocity saturation at a lower value

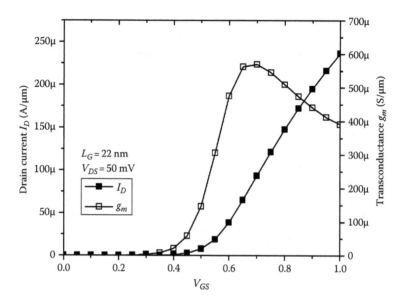

FIGURE 13.5
Illustration of the mobility degradation effect of the drain current and transconductance parameter of a short-channel MOS transistor.

of the applied drain bias. At a higher lateral electric field, the carrier velocity is estimated through a nonlinear relation as follows:

$$v_d = \frac{\mu_s \xi_y}{\left(1 + \dfrac{\xi_y}{\xi_{sat}}\right)} \tag{13.3}$$

where
ξ_y is the lateral electric field
ξ_{sat} is the critical electric field

When the field strength is comparable to or greater than ξ_{sat}, the velocity saturation becomes important. The velocity saturation effect reduces the drain current by a factor $1 + V_{DS}/\xi_{sat}L$. This factor becomes unity, (i.e., this effect becomes negligible) when either V_{DS} is small or the channel length is large. In the presence of the velocity saturation effect, the drain current saturation voltage is expressed as

$$\frac{1}{V_{DSat}} = \frac{m}{V_{GS} - V_T} + \frac{1}{\xi_{sat}L} \tag{13.4}$$

where m is the body-effect coefficient

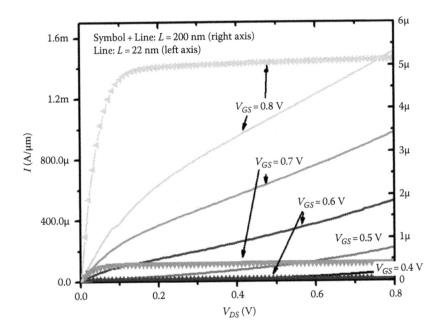

FIGURE 13.6
Drain current characteristics of a long- and a short-channel MOS transistor, illustrating the velocity saturation effect.

For a very-short-channel case, $\xi_{sat} \ll V_{GS} - V_T$, and simple mathematics shows that the saturation drain current is linearly proportional to $V_{GS} - V_T$ rather than being quadratic as for a long-channel case. The velocity saturation effect is illustrated in Figure 13.6 through its effect on the drain current. The square law dependence of the drain current on the gate-to-source voltage for a long-channel transistor is apparent from the spacing of the different curves. On the other hand, such dependence is linear for the short-channel MOS transistor.

13.6 Subthreshold Conduction and OFF-State Leakage Current

Depending on the magnitudes of the applied gate and drain voltages, a MOS transistor operates in either the linear or the saturation region in order to get a reasonable value of the drain current. However, if the applied gate voltage is reduced below the threshold voltage, a nonnegligible value of drain current flows for several tenths of a volt below the threshold voltage. This is because the inversion charges do not vanish abruptly. The subthreshold region immediately below the threshold voltage, where $\phi_F \leq \psi_s \leq 2\phi_F$, is referred to as the weak inversion region. The drain current that flows when $V_{GS} < V_T$ is referred to as the subthreshold current. In this region of operation, the drain current is dominated by the diffusion motion of the carriers. The weak inversion current/subthreshold leakage current can be explored by the following expression:

$$I_{DS} = \mu_{eff} C_{ox} \frac{W}{L} (m-1) \left(\frac{k_B T}{q}\right)^2 e^{q(V_{GS}-V_T)/mk_B T} \left(1 - e^{-qV_{DS}/k_B T}\right) \qquad (13.5)$$

where
 μ_{eff} is the effective mobility of the inversion carriers considering horizontal and lateral electric fields
 m is the body-effect coefficient

The subthreshold current is independent of the drain voltage once V_{DS} is larger than a few $k_B T/q$. An important parameter that measures the dependence of the subthreshold leakage current on the gate voltage is subthreshold swing parameter S, which is the inverse of the slope in the subthreshold region. This is defined as the gate voltage change needed to induce a drain current change of one order of magnitude. This is expressed as follows:

$$S = \left(\frac{d(\log_{10} I_{DS})}{dV_{GS}}\right)^{-1} = 2.3\frac{mkT}{q} = 2.3\frac{kT}{q}\left(1 + \frac{C_{dm}}{C_{ox}}\right) \qquad (13.6)$$

where C_{dm} is the depletion layer capacitance/unit area.

The typical value of S is 70–100 mV/decade. For the ideal case, $C_{ox} \to 0$, where $m \approx 1$, $S = 60$ mV/decade. If the Si-SiO$_2$ interface trap density is high, the subthreshold slope may degrade further.

The OFF-state current of a transistor is the drain current when the gate voltage is zero (drain voltage is usually considered to be high). For a 90 nm device, the major component of the OFF-state leakage current is the subthreshold drain current leakage. However, contributions of the junction leakage and the gate leakage are increasing. As seen from (13.5), the OFF-state leakage current is related to the threshold voltage, which in turn is modulated by the DIBL effect. A higher threshold voltage leads to a reduced OFF-state leakage current whereas a low value of the threshold voltage enhances this current. The OFF-state leakage current increases at the high drain bias because of the DIBL effect.

The OFF-state current can be measured from the I_D-V_G characteristics of a MOS transistor, as $I_{OFF} = I_{DS}$ @ $V_{DS} = V_{DD}$; $V_{GS} = 0$. We observe from Figure 13.7 that the OFF-state current significantly increases as the drain voltage is made high, because of the DIBL effect. The subthreshold slope for $V_{DS} = 50$ mV is calculated to be 116 mV/decade and that for $V_{DS} = 1$ V is calculated to be 296 mV/decade.

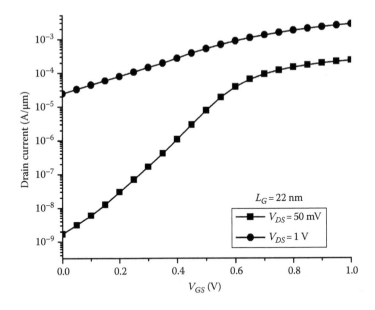

FIGURE 13.7
Illustration of OFF-state current.

13.7 Gate Leakage Current and Gate Oxide Thickness

The scaling down of the gate oxide thickness leads to a current component through the oxide. This current is caused by the carriers tunneling through the insulator region and is referred to as the gate leakage current. The gate leakage current increases exponentially as the oxide thickness is scaled down. In scaled MOS transistors, the oxide thickness is around 0.7–1.2 nm. This thickness comprises only a few layers of atoms and is approaching fundamental limits. The continued reduction of the thickness of the conventional oxides results in reliability degradation and unacceptable leakage current.

13.8 Solution Approaches

In the earlier sections of this chapter, the various nanoscale challenges of a MOS transistor are discussed in short. In the following section, we present some important approaches for overcoming the challenges. These are broadly classified into three categories: (1) channel engineering, (2) gate stack engineering, and (3) alternative device architectures. The fundamental idea for reducing the short-channel effects in a MOS transistor is to reduce (1) the oxide thickness, (2) the source/drain junction depth, and (3) the thickness of the depletion region.

13.8.1 Channel Engineering

In order to suppress the short-channel effects of the MOS transistors, various approaches like a locally high-doping concentration near the source/drain junctions and nonuniform doping in the channel from the interface toward the substrate have been implemented. These are collectively referred to as channel engineering.

13.8.1.1 Halo/Pocket Engineering

For an n-channel MOS transistor, more highly p-type-doped regions near the source and drain regions of the channel are sometimes developed. This increases the average doping concentration and leads to a threshold voltage roll-up. These regions with increased doping concentrations are commonly called halo/pocket. This is illustrated in Figure 13.8. The halo/pocket doping placed near the source/drain reduces the thickness of the depletion region of the source/drain to the channel junctions. With the reduction of the charge-sharing effects, the degradation of the threshold voltage with the lowering of the channel length is reduced. This also reduces the DIBL. The implant can be either symmetrical or asymmetrical with respect to the source and the drain. The pocket implant technology is very promising in the effort to tailor the short-channel performance of nanoscale MOS transistors. However, it has been found that the improvement of the short-channel immunity is traded-off with the degradation of the current drive, junction capacitance, and body effect. Therefore, careful tradeoffs need to be made between the short-channel effect immunity and other device electrical parameters.

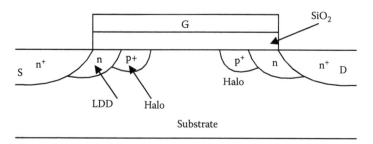

FIGURE 13.8
Schematic diagram of the n-channel MOS transistor with double halo implant.

13.8.1.2 Retrograde Channel Engineering

The retrograde channel engineering technique employs a vertically non-uniform doping profile with the channel surface (just below the interface) very lightly doped and the deeper channel region very heavily doped. This is shown in Figure 13.9. The low surface concentration increases the inversion carrier mobility by reducing the channel impurity scattering. On the other hand, the high-doped subsurface layer helps to reduce the short-channel effects. With an appropriate choice of the thickness of the low-doped region near the channel, the electron mobility is enhanced and the threshold voltage is reduced while maintaining good electric behavior versus short-channel effects.

13.8.2 Gate Stack Engineering: High-κ Metal-Gate Technology

Reduction of the gate leakage current is the key motivation for the replacement of SiO_2 with alternative gate dielectrics having high value of κ (permittivity) and properties as close to SiO_2 as possible. The requirement is, therefore, to find a substitute dielectric that gives the same capacitance with higher insulator thickness. Some essential properties required for a good candidate material to be used as a gate insulator are a high value of permittivity, good bonding capability with Si, compatibility of SiO_2 and metals

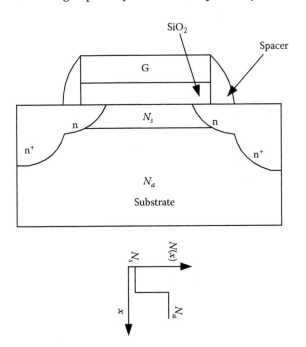

FIGURE 13.9
Schematic diagram of an n-channel MOS transistor with a retrograde channel profile.

with MOS processing, thermodynamic stability to be compatible with high-temperature processing, and good interface quality such that the number of electrically active defects are small. The thickness of the high-κ dielectric insulator is obtained from the following relationship:

$$t_{ox} = \frac{\kappa_{SiO_2}}{\kappa_d} t_d \qquad (13.7)$$

where
  κ_d is the relative permittivity of the dielectric material
  t_d is the gate dielectric thickness
  κ_{SiO_2} is the relative permittivity of SiO_2
  t_{ox} is known as the equivalent oxide thickness

Some popular materials used as an insulator in the current-day CMOS technology include ZrO_2, HfO_2, Si_3N_4, and $HfSiO_4$.

The selection of the right metal to be used as the gate electrode for the high-κ dielectric is very important. The combination of a high-κ dielectric and a poly-Si gate is not suitable for HP logic applications, since the resulting high-κ/poly-Si transistors have high threshold voltages and degraded channel mobility. The reason is that the Fermi level gets pinned at the interface, which is most likely due to the formation of defects at the interface. In addition, the surface phonon scattering gets enhanced, which also degrades the inversion carrier mobility. With high-κ gate dielectric, metal gates are used in the present-day CMOS technology. The effect of phonon scattering is reduced, which results in an improved carrier mobility. In order to achieve the desired value of the threshold voltage, metal gates with appropriate work functions are to be used. For the traditional planar CMOS applications on the bulk MOS technology, a p+ metal work function is needed for the p-channel transistor while an n+ metal work function is required for the n-channel transistor. A midgap-work-function gate (such as TiN) is symmetrical for n-channel and p-channel MOS transistors. The work function value is such that the Fermi level is placed at the midgap of the silicon substrate. With this, the CMOS processing scheme is simple, since only one mask and one metal is required for the gate electrodes.

13.9 Alternative Device Architectures

As mentioned earlier, in order to scale down the traditional bulk CMOS structure to the 10 nm gate limit, heavy channel doping is required to control the short-channel effects. A serious limitation associated with such a concept is that the inversion carrier mobility is severely degraded due to the impurity scattering and an increased transverse electric field. Above a concentration of ~2×10^{18} cm^{-3}, the inversion carrier mobility is seriously affected by impurity scattering. In addition, effects of the phonon and the interface scattering increase. The increased depletion charge results in a larger depletion capacitance and subthreshold slope. A large channel doping also inevitably enhances the band-to-band tunneling leakage between the body and the drain. The high channel doping increases the random dopant fluctuation effect, which critically affects the device-to-device statistical fluctuation of the threshold voltage.

The important requirements for a high-performance integrated circuit design are low V_T and low I_{OFF}, a small value of subthreshold slope S, and less variations of S and V_T. At small channel lengths, the gate control over the channel barrier is significantly reduced, competing with the drain terminal such that the transistor behaves like a resistor (see Figure 13.7) rather than a switch. The gate control over the channel can be enhanced by reducing the thickness of the gate oxide, that is, by making the gate closer to the channel. However, the drain could still have more control than the gate along other leakage current paths that are some distance below the Si surface. Leakage paths located far from the gate are responsible for the short-channel transistor leakage because they are at a distance from the gate (even with zero oxide thickness) where the potential barriers can be easily lowered by the drain. In order to address this growing leakage power problem, alternative device architectures such as an ultrathin body silicon-on-insulator (UTB-SOI) and FinFET have been proposed.

13.10 Ultrathin Body SOI Transistor

The schematic structure of an UTB-SOI transistor is shown in Figure 13.10. The transistor is built on a thin Si film on an insulator (SiO$_2$). Since the Si film is very thin (~5 nm), there does not exist any leakage path that is very far from the gate. The worst-case leakage path is along the bottom of the Si film. Therefore, the gate can effectively suppress the leakage. The subthreshold leakage current is reduced as the Si film is made thinner. The UTB-SOI MOS transistor has no deep silicon for the drain to affect. The drain coupling is eliminated by using the thin silicon film to build the channel. Scaling the gate length of UTB-SOI devices below 25 nm while maintaining good short-channel control requires silicon thickness of less than one quarter of the gate length. The channel is lightly doped to reduce the dopant fluctuation effect and increase the carrier mobility. The body effect is eliminated as the body is fully depleted. One challenge imposed by the UTB structure is the large parasitic source/drain resistance due to its thinness. By using selective deposition, a raised source/drain structure can be created to reduce the parasitic resistance. This transistor requires the SOI substrates with Si film uniformity of ±0.5 nm or less than two silicon atoms so that a 5 nm ultrathin Si film will not have excessive thickness nonuniformity. The schematic diagram of

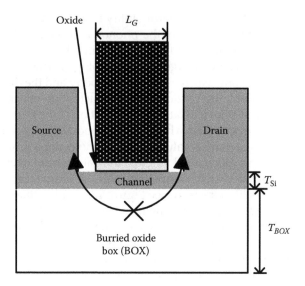

FIGURE 13.10
Schematic diagram of an UTB-SOI transistor with raised source and drain.

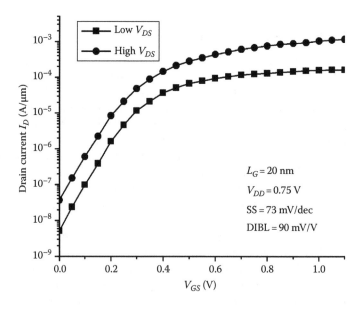

FIGURE 13.11
Typical I_D-V_G characteristics of an UTB-SOI transistor.

a typical UTB-SOI transistor is shown in Figure 13.10. The I_D-V_G characteristics of an UTB-SOI transistor with the buried oxide thickness T_{BOX} = 20 nm, channel thickness T_{Si} = 6 nm, and gate length L_G = 20 nm with strained SiGe channel is shown in Figure 13.11. The subthreshold slope is calculated to be 73 mV/decade and the DIBL coefficient is 90 mV/V. Samsung Electronics Co., Ltd. and ST Microelectronics have announced on May 2014 about the signing of a comprehensive agreement on a 28 nm fully depleted UTB-SOI technology for a multisource manufacturing collaboration.

13.11 Multigate FET

The second approach for eliminating the deeply submerged leakage paths is to provide gate control from more than one side of the channel. This is referred to as the multigate approach. The cross-sectional structure of a double-gate (DG) MOS transistor is shown in Figure 13.12. The Si film is thin so that no leakage path is far from either of the gates. The leakage current is very efficiently suppressed by both the gates. In a DG transistor, the thin silicon film is under a nearly volume-inversion condition, due to which the conduction of carriers is across the entire volume of the material.

The short-channel effects are controlled by the thin silicon film. Therefore, the gate length can be scaled down to 10 nm regime. The gate electrostatic integrity is better in a DG transistor because two gates are used to control the channel from both the sides. The subthreshold slope is steep in a DG transistor. In addition, low value of the transverse electric field and the negligible impurity scattering enhances the carrier mobility. This improves the drive current. The random dopant fluctuation effect is also eliminated. For a DG transistor, the

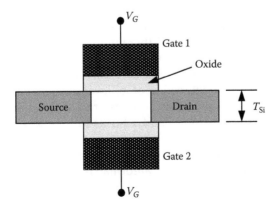

FIGURE 13.12
Schematic diagram of a double-gate MOS transistor.

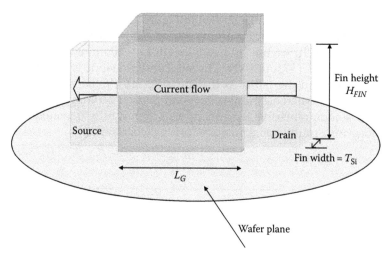

FIGURE 13.13
Schematic diagram of a fabricatable FinFET transistor.

threshold voltage is controlled by gate work function engineering. Although a DG structure has so many advantages, its fabrication process is complex.

The FinFET structure as shown in Figure 13.13 is what can be manufactured. The thin body is shaped like the fin of a fish. The fin width is denoted by T_{Si} and the fin height is denoted by H_{FIN}. The current flows parallel to the wafer plane. The channel is formed perpendicular to the plane of the wafer. The electrical channel width of the FinFET is given as $W = 2H_{FIN} + T_{Si}$. The length, height, and thickness are important parameters set by the foundry. If the fin is too thick, the major benefits of this device would be lost. However, fabrication of a very thin fin is a lithographically challenging task. The fins are constructed either on SOI or on bulk substrates. The fin is then coated with gate stack and is patterned similar to planar MOS technology. There are several variations of FinFET: tall FinFET, short FinFET, and nanowire FinFET. The tall FinFET has a large W and therefore large I_{ON}. The short FinFET is comparatively easier to fabricate. On the other hand, the nanowire FinFET gives the gate even more control over the silicon wire by surrounding it.

The typical I_D-V_G characteristics of a low leakage FinFET with gate length $L_G = 34$ nm, fabricated using 16 nm process technology is shown in Figure 13.14. Intel started using FinFET for its future commercial devices at 22 nm from 2011-2012.

BSIM-CMG is a compact model for the class of common multigate FETs, developed by the University of California, Berkeley. The latest version, BSIMCMG108.0.0 was officially released on August 22, 2014. BSIM-IMG, an independent multigate model, has been developed to model the electrical characteristics of the independent DG structures like the Ultrathin Body and BOX SOI transistors (UTBB).

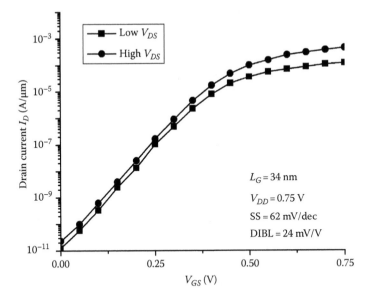

FIGURE 13.14
Typical *I-V* characteristics of a FinFET transistor.

13.12 Epitaxial Delta-Doped Channel (EδDC) MOS Transistor

Power dissipation has become the primary architectural limiter for integrated electronic circuits. The present-day chips burn too much power. This limits the designers from adding more functionality. Until the last few years, the common approach for the reduction of power dissipation was the reduction of supply voltage because of the quadratic relationship between the two. In recent years, however, this approach is limited by the increasing process variation that is intrinsic to deep-submicron process technologies. Random dopant fluctuation, which results from device-to-device fluctuations of the number and placement of dopant atoms in the channel of a MOS transistor, is a major source of process variability in the advanced CMOS technologies.

The V_T variation worsens as the transistor sizes are reduced. Therefore, for many years the semiconductor industry stuck to the use of 1 V power supply. This excess overdrive wastes significant amount of energy.

The two device architectures discussed above are efficient for reduction of the random dopant fluctuation effect and improved electrostatic integrity. However, enhanced manufacturing and/or design cost restrict their use for low-cost system-on-chip applications. Therefore, the advanced planar bulk

MOS transistor is still the preferred choice for SoC applications, such as the embedded system applications.

Recently, an epitaxial delta-doped channel (EδDC) MOS transistor structure is proposed, which belongs to the category of planar MOS transistor, which reduces the V_T variation and improves electrostatic integrity. This enables the scaling of bulk CMOS technology and reduction of the supply voltage. The schematic diagram of the cross section of an EδDC MOS transistor is shown in Figure 13.15. Similar to retrograde channel engineering, the doping concentration of the channel region is graded into two layers: The first layer is lightly doped and the second layer is very highly doped. The lightly doped layer at the interface reduces the V_T variation and improves the inversion carrier mobility through the reduced impurity scattering. On the other hand, the very highly doped layer improves the gate electrostatic integrity. The mobile charge carriers present in this high-doped layer terminate most of the electric field lines originating from the drain and the source regions and thereby screen out these from penetrating into the channel region. This layer also reduces any punch-through phenomenon between the source and the drain. The source-drain extension regions are designed such that the band-to-band tunneling (BTBT) leakage current associated with the junction between the SD extension and the channel region is reduced. The source-drain region is raised. This reduces the BTBT leakage current associated with the deep source drain and the high-doped screening layer. With the application of the substrate bias, the depletion width does not significantly change. The substrate control over the channel is high so that a significant amount of the substrate depletion charge terminates on the gate rather than on the

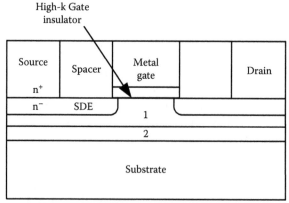

1: Lightly doped channel region

2: High-doped channel region

FIGURE 13.15
Schematic diagram of an n-channel EδDC MOS transistor along with the profile of the channel doping.

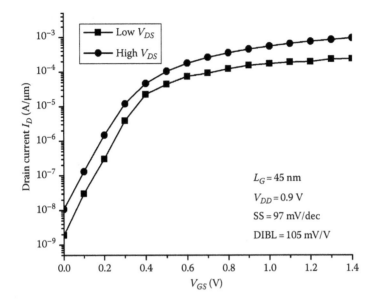

FIGURE 13.16
Typical *I-V* characteristics of an EδDC MOS transistor.

source and the drain. In an EδDC transistor, the subthreshold leakage current can be reduced by applying a reverse substrate bias because of its high substrate bias sensitivity.

The substrate bias characteristics of the EδDC transistor are better than those of the UTB-SOI and FinFET devices. The EδDC technology is compatible with the existing manufacturing and design flows, manufacturable in existing fabs, and compatible with legacy design IP and EDA tools and flows. A typical I_D-V_G characteristic of a EδDC transistor with gate length L_G = 45 nm is shown in Figure 13.16. The low-doped layer has a typical concentration of 2×10^{17} atoms/cm^3 and the high-doped layer has a typical concentration of around 3×10^{19} atoms/cm^3. The substrate concentration is typically 1×10^{18} atoms/cm^3.

Suggested Additional Readings

Bhattacharya, A.B., *Compact MOSFET Models for VLSI Design*, IEEE Press, John Wiley & Sons, Singapore, 2009.

Chau, R. et al., High-k/metal-gate stack and its MOSFET characteristics, *Proceedings of the IEEE Electron Device Letter*, 25(6), 408–410, June 2004.

Choi, Y.-K. et al., Ultrathin-body SOI MOSFET for deep-sub-tenth micron era, *Proceedings of the IEEE Electron Device Letter*, 21(5), 254–255, May 2000.

Frank, D.J., R.H. Dennard, E. Nowak, P.M. Soloman, Y. Taur, and H.-S.P. Wong, Device scaling limits of Si MOSFETs and their application dependencies, *Proceedings of the IEEE*, 89(3), 259–288, March 2001.

Hu, C.C., *Modern Semiconductor Devices for Integrated Circuits*, Pearson Education India, 2010.

International Technology Roadmap for Semiconductors, Process Integration, Devices and Structures, 2014, http://www.itrs.net/ Accessed on May 31, 2017.

Liu, Q. et al., High performance UTBB FDSOI devices featuring 20 nm gate length for 14 nm node and beyond, *Proceedings of IEEE IEDM Conference*, Washington, DC, 2013, pp. 9.2.1–9.2.4.

Ma, S.T. and J.R. Brews, Comparison of deep-submicrometer conventional and retrograde n-MOSFET's, *Proceedings of IEEE Transactions on Electron Devices*, 47(8), 1573–1579, August 2000.

Pandit, S., Nanoscale MOSFET: MOS transistor as basic building block, in *Introduction to Nano: Basics to Nanoscience and Nanotechnology*, Eds.: A. Sengupta and C.K. Sarkar, Springer Verlag, Berlin, Germany, 2015.

Pandit, S., C. Mandal, and A. Patra, *Nano-Scale CMOS Analog Circuits: Models and CAD Techniques for High-Level Design*, CRC Press, Boca Raton, FL, 2014.

Rogenmoser, R. and L.T. Clark, Reducing transistor variability for higher-performance lower-power chips, *Proceedings of the IEEE Micro*, 33(2), 18–26, March–April 2013.

Sengupta, S. and S. Pandit, Channel profile design of DC MOSFET for high intrinsic gain and low mismatch, *IEEE Transactions on Electron Devices*, 632, 551–557, February 2016.

Wu, S.-Y. et al., A 16 nm FinFET CMOS technology for mobile SoC and computing applications, *Proceeding of the IEEE IEDM Conference*, 2013, Washington, DC, pp. 9.1.1–9.1.4.

Yeo, Y.-C., T.-K. King, and C. Hu, MOSFET gate leakage modeling and selection guide for alternative gate dielectrics based on leakage considerations, *Proceeding of the IEEE Transactions on Electron Devices*, 50(4), 1027–1035, April 2003.

Yu, B., C.H.J. Wann, E.D. Nowak, K. Noda, and C. Hu, Short channel effect improved by lateral channel engineering in deep submicronmeter MOSFETs, *Proceedings of the IEEE Transactions on Electron Devices*, 44(4), 627–634, April 1997.

14

Nanotechnology Applications in Electron Devices

Sarosij Adak, Arghyadeep Sarkar, and Sanjit Kumar Swain

CONTENTS

14.1 Introduction

In the last three decades, microelectronics research activities have registered a nearly exponential growth. This continuous development has led to the miniaturization of devices and integrated circuits, which has consequently increased the circuit density and complexity of the systems at lower costs. However, the critical device size for commercial products is already approaching the sub 20 nm dimensions. Due to this downscaling, it becomes very difficult to maintain the necessary device performance because of significantly enhanced in short-channel effects. Presently, the device drive current has increased for fast switching. This enables an exponential increase in the leakage current and causes excessive standby power dissipation. Moreover, the physical limits and the ultimate performances of silicon MOSFETs, in terms of power handling, high frequency, maximum operating temperature, and breakdown voltage, have already reached their optimum values. Thus, it is indispensable to investigate new channel materials and improved device structures that would present us with energy-efficient solutions, high switching speeds, high power-handling capacity, high-frequency operation, maximum operating temperature, high breakdown voltage, and low power consumption.

From the new channel material perspective, high-mobility III–V semiconductors having significant transport advantages are extensively being researched as alternative channel materials for upcoming highly scaled devices. The III–V compound semiconductor binaries such as GaAs, GaN, and InP; ternaries such as InGaAs, AlGaAs, AlGaN, and AlInN; and quaternaries such as InAlGaN, InAlGaAs, and GaInPAs are widely studied for enhancing device performance. The GaAs-based compounds have much higher mobility than the silicon-based ones, and they are thus suitable for high-speed operations. On the other hand, III-nitride compounds have a higher bandgap than that of silicon, enabling wide-temperature, high-power, and high-frequency operations. The III-nitride compounds also offer higher breakdown voltage. The heterostructure devices using III–V compound semiconductors consist of two or more layers of different bandgaps grown one above the other. Such heterostructure leads to the formation of a high-mobility quantum well at the heterointerface of the wide-bandgap and the narrow-bandgap semiconductor layers. Hence, charge carriers confined in these quantum wells can move only in two dimensions (2D) and have very high mobility, and they are thus termed two-dimensional electron gas (2DEG).

The selection of the heterostructure system used in such devices can be done from a wide range of options such as AlGaN/GaN, AlInN/GaN, InGaAs/InP, and AlInAs/GaAs. Furthermore, in each of these heterostructures, wide variation in different performance parameters is possible by varying the mole fraction. Thus, with respect to the wide range of material options, the selection would depend on the desired properties for a particular set of applications. Along with the mastery of growth, III-nitride alloys are attractive for electronic applications due to their unique property of obtaining large polarization fields. The inbuilt polarization promotes the formation of heterostructures with high 2DEG carrier density, possessing high mobility at the heterointerface. The GaN-based heterostructures (such as the traditional AlGaN/GaN), having high breakdown voltage and large carrier density with high mobility, are ideal for high-frequency and high-power applications. The heterojunction in the AlGaN/GaN and AlInN/GaN confines the electrons to a triangular quantum well. The spatial separation between the channel carriers and the ionized donors in AlGaN/GaN and AlInN/GaN heterostructures helps to minimize Coulomb scattering.

An electron hall mobility in excess of 2000 cm²/V at room temperature has been demonstrated in the 2DEG channel. The existence of strong polarization fields within the GaN-based heterostructures offers new possibilities for device designs, which are not accessible in the conventional GaAs- and InP-based III–V semiconductors.

14.2 Material Structure and Polarization of Wurtzite GaN-Based Devices

GaN- and III-nitride-based semiconductors can exist under various crystal structures, such as wurtzite, zinkblende, and rock-salt [1,2]. The wurtzite structure is the most thermodynamically metastable structure at ambient conditions (WZ or 2H). Thus, the majority of the AlGaN/GaN devices are grown on the wurtzite phase. The wurtzite structure has a hexagonal unit cell and contains two intercepting hexagonal closed-packed sublattices. Therefore, it is defined by two lattice parameters, a_0 and c_0, having an ideal ratio $c_0/a_0 = \sqrt{8/3} \approx 1.633$, as shown in the stick and ball representation of Figure 14.1. Each sublattice is made by one type of atoms, which is shifted with respect to the other along the c-axis by the amount $u_0 = 3/8$ in fractional coordinates [3]. Due to the different metal cations, the bond lengths and the resultant c_0/a_0 ratios of the common III-nitride compounds and their alloys are different [3].

The wurtzite GaN-based semiconductors have a polar axis due to the lack of inversion symmetry in the <0001> direction. The difference in electro negativity (here given in Pauling's scale) between the Gallium (1.81) and/or

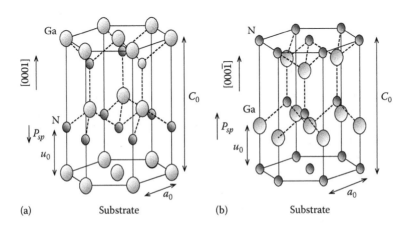

FIGURE 14.1
Schematic of the wurtzite (a) Ga-face and (b) N-face GaN.

Aluminum (1.61) and/or Indium (1.78) atoms, and the Nitrogen (3.04) atom, leads to group III-nitrides possessing high ionicity of the metal-nitrogen covalent bond [3]. By virtue of this electronic charge redistribution, inherent in the crystal structure, the group III-nitrides show a very strong polarization, referred as spontaneous polarization (*Psp*) [4]. Owing to the wurtzite structure, GaN and III-nitrides can have different polarities due to an uneven charge distribution between the neighboring atoms in the lattice. Figure 14.1 shows the two possible polarities: in the cation-face (Ga-face) structures, the polarization field points away from the surface to the substrate, while in the anion-face (N-face) structures the direction of the polarization field is inverted [3].

14.3 Polarization

The origin of polarization effect in the case of the GaN wurtzite crystal structure is mainly due to its non-centrosymmetry. The theoretical observation by King-Smith and Vanderbilt [5] elaborates the detailed analysis of the polarization effects quantitatively in the GaN-based heterostructure. This polarization effect facilitates the GaN-based HEMT and MOS-HEMT structures to be fabricated with a heterostructure having intrinsic layers. The 2DEG density in these devices was found to have a strong affinity to the thickness as well as the composition of the AlGaN barrier layer in the AlGaN/GaN HFET. This observation gives the idea to explain the origin of the 2DEG carriers in an undoped AlGaN/GaN HFET. The carriers in the channel are believed to have originated from the surface donor states present at the AlGaN surface. Broadly, there are two types of polarization effects that take place in a GaN-based heterostructure: spontaneous polarization and piezoelectric polarization. Both of these effects are explained in the following sections in details.

14.3.1 Spontaneous Polarization

This type of polarization depends on the polarity of the crystal structure. Spontaneous polarization, that is, polarization at zero strain, is basically due to bonds between the cation (Ga) and the anion (N) sites being noncentrosymmetric along the *c*-direction (0001) in a GaN crystal structure. We provide a simplified understanding of the polarization effects. The polarization vector present in the GaN crystal can be explained by the tetrahedral stick and ball diagram as shown in Figure 14.2a. The directions of polarization in this structure are defined as such due to the electron clouds being closer to the N atoms. This difference in the electro negativity together with the non-centrosymmetrical structure creates a polarization dipole between the cation (Ga) and the anion (N) sites in the GaN. The horizontal component

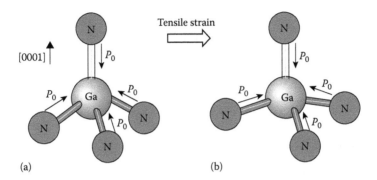

FIGURE 14.2
Tetrahedral stick and ball representation of GaN under (a) no strain and (b) tensile strain.

of polarization is usually assumed to have canceled out and only the vertical component of spontaneous polarization in the three diagonal bonds has been taken into account. The presence of a polarization dipole alone does not constitute a polarization charge existing at the interface of the heterostructure. Instead, the difference in polarization across the interface of the two layers (e.g., AlGaN/GaN) will result in polarization charge to be present at the interface of the heterostructure. Depending on the composition of the adjacent layers in the heterostructure, the difference in electronegativity across the heterointerfaces can be very different, thus resulting in a different polarization. This difference in polarization between two adjacent layers will result in a polarization charge across the heterostructure. The polarization charges due to spontaneous polarization are not always well defined as it requires polarization between two phases to be connected by a pseudomorphic boundary [6].

14.3.2 Piezoelectric Polarization

The spontaneous polarization exists in an unstrained nitride-based heterostructure. However, unstrained nitride-based heterostructures are fairly uncommon. The difference in the lattice constant or the difference in the thermal coefficient of expansion between the adjacent layers (e.g., GaN and AlGaN) would result in either a compressive or a tensile strain [6]. By virtue of the non-centrosymmetrical nature of the nitride-based crystal structure, the compressive or the tensile stress would alter the polarization in these structures. This can be explained with the tetrahedral stick and ball figure of GaN (Figure 14.2b). Due to the tensile strain in the GaN layer, there is a reduction of the vertical polarization component from the three diagonal bonds, and thus it would alter polarization. This would result in a difference of the polarization, which will eventually result in a polarization charge confined at the heterostructure interface. In contrast with the spontaneous polarization, where the polarization depends on the type and composition of the alloy, the piezoelectric polarization requires additional information such as strain, piezoelectric coefficients, and elastic constants, all of which might be directly or indirectly affected by the composition of nitride alloys. The resultant polarization is the combined effect of spontaneous and piezoelectric polarizations.

14.4 Fabrication Steps of AlGaN/GaN-Based HEMT Devices

The fabrication methods related to AlGaN/GaN heterostructure-based HEMT devices are considered to be better optimized with the sapphire/silicon carbide/silicon-based substrates. Sapphire is one of the most cost-effective materials for GaN epitaxial layer growth, but at the same time heat may not be proficiently channelized through the sapphire substrate due to its poor thermal conductivity. Consequently, Silicon (Si) and silicon carbide (SiC) are considered to be better with respect to the sapphire substrate for good thermal conductivity. In this case, the lattice constant of the substrates is close to GaN, but the growth process of GaN on these substrates is found to be not so easy and cost effective. Figure 14.3a represents the AlGaN/GaN heterostructure which is grown by the process of metal organic chemical vapor deposition (MOCVD). The AlGaN/GaN HEMT structure having an undoped AlN layer is sandwiched between the AlGaN and GaN layers to reduce the impurity scattering due to 2DEG in the channel electrons at the interface of AlGaN and GaN. The thickness of the unintentionally doped GaN buffer layer is generally in the range of 1–5 μm, and the thickness of the AlGaN barrier layer is 5–30 nm depending upon the application area. The mole fraction of aluminum in the AlGaN barrier is optimized in the order of 0.25 to have adequate confinement of the 2DEG electron by suitable conduction band offset and also to decrease the interfacial defects at the interface of the AlGaN barrier layer and the GaN buffer layer. Figure 14.3 indicates an outline of the AlGaN/GaN HEMT fabrication process.

14.5 Basic Operation Principle of AlGaN/GaN-Based HEMT Devices

AlGaN/GaN HEMT is a three-terminal device, where the current flowing between the drain and the source is restricted by the space charge, which can be changed by using the voltage to the gate terminal. The flow of current between drain and source is through the 2D conducting channel, which is formed by the 2DEG. The 2DEG is formed at the interface between the GaN

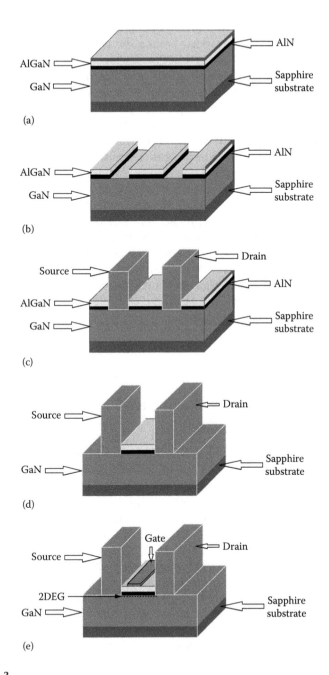

FIGURE 14.3
Schematic of basic steps for AlGaN/GaN HEMT fabrication: (a) AlGaN/GaN heterostructure grown by MOCVD; (b) recess etching for drain and source region, ohmic contact (Ti/Al) deposition, and annealing (700°C–850°C); (c) mesa etching; (d) gate recess etching; (e) gate metal stack (Schottky contact) deposition.

and the strained AlGaN barrier layers due to the difference in the spontane-ous and piezoelectric polarizations among the two crystals, and a conduc-tion band offset is formed on the GaN buffer layer which looks almost like 2D sheet charge. The quality and quantity of the 2DEG mainly depend on the thickness of the AlGaN barrier layer, Al mole fraction, substrate mate-rial, and growing method. After applying a positive drain voltage, the drain current flow through the 2DEG will start due to the potential difference between drain and source terminals. The drain current magnitude is con-trolled by the voltage applied to the gate terminal. When the gate terminal voltage increases in the negative magnitude, then the space charge under the gate spreads toward the 2D channel with electrons, and after reaching the 2D channel this starts to deplete beneath the gate area. As a result, the drain cur-rent reduces until the channel is pinched-off. When the applied gate terminal voltage is higher than the pinch-off voltage, then the electrons flow between the source and the drain terminal and, increasing the drain terminal volt-age, the drain-to-source (I_{DS}) current increases in a linear fashion up to a cer-tain value. After this certain drain-to-source current value, the current in the device is about to saturate. The saturated drain-to-source current depends upon the concentration of the 2DEG.

14.6 Applications of GaN-Based HEMT

14.6.1 High-Power Applications

The GaN material is basically a binary III–V direct bandgap semiconductor, having a wide bandgap of 3.4 eV, enabling these devices to operate at higher temperatures. As a result of this GaN power, HEMT devices require less cool-ing arrangements and fewer numbers of high-value processing steps related with complex structures designed for maximum heat removal [6]. Moreover, the GaN-based devices offer a higher breakdown voltage, and their criti-cal breakdown field is about 4 MV/cm as compared to those of GaAs and Si, which are 0.4 and 0.2 MV/cm, respectively [7]. Furthermore, there are several structural modifications which can improve the breakdown voltage as compared to the basic GaN-based HEMTs. A field plate is one of them. There are different types of field plates: gate field plate, source field plate, and multiple field plate [8–10]. Field plate length optimization is very much required to get the maximum breakdown voltage. A field plate is nothing but a metal electrode. When it is connected with the gate and the source then it is called gate field plate and source field plate, respectively. The introduction of field plate tends to distribute the electric field more along drain end rather than concentrate in the gate region. This phenomenon helps to widen the depletion region under the gate terminal. For this reason, the electric field is distributed more consistently between the gate and the drain region. J. G. Lee et al. used the optimization of a dual field plate in AlGaN/GaN on Si to achieve the breakdown voltage of 1590 V [11].

Additionally, the breakdown voltage can be further increased by using a back barrier [12] having a wider bandgap property and optimizing the gate-to-drain distance of the AlGaN/GaN HEMT [13].

14.6.2 Mixed-Signal Applications

Both normally on and normally off devices are very much required for the mixed-signal application. It is known that due to the presence of 2DEG in the channel of GaN-based HEMTs are generally normally on devices so it is very difficult to use the GaN-based HEMTs for a mixed-signal system on chip applications. After applying some techniques and structural modifications in the GaN-based HEMTs, the normally on devices turn into the normally off mode. Recess gate, n++ GaN cap layer, and piezo neutralization techniques [14–16] are the methods which can help the normally on devices to turn into normally off devices. Y. Kong et al. and Y. Tang et al. have demonstrated the AlGaN/GaN and InAlN/AlN/GaN enhancement and depletion mode HEMTs, respectively, which can be used for a mixed-signal system on chip applications [17,18].

14.6.3 RF Applications

GaN-based HEMT devices have become one of the interesting candidates for very-high-frequency applications due to their very high carrier density and high carrier velocity ($V_{peak} \sim 2.5 \times 10^7$ cm/s) in the GaN channel. The cut off frequency (f_t) and the maximum cut off frequency (f_{max}) are the major figures of merit for the Rf applications. These figures of merit can be further improved by scaling the device gate length and using a wide-bandgap back-barrier compound semiconductor material. M. L. Schuette et al. have investigated 27 nm gate-length InAlN/AlN/GaN HEMTs and obtained 348 GHz f_t and 337 GHz f_{max} for the normally off mode and 303 GHz f_t and 279 GHz f_{max} for the normally on mode HEMT devices [19]. These results are higher than those of 30 nm gate-length devices having 220 GHz f_t and 245 GHz f_{max} for the normally off mode and 192 GHz f_t and 221 GHz f_{max} for the normally on mode HEMT devices, which have been demonstrated by B. Song et al. [20]. Furthermore, D. S. Lee et al. have observed that after using a wide-bandgap AlGaN back barrier in the 65 nm gate-length InAlN/GaN HEMT, the f_t has enhanced from 195 to 210 GHz [21]. By the downscaling of the gate length of the GaN-based HEMT devices and by using the back barrier, the major Rf figures of merit are improved, which is highly favorable for Rf applications.

14.6.4 Sensor Applications

The higher bandgap, high chemical and thermal stability, nontoxicity to the living cells, and high sensitivity of the AlGaN/GaN HEMT devices are prospective for biosensor applications. R. Thapa et al. have confirmed level-free identification of DNA hybridization with the help of an AlGaN/GaN HEMT-based transducer having a biofunctionalized gate terminal [22]. H. H. Lee et al. have found a real-time and label-free detection method of the C-reactive protein using two AlGaN/GaN HEMT devices, that is, biosensing HEMT and reference HEMT [23]. AlGaN/GaN HEMT devices are also more suitable for biosensing applications as compared to conventional detection methods (enzyme-linked immunosorbent assay, immunoturbidimetry, and visual agglutination) of biological molecules as they take less time and are cost effective.

References

1. Trampert, A., O. Brandt, and K. H. Ploog, Crystal structure of group III nitrides. *Semiconductors and Semimetals* 50 (1997): 167–192.
2. Morkoç, H., *Handbook of Nitride Semiconductors and Devices, Materials Properties, Physics and Growth*, Vol. 1. John Wiley & Sons, Berlin, Germany, 2009.
3. Treidel, E. B., GaN-based HEMTs for high voltage operation design, technology and characterization, PhD dissertation, Technical University of Berlin, Berlin, Germany, 2012.
4. Sze, S. M., and K. Ng. Kwok, *Physics of Semiconductor Devices*. John Wiley & Sons, New York, 2006.
5. King-Smith, R. D. and D. Vanderbilt, Theory of polarization of crystalline solids. *Physical Review B* 47(3) (1993): 1651.
6. Pearton, S. J. et al., Fabrication and performance of GaN electronic devices. *Materials Science and Engineering: R: Reports* 30(3) (2000): 55–212.
7. Bandić, Z. Z. et al., High voltage (450 V) GaN schottky rectifiers. *Applied Physics Letters* 74(9) (1999): 1266–1268.
8. Karmalkar, S. and U. K. Mishra, Enhancement of breakdown voltage in AlGaN/GaN high electron mobility transistors using a field plate. *IEEE Transactions on Electron Devices* 48(8) (2001): 1515–1521.
9. Saito, W. et al., High breakdown voltage AlGaN-GaN power-HEMT design and high current density switching behavior. *IEEE Transactions on Electron Devices* 50(12) (2003): 2528–2531.
10. Xing, H. et al., High breakdown voltage AlGaN-GaN HEMTs achieved by multiple field plates. *IEEE Electron Device Letters* 25(4) (2004): 161–163.
11. Lee, J.-G. et al., High breakdown voltage (1590 V) AlGaN/GaN-on-Si HFETs with optimized dual field plates. *CS MANTECH Conference*, Boston, MA, April 23–26, 2012.
12. Lee, H.-S. et al., 3000-V 4.3-InAlN/GaN MOSHEMTs with AlGaN back barrier. *IEEE Electron Device Letters* 33(7) (2012): 982–984.

13. Hasan, Md. T. et al., Current collapse suppression by gate field-plate in AlGaN/GaN HEMTs. *IEEE Electron Device Letters* 34(11) (2013): 1379–1381.

14. Wang, R. et al., Gate-recessed enhancement-mode InAlN/AlN/GaN HEMTs with 1.9-A/mm drain current density and 800-mS/mm transconductance. *IEEE Electron Device Letters* 31(12) (2010): 1383–1385.

15. Kuzmik, J. et al., Proposal and performance analysis of normally off GaN++/InAlN/AlN/GaN HEMTs with 1-nm-thick InAlN barrier. *IEEE Transactions on Electron Devices* 57(9) (2010): 2144–2154.

16. Ota, K. et al., A normally-off GaN FET with high threshold voltage uniformity using a novel piezo neutralization technique. *2009 IEEE International Electron Devices Meeting (IEDM)*, Baltimore, MD. IEEE, 2009.

17. Kong, Y. et al., Monolithic integration of E/D-mode AlGaN/GaN MIS-HEMTs. *IEEE Electron Device Letters* 35(3) (2014): 336–338.

18. Tang, Y. et al., High-performance monolithically-integrated E/D mode InAlN/AlN/GaN HEMTs for mixed-signal applications. *2010 IEEE International Electron Devices Meeting (IEDM)*, San Francisco, CA. IEEE, 2010.

19. Schuette, M. L. et al., Gate-recessed integrated E/D GaN HEMT technology with $f_T/f_{max} > 300$ GHz. *IEEE Electron Device Letters* 34(6) (2013): 741–743.

20. Song, B. et al., Monolithically integrated E/D-mode InAlN HEMTs with $f_t/f_{max} > 200/220$ GHz. *2012 70th Annual Device Research Conference (DRC)*, University Press, Austin, TX. IEEE, 2012.

21. Lee, D. S. et al., InAlN/GaN HEMTs with AlGaN back barriers. *IEEE Electron Device Letters* 32(5) (2011): 617–619.

22. Thapa, R. et al., Biofunctionalized AlGaN/GaN high electron mobility transistor for DNA hybridization detection. *Applied Physics Letters* 100(23) (2012): 232109.

23. Lee, H. H. et al., Differential-mode HEMT-based biosensor for real-time and label-free detection of C-reactive protein. *Sensors and Actuators B: Chemical* 234 (2016): 316–323.

15

Micro/Nanoelectromechanical Systems

Atanu Kundu

CONTENTS

15.1 Introduction

A nanoelectromechanical system (NEMS) is a nanoscale version of the microelectromechanical system (MEMS), which has both an electrical and a mechanical component [1]. One nanometer is a billionth of a meter, or 10^{-9} meter. To visualize the nanometer dimension, it can be compared with the human hair. Normally, the diameter of human hair is around 100 μm; it has to be divided 100 times equally to get 1 μm, and that 1 μm has to be further divided 1000 times equally to get 1 nm.

NEMS/MEMS are very small devices or group of devices that can integrate both mechanical and electrical components within a space of a few nano/micrometers. NEMS can be constructed on one chip that contains one or more mechanical components and electrical circuitry for inputs and outputs having dimensions in nanometers. NEMS belongs to a much broader category of devices called transducers, whose operation involves interplay between different forms of energy [1].

Different types of transducers are electromagnetic, electrochemical, thermoelectric, radio acoustic, and electrochemical.

A major application of MEMS/NEMS is as sensors. The primary sensors are pressure sensors, chemical sensors, and inertial sensors. The basic applications of an inertial sensor include the accelerometer and the gyroscope. In addition to sensors, MEMS/NEMS mechanical parts consist of pumping devices, gear trains, movable mirrors, miniature robot, tweezers, tools, and laser. These devices have found numerous applications in the following fields [2–5]:

- Biomedical
- Optical communication
- Wireless network
- Aerospace
- Consumer products
- Gaming application
- Automotive industry
- Information technology

15.2 Sensor and Actuators

A sensor [6–8] is an object whose purpose is to detect events or changes in its environment and provide a corresponding output. An actuator [9–12] is a component of machines that is responsible for moving or controlling their mechanism or system.

15.3 How an Electromechanical System Works

M/NEMS usually consist of two parts: sensors and actuators. The sensor part collects information from its surroundings and the actuator executes the given commands by movements of different parts. M/NEMS is constructed with mechanical as well as electrical components. The electrical elements in M/NEMS process data while the mechanical elements act in response to that data; hence, the two elements form a complete system-on-a-chip (SoC) integrated circuitry. The sensor gathers information from the environment by measuring mechanical, thermal, biological, chemical, optical, mass, gravity, and magnetic phenomena and also converts this information about the physical quantities into electrical signals. After sensing these physical quantities, the data/information is collected and analyzed and then decisions are made. The electronic part of the system processes the information derived from the sensors and then directs the actuators to respond by moving, positioning, regulating, pumping, and filtering, in order to control the environment for some desired outcome or purpose, which is called the actuation operation. In electromechanical systems, mechanical and electrical elements work as independent devices and complete the entire work. The entire system is fabricated using the basic fabrication techniques and materials of nanoelectronics.

15.4 History of Electromechanical Systems

The automotive industry was one of the first industries that commercially embraced MEMS devices in the 1990s with the accelerometer, an M/NEMS device used to sense a sudden change in velocity, causing the activation of airbags at the time of a crash. Other applications in the automotive industry are monitors for tire pressure and fuel pressure, sensors for collision avoidance, crash sensors for air bag safety systems, sensors for skid detection, smart suspension for sport utility vehicles to rollover risk, automatic seat belt restraints, door locking, vehicle security, headlight leveling, and navigation.

In 1785, Charles-Augustin de Coulomb [13,14] designed an electromechanical system to measure electrical charge. Coulomb measured the magnitude of the electric forces between charged objects using the torsion balance that he invented, from which the now-famous Coulomb's law was derived. Coulomb's torsion balance consist of two spherical plates, where, the difference of charge is converted to an attractive/repulsive force. As time moved forward, the electromechanical system became smaller and more sophisticated due to numerous technological breakthroughs listed in Table 15.1.

15.5 NEMS and Scaling Effects

In the electromechanical systems, if the component size is in the order of micrometers, it is called a MEMS and if the components are in the order of nanometers, it is called a NEMS. Ultrasmall silicon-based NEMSs fail to achieve high-quality factors because of the dominance of the surface effect [16], such as surface oxidation [17], reconstruction, and thermoelastic damping [18]. However, processes such as electron beam lithography and micromachining now enable semiconductor nanostructures to be fabricated below 10 nm, which

TABLE 15.1

Technological Breakthroughs

1940	Development of pure semiconductors (Ge or Si) during the World War II.
1947	Invention of the point-contact transistor that heralded the beginning of the semiconductor circuit industry.
1949	Ability to grow pure single-crystal silicon improved the performance of semiconductor transistors. However, their cost and reliability were not completely satisfactory.
1953	Silicon oxidation.
1956	Nobel Prize for the invention of transistor.
1958	Jack Kilby was awarded Nobel Prize for the invention of Integrated Circuits (IC).
1959	Prof. Richard P. Feynman's pivotal lecture "There's plenty of room at the bottom" [15]. In that lecture, Prof. Feynman described the enormous amount of space available on the micro level. An entire encyclopedia could be written on the head of a pin. Thus, he described the enormous potential of micro- and nanofabrications.
1960	Invention of the planar batch fabrication process tremendously improved the reliability due to the invention of planar MOSFET at the same time.
1964	Resonant gate transistor with meal-beam micromachining, produced by Nathenson at Westinghouse, was the first engineered batch-fabricated MEMS device.
1970	Invention of the microprocessor, which found many applications that have been responsible for transforming our society, drove the higher demand.
1970–1980	MEMS commercialization was started and the first micromachined accelerometer was developed.
1982	Invention of the scanning tunneling microscope.
1984	Joint fabrication of MEMS and integrated circuits using the polysilicon surface micromachining process.
1986	Invention of the atomic force microscope.
1991	Discovery of the carbon nanotube and a fully integrated single-chip accelerometer by Analog Devices.
1996	Technique for producing carbon nanotubes of uniform diameter.
2008	One billion sensors produced by a single company (Bosch).
2010	Number of MEMS devices and applications continually increases.

enables the fabrication of NEMSs. However, there are other challenges to get fully realized NEMS devices, such as (1) communicating signals from nanoscale devices to the macroscopic world, (2) understanding and controlling mesoscopic machines, and (3) reproducing routine nanofabrication methods [19,20].

Components of the NEMS devices deflect or vibrate within a very small range during their operation. The deflection of a doubly clamped beam varies linearly with an applied force, and the amount of displacement is a few percent of the thickness of that nanodevice. Building a transducer that is sensitive enough to allow information to be transferred accurately and with great precision by sensing this small displacement is a challenge. Another difficulty is that the natural frequency of this motion is inversely proportional with size or mass. Hence, it increases with the decreasing size. The natural frequency is given as $\omega = \sqrt{k/m}$ and $=2\pi/\omega$; where k is the stiffness of the spring(Newton/meter), m is the mass (kg), and ω is the radian frequency. $\omega^2 = k/m$, which gives us $f = 2\pi\sqrt{k/m}$, where f is the natural frequency in hertz (cycle/s). Therefore, the ideal NEMS transducer should be capable of sensing displacements in the range of 10^{-15}–10^{-12} meters and of vibrating in the frequency of several gigahertz. Satisfying these two requirements simultaneously proves to be a very challenging task. Nanoscale electrode capacitances are of about 10^{-18} (farad) and less [21–23].

$$C = \varepsilon_r \varepsilon_0 \frac{A}{d}$$

C is the capacitance in farad.
A is the area overlapped by two plates.
ε_r is the relative static permittivity or dielectric constants of the materials between the plates; for vacuum, $\varepsilon_r = 1$. Table 15.2 denotes the permittivity of various elements.
ε_0 is the vacuum permittivity.

$$\varepsilon_0 \triangleq \frac{1}{C_0 \mu_0} = 8.8541878176...\times10^{-12} \text{ F/m}$$

C_0 is the speed of light
μ_0 is the vacuum permittivity
d is the distance between the plates

TABLE 15.2

Relative Permittivity of Materials

Material	ε_r
Vacuum	1 (by definition)
Air	1.0
Teflon	2.1
Polyethylene	2.25
Polyimide	3.4
Polypropylene	2.2–2.36
Polystyrene	2.4–2.7
Carbon disulfide	2.6
Mylar	3.1
Paper	3.85
Electroactive polymers	2–12
Mica	3–6
Silicon dioxide	3.9
Silicon nitride	7–8

As the changes in distance are in the range of a fraction of nanometer, the capacitance altered in the device motion is very small, which leads to unavoidable parasitic capacitance [22,23] that dominates in these nanodevices. Besides, optical methods like simple beam deflection or fiber-optic interferometry technique [24] fail due to the diffraction limit used in the scanning probe microscope to detect the deflection of the probe, so it is not applicable to measure the cross-section much less than the wavelength of the light.

One of the major challenges in achieving potentially high-quality NEMS devices are defects in bulk semiconductor material, interface and fabrication-induced surface damages [25–27]. These intrinsic effects can be minimized by carefully choosing the material, changing the device geometry, and changing the processing technology. Extrinsic effects also occur, such as resistance due to air during movement, clamping losses at the supports, and electrical losses mediated through the transducer [26–29]. The clamping and electrical losses can be minimized by careful engineering. In NEMSs, due to shrinking of the device size, the device properties are dominated by surfaces that can be minimized by ultrahigh-purity heterostructure. Nuclei have an intrinsic magnetic moment or spin, which can interact with an applied magnetic field. However, about 10^{14}–10^{16} nuclei are required to generate a measurable signal.

van der Waals force is present between molecules, atoms, and small particles owing to abrupt change in their electron clouds [30]. The surface-interaction-related force is called Casimir/van der Waals force, which occurs due to the electrostatic actuation [31]. It has been observed that molecules can attract each other at a moderate distance and repel each other at a close distance. This force-related measurement was accurately done by Theo Overbeek and Evert Verwaey in the 1940s, at Philips Research Laboratories in Eindhoven, Netherlands. The attractive forces, which were explained in the Quantum Mechanical Theory by Fritz London in 1930, are collectively called van der Waals forces [31–33]. However, when the separation between the molecules, atoms, or particles is below few nanometers, it becomes difficult to detect correctly by the theory predicted by London [33,34]. When the changes are faster than the time taken by the electromagnetic wave to propagate or the change in distance is few nanometers, it is essential that the special theory of relativity is taken into account. Hendrik Casimir explained how the delays modify the character of attractive forces, and the equation or the formula is denoted by the Casimir force, $F_c = ch\pi^2 A / (240L^4)$. Here, the force, F_c, depends on the plate size, A, and L is the distance between the plates, h is the Planck's constant, and c is the speed of light. When the separations between the plates are below 1 μm, the effect of Casimir force is present; otherwise, for larger distances, this force become negligible. As the NEMS devices having dimensions are in the order of 100 nm or sub 100 nm regime [35], it is also essential to consider this effect.

Besides, when device dimensions are in the order of the critical length of the interatomic distance of materials, the physical property of the material also changes. Therefore, the mechanical, electrical, thermal, and magnetic properties are dissimilar than the bulk property of a similar material [36,37].

During scaling, reduce in size causes changes in resistivity, thermal conductivity, mechanical properties and magnetic frustration. This provides limitations as well as opportunities [38–40]. For bulk material, the material property is independent of the size. However, for the device size in the order of nanoscale, the Young's modulus no more remains the same as the bulk value. Therefore, the mechanical property of the material changes due to the change in the Young's modulus.

15.6 Why Carbon Nanotubes, and Limitation of Si

The invention of transistor was revolutionary in the field of microelectronics technology. The vacuum tube used to require several hundred volts of anode voltage to operate, which required few watts of power, whereas transistors required only few millivolts of power. The driving force behind any modern electronic gadget is an integrated circuit (IC) containing a few billion (10^9) transistors. To enhance the functionality of an existing electronic gadget, it is desirable to put more number of transistors in the IC of the same chip area. For this reason the transistor size has to be scaled down day by day, with the increasing demand of complex portable device functionality. This leads to the aggressive scaling of the device size. The smaller transistor size will help to pack more and more devices in a given chip area; hence, the electronic gadgets will be smarter as well as cheaper. As the capacitance associated with the device will be reduced, the transistors will switch faster. The transistor will consume low power from the battery of the portable device and will also increase the functionality of the IC. After the inception of metal-oxide-semiconductor field effect transistor (MOSFET) in 1960 [41], it has been one of the most important devices in the semiconductor industry to be integrated in monolithic ICs to serve as a basic switching element for digital circuits and as an amplifier for analog applications. In the case of a MOSFET, the drain current equation in saturation is given in Equation 15.1

$$I_d = \left(\frac{\mu C_{ox}}{2}\right)\left(\frac{W}{L}\right)\left(V_{gs} - V_{th}\right)^2 \tag{15.1}$$

where
I_d is the drain current
μ is the mobility of the electron
C_{ox} is the oxide capacitance
W is the width of the channel
L is the length of the channel
V_{gs} is the gate to source voltage
V_{th} is the threshold voltage of the device

The drain current equation in Equation 15.1 is basically the function of technology ($\mu_n C_{ox}$), W/L, and Voltage ($V_{gs} - V_{th}$). Hence, drain current is represented as $I_d = f(\text{Technology}) \times f(\text{W/L}) \times f(\text{Voltage})$. Since, the mobility of an electron and the oxide capacitance cannot be changed once the device is fabricated following a particular technology, the technology parameter is constant once the device is fabricated. It signifies that if the technology and voltage parameters remain unchanged while W and L are reduced, keeping the ratio constant, the drain current, I_d, will also remain unchanged in spite of scaling the device. As a result, the MOSFET has been considered as a highly suitable transistor for VLSI circuits. The aggressive scaling of MOSFETs has started and the semiconductor industry has showcased a spectacular exponential growth in the number of transistors per IC for several decades, as predicted by Moore's law [42].

With the aggressive scaling of the MOSFET, the device channel length has reached below the 10 nm node. The scaling is essential for steady improvement in performances; however, as the channel length L is reduced to increase the number of components per chip, the so-called short channel effects (SCE) arise, which limit further scaling of the device dimensions. Due to the SCE, the device tends to lose control of the channel carriers when the source and drain are brought into close proximity. The source and drain can interact with each other ever in the off state when there is no conducting channel in between. Eventually, the gate terminal cannot turn off, and hence it increases the off-state leakage power. Therefore, power consumption of the device increases, and it is very difficult to use the device as there is no control over the gate. The device density is limited by the maximum allowable power dissipation. Therefore heat removal and processor cooling management is another important issue in this device dimension. The present MEMS/NEMS technology is dominated by silicon as it is compatible with the production for ICs and silicon-based micromachining technologies are much developed.

However, the silicon has certain intrinsic limitations as a material to work as a mechanical element. Silicon has poor mechanical properties like a low Young's modulus, weak fracture strength, a large coefficient of friction, poor thermal stability, and the surface stiction problem. Carbon-based materials show promise for MEMS/NEMS applications with their diverse and much better performance than silicon [43,44]. Carbon-based materials like diamond, carbon nanotube (CNT), graphene, SiC, and III–Vs nitrides possess excellent properties such as low mass, high Young's modulus, high thermal conductivity, and hydrophobic surface. This makes it very suitable for NEMS switches to microwave field application [45,46].

Among various materials for the fabrication of NEMSs, carbon nanostructures like diamond, graphene, and CNT have been considered as promising materials for a variety of applications due to their unique electronic and mechanical properties [47–49]. Smalley, Kroto, and their team discovered buckminsterfullerene, C_{60}, a new form of carbon in 1985. C_{60} was named

FIGURE 15.1
Buckyball (Buckminsterfullerine C_{60}) resembles a soccer-ball pattern.

buckminsterfullerene after the American architect Buckminster Fuller, who was famous for designing a large geodesic dome. They got the noble prize in 1996 in chemistry for their invention. C_{60} is a soccer-ball-like molecule, as shown in Figure 15.1, having exactly 1 nm diameter, made of pure carbon atoms bonded in hexagon and pentagon configurations [50,51].

It is worth mentioning that the Nobel prize in chemistry was given for figuring out how to make a carbon into a soccer ball. Why is it so important? It is an extremely stable carbon molecule with 720 protons, which means that it has to be made of exactly 60 carbon atoms with 12 protons each, and it does not react easily with other molecules, which is unusual because of a single carbon atom. It has whopping four spare electrons that can be used to make bonds with other atoms. The fact that this new molecule was not very reactive meant that each of these 60 carbon atoms had to have three of their electrons occupied with other carbons and only one electron free. This soccer-ball-like structure is the mixture of 12 pentagons and 20 hexagons in order for the cage to close completely. The free electron in each of the carbon atoms gives C_{60} a lot of flexibility and a high electrical conductivity, and even though it is very soft under normal conditions, C_{60} can be compressed between two diamond tips at 320,000 times atmospheric pressure to create a substance so hard it can dent diamond—the hardest substance on Earth. Buckyballs might soon find widespread use in medicine. When a molecule of Buckyball is attached to 12 molecules of nitrous oxide, the tiny structure can explode in a controlled reaction. Researchers call them Buckybombs, and eventually they could be used in individual cells to deliver medication or to destroy a tumor.

In 1991, Iijima discovered that a CNT is formed by rolling a sheet of graphene into a cylinder and the properties of CNTs depend on how the CNT is rolled [52]. CNTs can be geometrically considered as sheets of sp²-bonded graphene rolled into a long hollow tubule.

CNTs are at least 100 times stronger than steel although only one-sixth as heavy as steel [53]. They can conduct heat and electricity far better than copper. A CNT can be as thin as a few nanometers yet be as long as hundreds of microns. There are different types of CNTs, although they are usually categorized as either single-walled (SWCNT) or multiwalled (MWCNT) [54,55]. An SWCNT is just like a regular straw, having only one layer. CNTs can also occur as multiple concentric cylinders of carbon atoms—as a collection of nested tubes of continuously increasing diameters—and are called MWCNTs.

15.7 Subthreshold Swing of Silicon-Based Transistor

The subthreshold swing is a measure of how sharply the drain current changes with the change of gate voltage during switching from 0 V to the threshold voltage. The subthreshold swing is measured in mV/decade. The mV is the amount of incremental gate voltage required to change in the drain current by one decade. Due to fundamental thermodynamics constrains, the minimum value of the subthreshold swing at room temperature has to be 60 mV/decade. It is desired that during switching from 0 V to V_T, the drain current change should be more, with the minimum change of gate voltage. Therefore, the subthreshold swing value is desired to be as small as possible.

The subthreshold current is considered to be leakage current between the source and drain when the device is below the threshold voltage or in a weak inversion region. The drain current, I_D, of MOSFETs consists of a drift and a diffusion current component. In the weak inversion region ($V_{gs} < V_{th}$), the current between the source and the drain is dominated by the diffusion current component.

The subthreshold drain current is dominated by diffusion current and can be expressed by Equation 15.2 [56–59]:

$$I_{off} = \mu_n C_{ox}\left(\frac{W}{L}\right)\left(\frac{\sqrt{2q\varepsilon_{si}N_a}}{2\sqrt{\Psi_s}}\right)\left(\frac{kT}{q}\right)^2\left(\frac{n_i}{N_A}\right)^2\left(e^{\frac{q}{kT}\cdot\frac{(V_{GS}-V_{th})}{m}}\right)\left(1-e^{-\frac{q}{kT}V_{DS}}\right) \quad (15.2)$$

where

$$\Psi_s = \frac{(V_{GS}-V_{th})}{m}$$

$$m = 1+\frac{C_{dep}}{C_{ox}}$$

Body effect coefficient, $\gamma = \sqrt{2q\varepsilon_{si}N_a}$

kT/q is called thermal voltage

k is Boltzmann constant = 1.38×10^{-23} J/K

q is the charge of an electron = 1.602×10^{-19} C

It is evident from Equation 15.2 that to achieve low I_{off}, that is, to minimize off-state leakage current, channel length L has to be increased, which is against the device scaling. It is also evident from Equation 15.2 that the increase of threshold voltage V_{th} will decrease I_{off}; however, a threshold voltage rise will degrade the device performance. Another parameter in this equation, substrate doping, N_A, can minimize I_{off} as it is proportional to the off-state leakage current, but the substrate needs to be kept undoped to avoid nonuniform dopant distribution in sub 100 nm MOSFETs. High substrate doping will lead to increased SCEs, like impurity scattering.

Subthreshold slope (SS) is a measure of the ability to turn off a device, with a value typically of 60–100 mV per order of magnitude in the current change, normally expressed as 60–100 mV/decade. The lower the SS, the steeper the curve with V_{gs} and the leakage will be lower. The SS is considered as $(d(\log_{10}I_{off})/dV_{GS})^{-1}$. Therefore, performing the partial differentiation, the SS becomes

$$SS = \left(\frac{\partial\left(\log_{10}I_{off}\right)}{\partial V_{GS}}\right)^{-1} = 2.3\frac{kT}{q}\left(1+\frac{C_{dep}}{C_{ox}}\right) \tag{15.3}$$

$m = 1+ (C_{dep}/C_{ox})$, varies from 1.1 to 1.4 and kT/q = 26 mV.

Hence, subthreshold swing, SS (minimum), becomes $2.3 \times 1.1 \times 26 \approx 65.78$.

Therefore, the SS is always greater than ~60 mV/decade at room temperature. The oxide capacitance, $C_{ox}=\varepsilon A/t_{ox}$ and it is evident from Equation 15.3 that a sharper subthreshold swing or subthreshold swing minimization is possible by the use of lesser gate oxide thickness, t_{ox}, which causes larger C_{ox}. Therefore, to attain an improved subthreshold swing, the VLSI industry started to scale down the oxide thickness. The shrinking of oxide thickness leads to the gate tunneling current, which is another roadblock to minimize off-state leakage current.

15.8 Post-Silicon Material for CMOS and NEMS

The subthreshold leakage power increases significantly for a silicon-based MOS transistor as the device reaches the sub-100 nm regime. The subthreshold leakage power was considered insignificant for the earlier generations of ICs; however, for the advanced processors, the number of transistors are in the order of few billions (10^9) [60,61]. Though the subthreshold current is in the order of nano- or picoampere, the total leakage current still reaches the ampere range, which is the matter of concern.

Additionally, this leakage current is strongly temperature dependent. There is a strong coupling between temperature and leakage current, which further increases the total power dissipation of the circuit [62–64].

For a bulk MOSFET, the subthreshold swing cannot be lower than the 60 mV/decade due to the fundamental limits. This fundamental limit is one of the major shortcomings of CMOS devices, and alternative technologies need to be explored in order to achieve a lower subthreshold swing. The theoretical value of the subthreshold swing is 60 mV/decade, but practically all CMOS-based transistors (bulk, fully-depleted SOI [FDSOI] and FinFET) have SSs higher than *60 mV/decade*. The alternative devices like tunneling CNT transistors (T-CNFET), nanowire-based devices (NW-FET), and impact-ionization-based MOS (I-MOS) have the subthreshold swing of *40, 35,* and *8.9 mV/decade*, respectively. However, it has been experimentally shown that an electromechanical device shows subthreshold swing of *2 mV/decade* due to having physically movable gate, by which the source and drain can be separated, as shown in Figure 15.2 [65–67].

Therefore, major emphasis is given on these devices to design transistors for IC fabrications. Diamond, CNT, and graphene are highly suitable materials for designing these electromechanical devices due to their low mass, excellent chemical stability, and high mechanical stiffness. Besides, these materials possess high thermal conductivity and high saturation current density. These carbon-based materials have hydrophobic surfaces, by which the surface stiction problem between the contacts can be avoided [68].

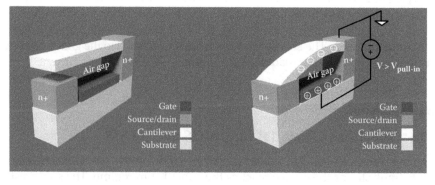

FIGURE 15.2
Basic operation of a suspended-gate NEM FET working as an electrostatic switch.

15.9 Graphene NEMS

Graphene is recently shown to have made numerous scientific and technological breakthroughs with novel nanodevice applications. Graphene is defined as a flat mono layer of sp^2-bonded carbon atoms, which is the basic building block for graphitic materials. The conventional theory of thermodynamics says that "freestanding" graphene is unstable and it is required to roll into a nanotube or other curved structure.

Graphene is a zero bandgap semiconductor material, having massless charge carriers. Graphene possess high carrier mobility with high carrier density, even at a very high room temperature, for its extraordinary electronic band structure. Besides, graphene has high thermal conductivity, ultrahigh stiffness, and hydrophobic surface. Bulk graphite is highly anisotropic, which is highly suitable for nanomachining of devices having a high aspect ratio. The Young's modulus of graphite is as high as 920 GPa, which is very high when compared to silicon's Young's modulus at 130 GPa. The breaking strength of graphene is 200 times greater than steel. These properties make graphene a potential candidate for NEMS [69–71].

The thermal conductivity of graphene (48 W/cm K) is also much higher compared to that of silicon (1.5 W/cm K), and its intrinsic thermal conductivity is independent of the number of the layers. The preparation method determines the electrical properties of graphene. Graphene scratched from bulk graphite has superior electrical properties compared to the other semiconductor materials. Measurements have revealed that it has higher electron mobility (>200,000 cm^2/V s) at room temperature than has silicon (1450 cm^2/V s). Graphene's current carrying capacity is also around thousand times greater than that of Cu [44,72].

15.10 Diamond NEMS

One of the more interesting prospects in the semiconductor world is diamond, for many of its properties are superior to those of silicon. Diamond is a wide bandgap semiconductor material of 5.5 eV, and hence it is highly suitable for high-frequency, high-power, and high-temperature devices [73].

Diamond has also very high breakdown field (10^7 V/cm) and the maximum thermal conductivity in materials (20 W/cm K). The diamond p–n junction diode shows rectification even at 370 K, which implies that it is possible to use diamond as a semiconductor in high-temperature conditions [74].

For its wide bandgap, diamond is intrinsically insulating; therefore, additional insulators such as SiO_2 can be avoided in diamond devices, either electronically or electromechanically. In order to make diamond conducting, it requires a doping process. Boron is used as a p-type dopant and nitrogen is used as an n-type dopant. The doping process is similar to the silicon doping process; however, it requires more time compared to the silicon doping because it is very difficult to diffuse a dopant through the strongly bonded diamond structure.

The electron and hole mobilities in a diamond are 4500 and 3800 cm^2/V s, respectively, which are quite higher than silicon. The saturation velocity is also very high in diamond: it is 2.5×10^7 and 10^7 cm/s for electrons and holes, respectively. This makes diamond excellent for electronic devices under high electric fields [72,75].

The diamond's Young's modulus is of 1200 GPa, which is the highest among the carbon-based materials. Therefore, diamond has the potential to work as the best semiconductor material for semiconductor, photonic, and MEMS/NEMS applications. Besides, diamond has unique physical, chemical, thermal, and mechanical properties due to its strong and short covalent bond (sp^3) between the carbon atoms [72,75].

Movable or suspended structures formation is possible by micro- or nanomachining of diamond. However, due to the lack of a solid carbon oxide on the diamond surface and its high resistance to chemical attack, bulk machining of diamond is quite challenging.

15.11 Carbon Nanotubes NEMS

CNT is the cylindrical structure formed by a rolled-up graphene sheet, with a length up to 1 µm. Due to its unique configuration; nano size; and excellent electrical and mechanical properties like high stiffness, remarkable thermal conductivity, low dimension, and low mass, CNT has become the most attractive material for NEMS applications [54].

Due to an application of stress, however, NTs get deformed and consequently changes take place in their bandgap. The changes in the bandgap transfer CNT sometimes from the semiconducting to the metallic type. This property of CNT can be utilized for various electromechanical sensor devices. It is feasible to design different CNT-based NEMS devices such as atomic-resolution mass sensors, displacement sensors, and high-frequency (GHz) resonators [54,76,77].

A CNT is mainly of two types: Metallic CNT and Semiconducting CNT. The metallic CNT is metallic in nature, having a bandgap of approximately zero, and MWCNTs, consisting of multiple graphene layers like concentric cylinders, are mostly metallic in nature. MWCNTs are mostly used for

TABLE 15.3

Properties of Carbon-Based Materials Compared with Si and Other Materials

Properties	Si	SWCNT	Graphene	Diamond
Density (g/cm³)	2.33	1.3–1.4	>1	3.52
Young's modulus (GPa)	130	~1,000	~1,000	1200
Thermal conductivity (W/cm K)	1.5	35	48	24
Bandgap (eV)	1.1	0–2	~0	5.5
Electron mobility (cm²/V s)	1450	100,000	200,000	4500
Hole mobility (cm²/V s)	480	4,000	>100,000	3800
Breakdown field (MV/cm)	0.3	—	—	10
Dielectric constant	11.8	—	—	5.5

interconnection inside a chip that offers negligible resistance. An SWCNT is usually semiconducting in nature. SWCNTs have a narrow energy bandgap of 0.3 eV and above and are used for manufacturing of active electronics devices. The thermal conductivity of SWCNT (35 W/cm K) is also much higher compared to that of silicon (1.5 W/cm K), and the intrinsic thermal conductivity is independent of the number of the layers. The thermal conductivity of CNT increases with increasing length in the range of 0.5–7 μm [78]. Electron mobility in CNT is 100,000 cm²/V s at room temperature, whereas silicon's electron mobility is 1,450 cm²/V s. The conductivity of CNTs can be even better than Cu [79,80] and it can withstand very high current densities (10^9 A/cm²).

The resistance, R, of a CNT is described as $h/(2e^2M)$, where h is the Planck's constant, e is the charge of electron, and M is the number of modes in the CNT with energies between the Fermi levels of the electrodes. The reason of high stiffness of a CNT is the sp^2 bonding between carbon–carbon atoms and the accurate Young's modulus is not yet defined. The Young's modulus can vary from 320 to 5500 GPa in various conditions [72,80]. Table 15.3 represents the properties of carbon-based materials compared with Si and other materials.

15.12 Switches and Hybrid NEMS–CMOS Integrated Circuits

Diamond MEM/NEM switches have the following advantages:

1. High thermal conductivity, allowing high-power and high-temperature operations.
2. A thermally stable Young's modulus with high value.
3. Low or no stiction between the contact surfaces.

Therefore, diamond MEM/NEM switches are expected to be highly reliable.

Although, the NEMS device features an exceptionally low OFF current, a compromise is made in the value of ON current. NEMS devices, unlike CMOS transistors, offer a very low ON current, which limits its utility as a component in an SoC. This necessitates the need of a design that incorporates the off-state characteristics of a NEMS switch with the high I_{ON} capabilities of CMOS technology. Succeeding to do so will minimize leakage and expedite switching while maintaining higher I_{ON}, which is required for analog applications. Hence, integration of CMOS and NEMS devices will prove to have significant advantages for the semiconductor industry.

CNT switches, on account of their high mechanical stiffness, can work in extremely high resonant frequencies of the gigahertz range. With a Young's modulus as high as 1 TPa [44,54,72–76], SWCNTs are found to be highly elastic in nature. NEMS based on CNTs are attractive for numerous reasons. For example, electrostatic switches as shown in Figure 15.2 can be designed where SWCNTs are suspended over a ground electrode. The CNT, on experiencing a potential difference between itself and the ground, deflects. When the potential is large enough, a contact is established between the tube and the ground electrode.

15.13 Conclusion

With benefits such as high mechanical stiffness, thermal conductivity, hydrophobic surface, and super electronic properties, carbon materials prove to be an excellent choice for MEMS/NEMS applications. Carbon MEMS/NEM switches also exhibit impressive chemical stability, reliability, and high-speed switching characteristics. Also, unlike Si-based switches, carbon MEMS/NEMS avoid stiction and abrasion.

Since the discovery of CNTs, several researches have been extensively carried out. However, some issues still remain to be resolved, most important of which is the repeatability of CNTs. The growth of CNTs is still needed to be optimized so that the structures can be reproduced consistently. Some challenging aspects involve the control over nanotube type (semiconducting or metallic), placement in desired positions, chirality, and diameter [54,74–80].

Another challenge stems from the fact that CNTs are extremely tiny, which makes them difficult to be integrated on a precise location for forming reliable contacts [81,82]. Although individual CNTs can be manipulated, the most direct way to do so is by using a scanning probe microscope [83]. This method is used to demonstrate prototypes and involves a very time-consuming process. The difficulty related to the manipulation of individual nanotubes results in poor repeatability. Condensed CNT aggregations, like membranes or "wafer," were utilized to improve device performance.

Polycrystalline and nanocrystalline diamond switches also suffer from similar problems, such as reproducibility, reliability, and controllability. This is due to the grain boundaries, stress, and impurities present in the material. Nevertheless, further research is encouraged in spite of the troubles relating to the processing of single-crystal diamond since ultra-nanocrystalline diamond shows promising results. To achieve ideal performance, efforts should be aimed at

1. Developing bulk and surface machining technique for batch production of suspended structures
2. Creating compatible device concepts with the batch production process, and
3. Large-area growth of single-crystal diamond

Attempts have been made recently in (1) and (3), and the world's first diamond NEM switch has been demonstrated [84]. NEMS can also withstand high acceleration without any breakage or significant disturbance. Note that the required proof mass should be present in the material.

For MEMS switches, low switching speeds restrict their usage to places where high speed is not needed, such as in radio-frequency (RF) applications. Additionally, large switching voltages of around 30–70 V necessitates the use of voltage up-converter components. Moreover, the popularity of MEMS has been impacted by its reliability issues. Downscaling devices to NEMS removes the switching speed and actuation voltage problems. Also, mechanical resonances in the gigahertz regime have been observed and reported.

References

1. Witkamp, B., M. Poot, and H.S. van der Zant, Bending-mode vibration of a suspended nanotube resonator, *Nano Letters*, 6(12), 2904–2908, 2006.
2. Cimalla, V., J. Pezoldt, and O. Ambacher, Group III nitride and SiC based MEMS and NEMS: Materials properties, technology and applications, *Journal of Physics D: Applied Physics*, 40(20), 6386, 2007.
3. Leondes, C.T., *Mems/Nems: (1) Handbook Techniques and Applications Design Methods, (2) Fabrication Techniques, (3) Manufacturing Methods, (4) Sensors and Actuators, (5) Medical Applications and MOEMS.* Springer Science & Business Media, Berlin, Germany, 2007.
4. Gammel, P., G. Fischer, and J. Bouchaud, RF MEMS and NEMS technology, devices, and applications, *Bell Labs Technical Journal*, 10(3), 29–59, Fall 2005.
5. Imboden, M. et al., High-speed control of electromechanical transduction: Advanced drive techniques for optimized step-and-settle response of MEMS micromirrors, *IEEE Control Systems*, 36(5), 48–76, Oct. 2016.

6. Iontchev, E., R. Kenov, R. Miletiev, I. Simeonov, and Y. Isaev, Hardware implementation of quad microelectromechanical sensor structure for inertial systems, In *Proceedings of the 2014 37th International Spring Seminar on Electronics Technology*, Dresden, Germany, 2014, pp. 417–420.

7. Kaasalainen, J. and A. Manninen, DC voltage reference based on a square-wave-actuated microelectromechanical sensor, *IEEE Transactions on Instrumentation and Measurement*, 60(7), 2506–2511, July 2011.

8. Efremov, G.I. and N.I. Mukhurov, Rod-type microelectromechanical sensor, In *Proceedings of the International Semiconductor Conference, 2001 (CAS 2001)*, Sinaia, Romania, 2001, pp. 363–366, Vol. 2.

9. Jain, A. and M.A. Alam, Extending and tuning the travel range of microelectromechanical actuators using electrically reconfigurable nano-structured electrodes, *Journal of Microelectromechanical Systems*, 22(5), 1001–1003, Oct. 2013.

10. Mehdizadeh, E. and S. Pourkamali, Deep submicron parallel scanning probe lithography using two-degree-of-freedom microelectromechanical systems actuators with integrated nanotips, *IET Micro & Nano Letters*, 9(10), 673–675, Oct. 2014.

11. Toshiyoshi, H. et al., Microelectromechanical digital-to-analog converters of displacement for step motion actuators, *Journal of Microelectromechanical Systems*, 9(2), 218–225, June 2000.

12. Kim, S. et al., Microelectromechanical Systems (MEMS) based-ultrasonic electrostatic actuators on a flexible substrate, *IEEE Electron Device Letters*, 33(7), 1072–1074, July 2012.

13. Coulomb, C.A., Recherches théoriques et expérimentales sur la force de torsion: & sur l'élasticité des fils de métal: Application de cette théorie à l'emploi des métaux dans les arts & dans différentes expériences de physique: Construction de différentes balances de torsion, pour mesurer les plus petits degrés de force: Observations sur les loix de l'élasticité & de la coherence (pp. 229–270), 1784.

14. Martínez, A.A., Replication of Coulomb's torsion balance experiment, *Archive for History of Exact Sciences*, 60(6), 517–563, 2006.

15. Feynman, R.P., There's plenty of room at the bottom, *Engineering and Science*, 23(5), 22–36, 1960.

16. Ke, C. and H.D. Espinosa (eds.), Nanoelectromechanical systems (NEMS) and modeling, *Handbook of Theoretical and Computational Nanotechnology*, American Scientific Publishers, CA, Vol. 1 (pp. 1–38), 2006.

17. Koochi, A., A. Kazemi, F. Khandani, and M. Abadyan, Influence of surface effects on size-dependent instability of nano-actuators in the presence of quantum vacuum fluctuations, *Physica Scripta*, 85(3), 035804, 2012.

18. Beni, Y.T., A. Koochi, and M. Abadyan, Theoretical study of the effect of Casimir force, elastic boundary conditions and size dependency on the pull-in instability of beam-type NEMS, *Physica E: Low-Dimensional Systems and Nanostructures*, 43(4), 979–988, 2011.

19. Dai, H., Carbon nanotubes: Opportunities and challenges, *Surface Science*, 500(1), 218–241, 2002.

20. Roukes, M.L., Nanoelectromechanical systems, In *Transducers' 01 Eurosensors XV*. Springer, Berlin, Germany, 2001, pp. 658–661.

21. Ekinci, K.L., Electromechanical transducers at the nanoscale: Actuation and sensing of motion in nanoelectromechanical systems (NEMS), *Small*, 1(8–9), 786–797, 2005.

22. Dambrine, G., A. Cappy, F. Heliodore, and E. Playez, A new method for determining the FET small-signal equivalent circuit, *IEEE Transactions on Microwave Theory and Techniques*, 36(7), 1151–1159, 1988.
23. Prégaldiny, F., C. Lallement, and D. Mathiot, A simple efficient model of parasitic capacitances of deep-submicron LDD MOSFETs, *Solid-State Electronics*, 46(12), 2191–2198, 2002.
24. Gangopadhyay, T.K., Prospects for fibre Bragg gratings and Fabry-Perot interferometers in fibre-optic vibration sensing, *Sensors and Actuators A: Physical*, 113(1), 20–38, 2004.
25. Pang, S.W., D.D. Rathman, D.J. Silversmith, R.W. Mountain, and P.D. DeGraff, Damage induced in Si by ion milling or reactive ion etching, *Journal of Applied Physics*, 54(6), 3272–3277, 1983.
26. Verhagen, E., S. Deléglise, S. Weis, A. Schliesser, and T.J. Kippenberg, Quantum-coherent coupling of a mechanical oscillator to an optical cavity mode, *Nature*, 482(7383), 63–67, 2012.
27. Jang, J.E., S.N. Cha, Y. Choi, G.A.J. Amaratunga, D.J. Kang, D.G. Hasko, J.E. Jung, and J.M. Kim, Nanoscale capacitors based on metal-insulator-carbon nanotube-metal structures, *Applied Physics Letters*, 87(26), 263103, 2005.
28. Jang, J.E., S.N. Cha, Y.J. Choi, D.J. Kang, T.P. Butler, D.G. Hasko, J.E. Jung, J.M. Kim, and G.A. Amaratunga, Nanoscale memory cell based on a nanoelectromechanical switched capacitor, *Nature Nanotechnology*, 3(1), 26–30, 2008.
29. Baker, D.W. and E.A. Herr, Parasitic effects in microelectronic circuits, *IEEE Transactions on Electron Devices*, 12(4), 161–167, Apr. 1965.
30. Margenau, H., Van der Waals forces, *Reviews of Modern Physics*, 11(1), 1, 1939.
31. Hamaker, H.C., The London—van der Waals attraction between spherical particles, *Physica*, 4(10), 1058–1072, 1937.
32. Casimir, H.B.G. and D. Polder, The influence of retardation on the London-van der Waals forces, *Physical Review*, 73(4), 360, 1948.
33. Dzyaloshinskii, I.E., E.M. Lifshitz, and L.P. Pitaevskii, General theory of van der Waals' forces, *Physics-Uspekhi*, 4(2), 153–176, 1961.
34. Van Oss, C.J., R.J. Good, and M.K. Chaudhury, The role of van der Waals forces and hydrogen bonds in "hydrophobic interactions" between biopolymers and low energy surfaces, *Journal of Colloid and Interface Science*, 111(2), 378–390, 1986.
35. Craighead, H.G., Nanoelectromechanical systems, *Science*, 290(5496), 1532–1535, 2000.
36. Weis, R.S. and T.K. Gaylord, Lithium niobate: Summary of physical properties and crystal structure, *Applied Physics A*, 37(4), 191–203, 1985.
37. Gottstein, G., *Physical Foundations of Materials Science*. Springer Science & Business Media, Berlin, Germany, 2013.
38. Durkan, C. and M.E. Welland, Size effects in the electrical resistivity of polycrystalline nanowires, *Physical Review B*, 61(20), 14215, 2000.
39. Vazquez-Mena, O., G. Villanueva, V. Savu, K. Sidler, M.A.F. Van Den Boogaart, and J. Brugger, Metallic nanowires by full wafer stencil lithography, *Nano Letters*, 8(11), 3675–3682, 2008.
40. Mengotti, E., L.J. Heyderman, A.F. Rodríguez, F. Nolting, R.V. Hügli, and H.-B. Braun, Real-space observation of emergent magnetic monopoles and associated Dirac strings in artificial kagome spin ice, *Nature Physics*, 7(1), 68–74, 2011.
41. Troutman, R.R., VLSI limitations from drain-induced barrier lowering, *IEEE Journal of Solid-State Circuits*, 14(2), 383–391, 1979.

42. Thompson, S. et al., MOS scaling: Transistor challenges for the 21st century, *Intel Technology Journal*, Q3, 1998.

43. Guisinger, N.P. and M.S. Arnold, Beyond silicon: Carbon-based nanotechnology, *MRS Bulletin*, 35(04), 273–279, 2010.

44. Zang, X., Q. Zhou, J. Chang, Y. Liu, and L. Lin, Graphene and carbon nanotube (CNT) in MEMS/NEMS applications, *Microelectronic Engineering*, 132, 192–206, 2015.

45. Li, D. and R.B. Kaner, Graphene-based materials, *Nature Nanotechnology*, 3, 101, 2008.

46. Stankovich, S., D.A. Dikin, G.H.B. Dommett, K.M. Kohlhaas, E.J. Zimney, E.A. Stach, R.D. Piner, S.T. Nguyen, and R.S. Ruoff, Graphene-based composite materials, *Nature*, 442(7100), 282–286, 2006.

47. Smalley, R.E., *Carbon Nanotubes: Synthesis, Structure, Properties, and Applications*, Vol. 80, M.S. Dresselhaus, G. Dresselhaus, and P. Avouris (eds.). Springer Science & Business Media, 2003.

48. Dresselhaus, M.S., G. Dresselhaus, J.-C. Charlier, and E. Hernandez, Electronic, thermal and mechanical properties of carbon nanotubes, *Philosophical Transactions of the Royal Society of London A: Mathematical, Physical and Engineering Sciences*, 362(1823), 2065–2098, 2004.

49. Krasheninnikov, A.V. and F. Banhart, Engineering of nanostructured carbon materials with electron or ion beams, *Nature Materials*, 6(10), 723–733, 2007.

50. Wudl, F., The chemical properties of buckminsterfullerene (C60) and the birth and infancy of fulleroids, *Accounts of Chemical Research*, 25(3), 157–161, 1992.

51. Hedberg, K. and L. Hedberg, Bond lengths in free molecules of buckminsterfullerene, C60, from gas-phase electron diffraction, *Science*, 254(5030), 410, 1991.

52. Iijima, S., Helical microtubules of graphitic carbon, *Nature*, 354(6348), 56–58, 1991.

53. Iijima, S., Carbon nanotubes: Past, present, and future, *Physica B: Condensed Matter*, 323(1), 1–5, 2002.

54. Baughman, R.H., A.A. Zakhidov, and W.A. de Heer, Carbon nanotubes—The route toward applications, *Science*, 297(5582), 787–792, 2002.

55. Hirsch, A., Functionalization of single-walled carbon nanotubes, *Angewandte Chemie International Edition*, 41(11), 1853–1859, 2002.

56. Wanlass, F.M. and C.T. Sah, *International Solid-State Circuits Conference, Digest*, University of Pennsylvania, New York. Lewis Winner, 1963.

57. Stockinger, M., Optimization of ultra-low-power CMOS transistors, PhD dissertation, Institute for Microelectronics, Vienna, Austria, 2000.

58. Taur Giustolisi, G., G. Palumbo, M. Criscione, and F. Cutri, A low-voltage low-power voltage reference based on subthreshold MOSFETs, *IEEE Journal of Solid-State Circuits*, 38(1), 151–154, 2003.

59. Taur, Y., X. Liang, W. Wang, and H. Lu, A continuous, analytic drain-current model for DG MOSFETs, *IEEE Electron Device Letters*, 25(2), 107–109, 2004.

60. Narendra, S., V. De, S. Borkar, D.A. Antoniadis, and A.P. Chandrakasan, Full-chip subthreshold leakage power prediction and reduction techniques for sub-0.18-μm CMOS, *IEEE Journal of Solid-State Circuits*, 39(3), 501–510, 2004.

61. Kim, N.S., T. Austin, D. Baauw, T. Mudge, K. Flautner, J.S. Hu, M.J. Irwin, M. Kandemir, and V. Narayanan, Leakage current: Moore's law meets static power, *Computer*, 36(12), 68–75, 2003.

62. Wright, P. and K.C. Saraswat, Thickness limitations of SiO_2 gate dielectrics for MOS ULSI, *IEEE Transactions on Electron Devices*, 37(8), 1884–1892, Aug. 1990.

63. Wilk, G.D., R.M. Wallace, and J.M. Anthony, High-k gate dielectrics: Current status and materials properties considerations, *Journal of Applied Physics*, 89(10), 5243–5275, 2001.

64. Gopalakrishnan, K., P.B. Griffin, and J.D. Plummer, I-MOS: A novel semiconductor device with a subthreshold slope lower than kT/q, In *International Electron Devices Meeting*, San Francisco, CA, 2002 (IEDM'02), Dec. 2002, pp. 289–292.

65. Dadgour, H.F. and K. Banerjee, Hybrid NEMS-CMOS integrated circuits: A novel strategy for energy-efficient designs, *IET Computers & Digital Techniques*, 3(6), 593–608, 2009.

66. Theis, T.N., (Keynote) In quest of a fast, low-voltage digital switch, *ECS Transactions*, 45(6), 3–11, 2012.

67. Cristoloveanu, S., J. Wan, and A. Zaslavsky, A review of sharp-switching devices for ultra-low power applications, *IEEE Journal of the Electron Devices Society*, 4(5), 215–226, 2016.

68. Zhang, L.L. and X.S. Zhao, Carbon-based materials as supercapacitor electrodes, *Chemical Society Reviews*, 38(9), 2520–2531, 2009.

69. Geim, A.K., Graphene: Status and prospects, *Science*, 324(5934), 1530–1534, 2009.

70. Kim, S.Y. and H.S. Park, Multilayer friction and attachment effects on energy dissipation in graphene nanoresonators, *Applied Physics Letters*, 94(10), 101918, 2009.

71. Chen, C., S. Rosenblatt, K.I. Bolotin, P. Kim, I. Kymissis, H.L. Stormer, T.F. Heinz, and J. Hone, Nems applications of graphene, In *2009 IEEE International Electron Devices Meeting (IEDM)*, Baltimore, MA, Dec. 2009, pp. 1–4.

72. Liao, M. and Y. Koide, Carbon-based materials: Growth, properties, MEMS/NEMS technologies, and MEM/NEM switches, *Critical Reviews in Solid State and Materials Sciences*, 36(2), 66–101, 2011.

73. Matsudaira, H., S. Miyamoto, H. Ishizaka, H. Umezawa, and H. Kawarada, Over 20-GHz cutoff frequency submicrometer-gate diamond MISFETs, *IEEE Electron Device Letters*, 25(7), 480–482, July 2004.

74. Okano, K., H. Kiyota, T. Iwasaki, Y. Nakamura, Y. Akiba, T. Kurosu, M. Iida, and T. Nakamura, Fabrication of a diamond pn junction diode using the chemical vapour deposition technique, *Solid-State Electronics*, 34(2), 139–141, 1991.

75. Klein, C.A. and G.F. Cardinale, Young's modulus and Poisson's ratio of CVD diamond, *Diamond and Related Materials*, 2(5–7), 918–923, 1993.

76. Chiu, H.Y., P. Hung, H.W.C. Postma, and M. Bockrath, Atomic-scale mass sensing using carbon nanotube resonators, *Nano Letters*, 8(12), 4342–4346, 2008.

77. Lassagne, B., D. Garcia-Sanchez, A. Aguasca, and A. Bachtold, Ultrasensitive mass sensing with a nanotube electromechanical resonator, *Nano Letters*, 8(11), 3735–3738, 2008.

78. Wang, Z.L., D.W. Tang, X.B. Li, X.H. Zheng, W.G. Zhang, L.X. Zheng, Y.T. Zhu, A.Z. Jin, H.F. Yang, and C.Z. Gu, Length-dependent thermal conductivity of an individual single-wall carbon nanotube, *Applied Physics Letters*, 91(12), 123119, 2007.

79. Tans, S.J., A.R. Verschueren, and C. Dekker, Room-temperature transistor based on a single carbon nanotube, *Nature*, 393(6680), 49–52, 1998.

80. Tans, S.J., M.H. Devoret, H. Dai, A. Thess, R.E. Smalley, L.J. Georliga, and C. Dekker, Individual single-wall carbon nanotubes as quantum wires, *Nature*, 386(6624), 474–477, 1997.

81. Vijayaraghavan, A., S. Blatt, D. Weissenberger, M. Oron-Carl, F. Hennrich, D. Gerthsen, H. Hahn, and R. Krupke, Ultra-large-scale directed assembly of single-walled carbon nanotube devices, *Nano Letters*, 7(6), 1556–1560, 2007.

82. Mahar, B., C. Laslau, R. Yip, and Y. Sun, Development of carbon nanotube-based sensors—A review, *IEEE Sensors Journal*, 7(2), 266–284, 2007.
83. Neuman, K.C. and A. Nagy, Single-molecule force spectroscopy: Optical tweezers, magnetic tweezers and atomic force microscopy, *Nature Methods*, 5(6), 491, 2008.
84. Liao, M., S. Hishita, E. Watanabe, S. Koizumi, and Y. Koide, Suspended single-crystal diamond nanowires for high-performance nanoelectromechanical switches, *Advanced Materials*, 22(47), 5393–5397, 2010.

Index

A

Absorption coefficient, 7, 77, 205
Acetone-sensing mechanism, 224
Acetone vapors, 232
Adhesive film technology (AFT), 237
AFM, *see* Atomic force microscopy
Agilent GUI data-logging
        software, 232
Agilent U34411A multimeter, 232
Algae, 137
Aniline, 251
Approximation method
    Rayleigh–Schrodinger perturbation
        method, 26–28
    variational method, 26
    Wentzel–Kramers–Brillouin
        method, 28–29
Arc-discharge technique, 252
Atomic force microscopy (AFM), 70,
        83, 176, 300
Atomic/molecular gas condensation
        process, 61–63

B

Bacteria
    Bacillus species, 134
        B. licheniformis, 133
        B. subtilis culture filtrate, 133
    Delftia acidovorans, 133
    palladium, 133
    Pseudomonas cells, 133
    Rhodopseudomonas capsulata, 133
    silver and gold nanoparticles, 134
Ball-and stick model, 243
Ball grid array (BGA), 234
Ballistic transport
    carrier transport, 96–97
    continuity equation, 97
    current density, 96
    Einstein relation, 97–98
    electrochemical force, 97–98
    Landauer formula, 98–100

1D channel, 98–99
    quasiballistic transport, 96–97
Band gap, 35, 88, 91, 198, 200, 205, 207,
        209, 245–246, 286, 292, 310–311
Band-to-band tunneling (BTBT), 277, 282
BGA, *see* Ball grid array
Biogenic nanoparticles, 132, 140, 161
Biological synthesis, of nanoparticles
    advantages, 132
    biogenic nanoparticles, 132
    definition, 131
    limitations, 142
Biosensing applications, 294
Black body, 7, 9
Bloch's theorem, 41
Bottom-top approach, 250
Breakdown voltage, 286, 292–293
Brust–Schiffrin technique, 111–112
Buckyball (buckminsterfullerene C_{60}),
        304–305

C

Carbon nanotubes (CNTs), 242, 303–306
    band structure, of nanoparticles, 42
    HEBM, 69
    NEMS, 310–311
        buckminsterfullerene, 304–305
        drain current equation, 304
        electronic gadget, 303
        integrated circuit, 303
        MOSFET, 303–304
        MWCNT, 306
        SCE, 304
        SWCNT, 306
        Young's modulus, 304
    PECVD, 56
    and Si limitation, 310–311
        buckminsterfullerene, 304–305
        drain current equation, 304
        electronic gadget, 303
        integrated circuit, 303
        MOSFET, 303–304
        MWCNT, 306

Milton Keynes UK
Ingram Content Group UK Ltd.
UKHW021629071024
449327UK00020BA/1253